Diversity, Distribution and Phylogeny of Vector Insects

Diversity, Distribution and Phylogeny of Vector Insects

Editors

Carlos Barceló
Ignacio Ruiz-Arrondo

MDPI

Basel • Beijing • Wuhan • Barcelona • Belgrade • Novi Sad • Cluj • Manchester

Editors
Carlos Barceló
University of the Balearic Islands
Palma
Spain

Ignacio Ruiz-Arrondo
San Pedro University Hospital-CIBIR
Logroño
Spain

Editorial Office
MDPI
St. Alban-Anlage 66
4052 Basel, Switzerland

This is a reprint of articles from the Special Issue published online in the open access journal *Diversity* (ISSN 1424-2818) (available at: https://www.mdpi.com/journal/diversity/special_issues/Diversity_Distribution_Phylogeny_Vector_Insects).

For citation purposes, cite each article independently as indicated on the article page online and as indicated below:

Lastname, A.A.; Lastname, B.B. Article Title. *Journal Name* **Year**, *Volume Number*, Page Range.

ISBN 978-3-7258-0443-6 (Hbk)
ISBN 978-3-7258-0444-3 (PDF)
doi.org/10.3390/books978-3-7258-0444-3

Cover image courtesy of Mikel A. González

Contents

diversity MDPI

Editorial

Diversity, Distribution and Phylogeny of Vector Insects

Carlos Barceló [1,*] and Ignacio Ruiz-Arrondo [2]

1 Applied Zoology and Animal Conservation Group, University of the Balearic Islands (UIB), Cra. Valldemossa Km. 7.5, 07122 Palma, Balearic Islands, Spain
2 Centre of Rickettsiosis and Arthropod-Borne Diseases, Hospital Universitario San Pedro-CIBIR, 26006 Logroño, La Rioja, Spain; iruizarr@gmail.com
* Correspondence: carlos.barcelo@uib.es

Mosquitoes (Fam. Culicidae), sand flies (Subfam. Phlebotominae), biting midges (Fam. Ceratopogonidae), black flies (Fam. Simuliidae) and stable flies (Fam. Muscidae) are groups of insects capable to transmit pathogens of public health and veterinary importance [1–5]. These pathogens include viruses, bacteria, protozoans and helminths that affect humans (e.g., Zika and dengue viruses), domestic and wild animals (e.g., Bluetongue and Equine infectious anemia virus), and agents that cause zoonotic diseases such as West Nile, Leishmaniasis and Onchocerciasis [6–10]. These insects not only transmit pathogens, but also are responsible for human and domestic livestock nuisances and allergies [3,4].

More than 700,000 humans die every year due to pathogens transmitted by vector insects worldwide, such as malaria and dengue [11–13]. Outbreaks have nowadays become increasingly frequent in certain regions. For example, in Europe, viruses such as West Nile and bluetongue occur in several countries, and leishmaniasis cases have increased during the last decade [2,7,14]. Control measures against these diseases are usually focused on the vector insect since, in most cases, vaccines are not available [15]. Knowing the distribution of the different vectors and understanding their biology, behaviors and genetic relationships could help humans to manage the control of vector diseases transmitted by them; therefore, the study of the diversity and phylogeny of vector insects in determined areas is crucial to understand the relationships between its hosts and the pathogens that can be transmitted, including aspects related to their epidemiology.

In this Special Issue, we included studies about the identification, diversity, distribution and population dynamics of mosquitoes, sand flies, biting midges and black flies in different countries, including a review of stable flies. In this context, this Special Issue includes 12 articles, of which 11 are research papers and one is a review. The group with the highest number of publications is mosquitoes, with six papers. Ruiz-Arrondo et al. (contribution 1) studied the species composition and population dynamics of Culicidae in three aquatic ecosystems located in a peri-urban area of a city in northern Spain. They observed that the different hydrological management practices of each environment could play a key role in determining the abundance of mosquito genera. In the same line, Rosa-Silva et al. (contribution 2) studied the effect of rain on the spatial distribution and abundance of container-breeding mosquitoes in Brazil, demonstrating that human occupation and rainfall impact interactions between invasive urban species such as *Aedes albopictus* and *Aedes aegypti*. and sylvatic species. Continuing with studies on the distribution of mosquitoes, this time in the state of Queretaro in Mexico, Ortega Morales et al. (contribution 3) identified 33 new records of mosquito species for this state, including two undescribed species. They morphologically and molecularly described one of them as *Shannoniana huasteca* n. sp. In turn, the study conducted by Ibáñez-Justicia et al. (contribution 4) includes novel information regarding malaria vector species from the *Anopheles maculipennis* complex in 161 locations (including overwintering sites) during the National Surveillance program in the Netherlands, with the first record of *Anopheles daciae* in the country. The study of Vanderheyden et al. (contribution 5) reported the distribution of *Culex pipiens* s.s and

Citation: Barceló, C.; Ruiz-Arrondo, I. Diversity, Distribution and Phylogeny of Vector Insects. *Diversity* **2024**, *16*, 95. https://doi.org/10.3390/d16020095

Received: 22 January 2024
Accepted: 30 January 2024
Published: 1 February 2024

Cx. torrentium in Belgium, using the COI and ACE2 loci. In the only study of this Special Issue carried out in Africa, phylogenetic analyses were conducted by Montalvo-Sabino et al. (contribution 6) on 65 mosquito species collected in South Africa, Mozambique and Angola, with new records for Afrotropical fauna. In addition, male genitalia were morphologically identified and associated with phylogenetic analysis.

Culicoides and black flies constitute the group of insect vectors with the second highest number of papers published in this Special Issue, with each having two publications. On the one hand, there is an ecological study of two of the main vector species of bluetongue in Europe, *Culicoides obsoletus* s.l. and *Culicoides imicola*, in which Barceló et al. (contribution 7) determined its nocturnal activity, showing different activity patterns according to the month and the time after the sunset. On the other hand, there is a molecular study in which Dähn et al. (contribution 8) designed PCR primers based on the COI gene to identify 21 biting midge species from the subgenus *Culicoides* in the Paleartic region. In addition, PCR assays were performed for the first time on recently described species of the Pulicaris group. Regarding the studies about black flies, Adler and Reeves (contribution 9) identified a pattern of north–south differentiation in the genus *Prosimulium* of western North America, and described a new species, *Prosimulium supernum*. Meanwhile, Kúdelová et al. (contribution 10) carried out the DNA barcoding of 25 black fly species from Slovakia, finding hidden diversity as well as shared barcode sequences among the studied species. A study on sandflies has also been published in this Special Issue. A new sand fly record for Spain, *Phlebotomus perfiliewi* s.l., was morphologically and phylogenetically described in González et al. (contribution 11), including a discussion about potential confusion with *Phlebotomus perniciousus*. Finally, Duvallet and Hogsette (contribution 12) reviewed several aspects of the genus *Stomoxys* sp., including worldwide diversity, distribution and genetic approaches to identify the origin and population dynamics of this genus.

Author Contributions: C.B. wrote the first draft of the manuscript. I.R.-A. edited the manuscript. All authors have read and agreed to the published version of the manuscript.

Acknowledgments: We want to thank the contributions of all authors involved in this Special Issue, the anonymous reviewers and the editorial staff of Diversity.

Conflicts of Interest: The authors declare no conflicts of interest.

List of Contributions

1. Ruiz-Arrondo, I.; Alarcón-Elbal, P. M.; Blanco-Sierra, L.; Delacour-Estrella, S.; de Blas, I.; Oteo, J. A. Species composition and population dynamics of Culicidae during their peak abundance period in three peri-urban aquatic ecosystems in Northern Spain. *Diversity* **2023**, *15*, 938. https://doi.org/10.3390/d15080938
2. Rosa-Silva, H.; Cardoso, J.G.; Reis-Júnior, R.; Corgosinho, P. H.; Faria, M. L.; Ribeiro, S. P., Abreu F. V. S.; Borges, M. A. Coexistence and spatial distribution of invasive and sylvatic container-breeding mosquitoes in city–forest ecotone within the Brazilian semi-arid. *Diversity* **2023**, *15*, 822. https://doi.org/10.3390/d15070822
3. Ortega-Morales, A. I.; Hernández-Triana, L. M.; Siller-Rodríguez, Q. K. The mosquitoes of Querétaro, Mexico: distribution, ecology, and discovery of *Shannoniana huasteca* n. sp. (Diptera: Culicidae). *Diversity* **2023**, *15*, 697. https://doi.org/10.3390/d15060697
4. Ibáñez-Justicia, A.; Smitz, N.; Blom, R.; Vanderheyden, A.; Jacobs, F.; Meganck, K.; Gombeer, S.; Backeljau, T.; Koenraadt, C. J. M.; Griep, J. S.; De Meyer, M.; Stroo, A. *Anopheles maculipennis* complex in the Netherlands: first record of *Anopheles daciae* (Diptera: Culicidae). *Diversity* **2022**, *14*, 636. https://doi.org/10.3390/d14080636
5. Vanderheyden, A.; Smitz, N.; De Wolf, K.; Deblauwe, I.; Dekoninck, W.; Meganck, K.; Gombeer, S.; Vanslembrouck, A.; De Witte, J.; Schenider, A.; Verlé, I.; De Meyer, M.; Backeljau, T.; Müller, R.; Van Bortel, W. DNA Identification and diversity of the vector mosquitoes *Culex pipiens* ss and *Culex torrentium* in Belgium (Diptera: Culicidae). *Diversity* **2022**, *14*, 486. https://doi.org/10.3390/d14060486

6. Montalvo-Sabino, E.; Abílio, A. P.; Guarido, M. M.; Valadas, V.; Novo, M. T.; Kampango, A.; Sousa, C. A.; Fafetine, J.; Venter, M.; Thompson, P. N.; Braack, L.; Cornel, A. J.; Parreira, R.; Gouveia de Almeida, A. P. Morphological and molecular characterization using genitalia and *CoxI* barcode sequence analysis of Afrotropical mosquitoes with arbovirus vector potential. *Diversity* **2022**, *14*, 940. https://doi.org/10.3390/d14110940
7. Barceló, C.; del Río, R.; Miranda, M. A. seasonal and nocturnal activity of *Culicoides* spp. (Diptera: Ceratopogonidae) adapted to different environments in the Balearic Islands. *Diversity* **2023**, *15*, 690. https://doi.org/10.3390/d15050690
8. Dähn, O.; Werner, D.; Mathieu, B.; Kampen, H. Development of conventional multiplex PCR assays for the identification of 21 West Palearctic biting midge taxa (Diptera: Ceratopogonidae) belonging to the *Culicoides* Subgenus *Culicoides*, including recently discovered species and genetic variants. *Diversity* **2023**, *15*, 699. https://doi.org/10.3390/d15060699
9. Adler, P. H.; Reeves, W. K. North–South differentiation of black flies in the Western cordillera of North America: a new species of *Prosimulium* (Diptera: Simuliidae). *Diversity* **2023**, *15*, 212. https://doi.org/10.3390/d15020212
10. Kúdelová, T.; Krčmárik, S.; Lužáková, I.; Bujačková, B.; Matická, K.; Kúdela, M. DNA barcoding of black blies (Diptera: Simuliidae) in Slovakia and its utility for species identification. *Diversity* **2023**, *15*, 661. https://doi.org/10.3390/d15050661
11. González, M. A.; Ruiz-Arrondo, I.; Gutiérrez-López, R.; Barceló, C.; Miranda, M. Á. (2023). First record of *Phlebotomus* (*Larroussius*) *perfiliewi* (Diptera: Psychodidae), vector of *Leishmania infantum* and phleboviruses, in Spain. *Diversity* **2023**, *15*, 400. https://doi.org/10.3390/d15030400
12. Duvallet, G.; Hogsette, J. A. Global diversity, distribution, and genetic studies of stable flies (*Stomoxys* sp.). *Diversity* **2023**, *15*, 600.

References

1. Schaffner, F.; Medlock, J.M.; Van Bortel, A.W. Public health significance of invasive mosquitoes in Europe. *Clin. Microbiol. Infec.* **2013**, *19*, 685–692. [CrossRef] [PubMed]
2. European Centre for Disease Prevention and Control. *The Spatial Relationship between the Presence and Absence of Leishmania spp. and Leishmaniasis, and Phlebotomine Sand Fly Vectors in Europe and Neighboring Countries*; ECDC: Stockholm, Sweden, 2023. [CrossRef]
3. Mullen, G.R.; Murphree, C.S. Biting midges (Ceratopogonidae). In *Medical and Veterinary Entomology*, 3rd ed.; Mullen, G.R., Durden, L.A., Eds.; Academic Press: Amsterdam, UK, 2019; pp. 213–236. [CrossRef]
4. Cupp, E.W.; Cupp, M.S. Black fly (*Diptera: Simuliidae*) salivary secretions: Importance in vector competence and disease. *J. Med. Entomol.* **1997**, *34*, 87–94. [CrossRef] [PubMed]
5. Patra, G.; Behera, P.; Das, S.K.; Saikia, B.; Ghosh, S.; Biswas, P.; Kumar, A.; Alam, S.S.; Kawlni, L.; Lalnunpuia, C.; et al. *Stomoxys calcitrans* and its importance in livestock. *Int. J. Adv. Agric. Res.* **2018**, *6*, 30–37.
6. Whiteman, A.; Loaiza, J.R.; Yee, D.A.; Poh, K.C.; Watkins, A.S.; Lucas, K.J.; Rapp, T.J.; Kline, L.; Ahmed, A.; Chen, S.; et al. Do socioeconomic factors drive *Aedes* mosquito vectors and their arboviral diseases? A systematic review of dengue, chikungunya, yellow fever, and Zika Virus. *One Health* **2020**, *11*, 100188. [CrossRef] [PubMed]
7. Jani, C.; Kakoullis, L.; Abdallah, N.; Mouchati, C.; Page, S.; Colgrove, R.; Chen, L.H. West Nile virus: Another emerging arboviral risk for travelers? *Curr. Infect. Dis. Rep.* **2022**, *24*, 117–128. [CrossRef] [PubMed]
8. Sick, F.; Beer, M.; Kampen, H.; Wernike, K. *Culicoides* biting midges—Underestimated vectors for arboviruses of public health and veterinary importance. *Viruses* **2019**, *11*, 376. [CrossRef] [PubMed]
9. Baldacchino, F.; Muenworn, V.; Desquesnes, M.; Desoli, F.; Charoenviriyaphap, T.; Duvallet, G. Transmission of pathogens by *Stomoxys* flies (*Diptera, Muscidae*): A review. *Parasite* **2013**, *20*, 26. [CrossRef] [PubMed]
10. Lakwo, T.; Oguttu, D.; Ukety, T.; Post, R.; Bakajika, D. Onchocerciasis elimination: Progress and challenges. *Res. Rep. Trop. Med.* **2020**, *11*, 81–95. [CrossRef] [PubMed]
11. World Health Organization. World Malaria Report 2021. Available online: https://www.who.int/teams/global-malaria-programme/reports/world-malaria-report-2021 (accessed on 11 April 2022).
12. World Health Organisation. Leishmaniasis. 8.1.22. Available online: https://www.who.int/news-room/fact-sheets/detail/leishmaniasis (accessed on 30 May 2022).
13. World Health Organisation. Dengue and Severe Dengue. 10.1.2022. Available online: https://www.who.int/news-room/fact-sheets/detail/dengue-and-severe-dengue (accessed on 30 May 2022).

14. Alzuheir, I.M. A review analysis of global Bluetongue virus-related research articles from 2000 to 2020. *Iraqi J. Vet. Sci.* **2022**, *36*, 341–347. [CrossRef]
15. Wilson, A.L.; Courtenay, O.; Kelly-Hope, L.A.; Scott, T.W.; Takken, W.; Torr, S.J.; Lindsay, S.W. The importance of vector control for the control and elimination of vector-borne diseases. *PLoS Negl. Trop. Dis.* **2020**, *14*, e0007831. [CrossRef] [PubMed]

diversity

MDPI

Article

Species Composition and Population Dynamics of Culicidae during their Peak Abundance Period in Three Peri-Urban Aquatic Ecosystems in Northern Spain

Ignacio Ruiz-Arrondo [1], Pedro María Alarcón-Elbal [2,3,*], Laura Blanco-Sierra [4], Sarah Delacour-Estrella [5], Ignacio de Blas [5] and José A. Oteo [1]

1 Centre of Rickettsiosis and Arthropod-Borne Diseases, Hospital Universitario San Pedro-CIBIR, 26006 Logroño, Spain; irarrondo@riojasalud.es (I.R.-A.); jaoteo@riojasalud.es (J.A.O.)
2 Laboratorio de investigación de Entomología, Departamento de Zoología, Facultad de Ciencias Biológicas (Bloque B), Universidad de Valencia, 46100 Burjasot, Spain
3 Department of Animal Production and Health, Veterinary Public Health and Food Science and Technology (PASAPTA), Facultad de Veterinaria, Universidad Cardenal Herrera-CEU, CEU Universities, 46115 Valencia, Spain
4 Centre d'Estudis Avançats de Blanes (CEAB-CSIC), 17300 Girona, Spain; lblanco@ceab.csic.es
5 Department of Animal Pathology, Faculty of Veterinary Sciences, Instituto Universitario de Investigación Mixto Agroalimentario de Aragón (IA2), Universidad de Zaragoza, 50013 Zaragoza, Spain; delacour@unizar.es (S.D.-E.); deblas@unizar.es (I.d.B.)
* Correspondence: pedro.alarcon@uv.es or pedro.alarconelbal@uchceu.es

Abstract: For a better understanding of the transmission cycles of mosquito-borne diseases, it is essential to explore the species composition and population dynamics, mainly during their peak abundance period. These investigations do not normally include peri-urban areas, despite their significant epidemiological interest. To address these gaps, an entomological survey was carried out in 2016 and 2017 in three aquatic ecosystems located on the outskirts of the city of Logroño, in northern Spain: the Iregua River and La Grajera (La Rioja) and Las Cañas Reservoirs (Navarra). Mosquitoes were captured using BG-Sentinel traps baited with CO_2 and BG-lure, as well as through the human landing collection method. In total, 6793 mosquito specimens were captured, representing 24 taxa within six genera. A specific PCR based on the ITS2 gene was used to differentiate members of the *Anopheles claviger* complex, and all individuals were identified as *An. claviger* sensu stricto. La Grajera had the most diverse culicid fauna, with 19 taxa, followed by Las Cañas (n = 15) and the Iregua River (n = 13). The composition and abundance of Culicidae varied across the aquatic ecosystem. We observed that the different hydrological management practices of each environment could play a key role in determining the abundance of mosquito genera. The overall risk of mosquito bites in the study area is expected to be relatively low and will depend on the freshwater ecosystem and the time of year.

Keywords: Culicidae; diversity; ITS2; mosquito; La Rioja; *Anopheles claviger* sensu stricto; Upper Ebro Valley

Citation: Ruiz-Arrondo, I.; Alarcón-Elbal, P.M.; Blanco-Sierra, L.; Delacour-Estrella, S.; de Blas, I.; Oteo, J.A. Species Composition and Population Dynamics of Culicidae during their Peak Abundance Period in Three Peri-Urban Aquatic Ecosystems in Northern Spain. *Diversity* **2023**, *15*, 938. https://doi.org/10.3390/d15080938

Academic Editor: Michael Wink

Received: 23 June 2023
Revised: 11 August 2023
Accepted: 15 August 2023
Published: 18 August 2023

1. Introduction

The arthropod-borne diseases represent a major risk to humans, livestock, pets and wildlife worldwide [1]. In Spain, there is a wide range of organisms that transmit diseases in our environment, mostly included in the so-called zoonoses [2]. Although reviewing the issue of arthropod-borne diseases in Spain is complex, it is unquestionable that since the beginning of the 21st century, the importance of Diptera has been extraordinary. The burden of some diseases such as bluetongue and leishmaniosis has led to an increase in studies on *Culicoides* biting midges (Ceratopogonidae) and phlebotomine sand flies (Psychodidae), respectively [3,4]. Black flies (Simuliidae) have also become very relevant pests in certain Spanish regions due to the annoyance caused by their painful bites [5].

Despite the above, mosquitoes (Culicidae) continue to be the most studied insects because of their major role as vectors. Invasive mosquito species and the pathogens they transmit represent a serious risk to both animal and public health [6], especially in areas with a Mediterranean climate, where these insects have a major impact throughout much of the year [7]. Therefore, studies of these vectors have not been addressed as much in climate regions such as the north-western, Cantabrian cornice, and inland regions, which are less suitable for their development, although a change in this regard has been observed in the last decade. In fact, since the detection of the Asian tiger mosquito, *Aedes albopictus* (Skuse, 1894), in the Basque Country in 2015 [8], interest in mosquitoes in northern Spain has increased. The recent finding of the invasive *Aedes japonicus* (Theobald, 1901) in Asturias in 2018 and its later detection in Cantabria and the Basque Country in 2019 and 2020, respectively [9,10], again highlighted the importance of conducting mosquito surveys in the northern regions of the country, focusing not only on specific invasive alien mosquitoes but also on native species.

Several studies on mosquitoes have been recently carried out in this part of the Iberian Peninsula, among which faunistic analysis predominated [11–13]. Specifically, since 2016, when a vector surveillance programme (mainly focused on culicids) was implemented in La Rioja region, 25 species have been identified [14–16]. However, these studies have not delved into seasonal population dynamics, which are essential to better understand mosquito-borne disease transmission cycles, identify sources of infection, and successfully control these vectors [17,18].

Most studies in Spain have been carried out in either natural areas or urban environments [19–21]. However, the urban-to-wild gradient is composed of distinct environmental habitats that influence the distribution, diversity, and abundance of mosquito species [20]. The rapid growth of urban areas in Spain has led to an increase in peri-urban habitats, which are areas where urban and rural environments overlap. However, the characteristics of peri-urban habitats differ from those of urban and natural habitats in many ways (e.g., fragmentation, pollution, noise, and light, among others) [22]. Peri-urban areas can provide freshwater ecosystems, such as riverbanks or wetlands, which have high value in terms of ecological diversity. They also enhance the human well-being of urban residents by providing a landscape for recreational activities in addition to their aesthetic value. As a negative counterpart to these aquatic habitats, there is the possibility of disturbance by haematophagous dipterans and their potential for disease transmission [23]. Therefore, it is necessary to monitor the presence of, among others, the members of the family Culicidae in these environments. In this respect, few Spanish studies have examined the mosquitoes that inhabit these particular environments [24–26]. The study of the mosquito fauna associated with peri-urban environments should be of increasing interest, as it is in these often-naturalised environments that the greatest number of encounters between vectors and humans occur.

We aimed to provide data on the species composition and population dynamics of mosquitoes during the peak abundance period in three nearby freshwater ecosystems located in peri-urban areas of Logroño, La Rioja, as well as to detect differences in the frequency of genera and relevant species in each location. Given the ecological aspects studied for the different species detected in the three peri-urban ecosystems, a succinct discussion about the possible public health implications of this mosquito species is provided.

2. Materials and Methods

2.1. Study Area

This study was conducted in the Autonomous Community of La Rioja (northern Spain) and a nearby area of the Chartered Community of Navarra (Figure 1). The entomological surveys were carried out in three sampling sites on the outskirts of the city of Logroño, the capital of the province of La Rioja, lying on the Upper Ebro Valley. This city is 385 m asl;

to the north are the Cantabrian Mountains, to the south the Iberian System. Regarding its climate, it is slightly continental, with quite cold, relatively rainy winters and hot, sunny summers but with some cool and rainy periods. The average temperature of the coldest month (January) was 5 °C; that of the warmest months (July, August) was 20.6 °C. Precipitation amounts to 633 mm per year. It ranges from 29 mm in the driest month (August) to 66 mm in the wettest ones (April, May) [27].

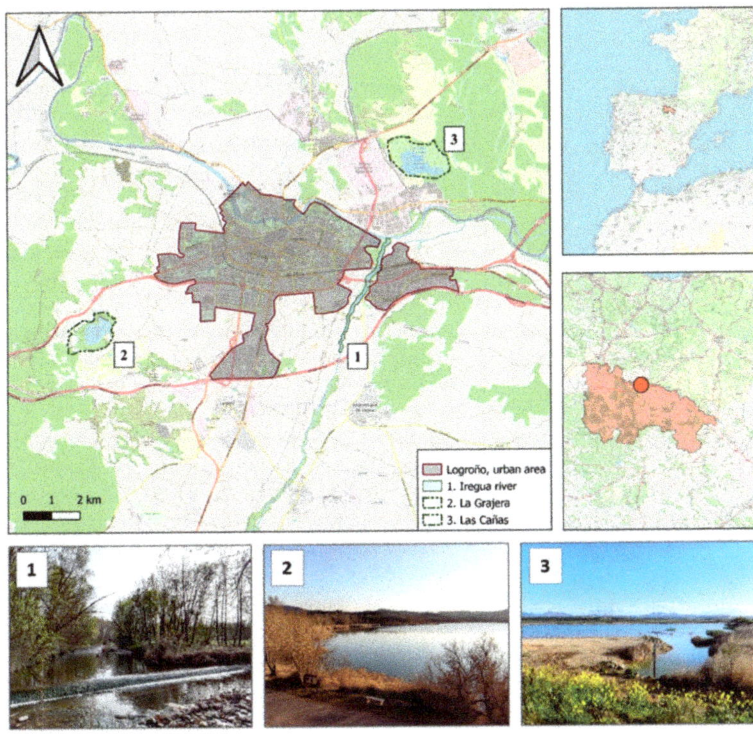

Figure 1. Location of the three peri-urban aquatic ecosystems sampled on the outskirts of the city of Logroño (red dot), northern Spain. (1) Iregua River; (2) La Grajera Reservoir; (3) Las Cañas Reservoir.

The first sampling freshwater habitat was the Iregua River, as it flows through Logroño, near its mouth in the Ebro River. From its source in the Sierra Cebollera to its mouth, this river crosses La Rioja from south to north over a length of 64 km, with a catchment area of 692 km^2 and an average flow rate of 3.32 m^3/s [28]. As it passes through the municipality, the river is mainly characterised by its permanently low flow; the natural vegetation is reduced to some gravel beaches and the slopes that delimit the flood channel. On the former, which were originally submerged, small sea bream and sea bream grow, as well as other shrubs and herbaceous plants of various species. The second peri-urban aquatic ecosystem was La Grajera, one of the few wetlands that exist in La Rioja. This place is currently protected and covers a total area of 87 ha, 32 ha with a permanent sheet of water supplied by a single inlet. It is the last enclave on the Camino de Santiago (Way of St. James) as it passes through Logroño, and it also has an educational classroom, a recreation area, a coffee shop and restaurants, paths around the lagoon, and a golf course with 18 holes [29]. The third and last aquatic environment was La Laguna de las Cañas, a nature reserve covering an area of 178 ha and one of the most important wetlands (80 ha) of Navarra, located in the municipality of Viana, just a few hundred metres from an industrial area of the city of Logroño. This reservoir collects runoff water from a catchment area of 6602 ha through a network of streams and ditches and can reach a volume of dammed water of around 0.7 hm^3 [30]. Las Cañas has several types of environments: crops, Mediterranean scrub in the surroundings, banks, flooded soils, small reed beds in the outer zone and large reed beds in the interior, open waters, islets, and trees, with a changing water level. These areas were chosen according to the presence of mosquitoes, waterflow, and ornithological richness and because they are located in peri-urban areas and frequently visited by people who want to enjoy contact with nature (Figure 1).

2.2. Mosquito Collection and Identification

Mosquitoes were collected from July to September 2016 and from May to September 2017. As can be observed, this study has primarily been carried out during the summer season, as this is the period when mosquitoes concentrate their activity in northern Spain.

The trapping effort was different in the case of the Iregua River, where it started a week later in 2016 and at the end of June in 2017. A total of four or five BG-Sentinel TM traps (BioGents GmbH, Regensburg, Germany) baited with BG-Lure® and 1 kg of CO_2 pellets were set once every two weeks in each location. In total, mosquito traps were run 223 nights during the study period, 81 in 2016 (18 in the Iregua River, 31 in La Grajera, and 31 in Las Cañas) and 143 in 2017 (32 in the Iregua River, 59 in La Grajera, and 52 in Las Cañas). The traps were set at dusk (19:00–21:00) and removed after dawn (8:00–9:00) the following day. The traps located in the two reservoirs were placed in five sampling stations along the entire perimeter of the aquatic ecosystem (separated between 400 and 500 m in La Grajera and 400 and 600 m in Las Cañas). In the case of the Iregua River, four traps were placed along the right bank of the river, covering 1 km of riverbed until it flows into the Ebro River. All sampling stations were selected on the basis of several criteria: covering different locations in the wetland; the presence of some vegetation cover in the vicinity; easy accessibility; and proximity to the aquatic ecosystems' water surface. La Grajera and Las Cañas are 1.7 km and 2 km away from the urban area of Logroño, respectively. The Iregua River separates two urban areas of Logroño, forming a green corridor between 100 and 200 m wide.

During the setting and removal of the traps, host-seeking females were captured by human landing collection (HLC) during the time the collector went from the vehicle to the exact location of the trap, set up the trap, and returned to the car (15 min). The collector exposed their legs and arms and caught landing mosquitoes with a hand-held mouth aspirator. Collection bags containing insects were stored at −80 °C until further processing.

For identification, the mosquitoes were separated from other insects, enumerated, sexed, and determined at the species level based on external morphological characters and male terminalia using the taxonomic keys of Schaffner et al. [31] and Becker et al. [32]. Males' genitalia were mounted on a slide using Hoyer's medium. A specific study on the molecular identification of different specimens of each species was previously carried out in La Rioja region, in which individuals of *Anopheles maculipennis* s.l. were analysed using the internal transcribed spacer 2 (ITS2) gene and identified as *Anopheles atroparvus* (Van Thiel, 1927) [15]. For this reason, we will henceforth refer to *An. atroparvus* in the present study. However, it had not been possible to determine which species of the Claviger complex were present in the region [15]. Therefore, the specific protocol of Kampen et al. [33] based on the ITS2 gene was used to differentiate between *Anopheles claviger* s.s. (Meigen, 1804) and *Anopheles petragnani* Del Vecchio, 1939. Each species generates PCR products of species-specific lengths: *An. claviger* s.s. at 269 bp and *An. petragnani* at 367 bp. The amplified products were sequenced in both senses using the BigDye R Terminator v3.1 Cycle Sequencing Kit (Applied Biosystems, Forest City, CA, USA) at the Sequencing Unit, Centre for Biomedical Research of La Rioja (CIBIR). The sequences were edited using BioEdit 7.2 software and compared with the sequences published in Kampen et al. [33] and deposited in GenBank.

2.3. Data Analysis

The number of mosquitoes was compared by zone with full-factorial general linear models (GLMs) including zone and period to discard the effect of seasonality. In order to detect differences in the frequency of genera and relevant species depending on the zone, similar GLMs were carried out including the number of mosquitoes as a weighted factor. In case of significant differences between zones, Duncan's *post hoc* test was carried out. A statistical analysis was performed using SPSS 19.0 for Windows (IBM, Armonk, NY, USA), setting the alpha error to 0.05. Figure 1 was created by QGIS V.3.0.1 (QGIS Development Team 2018, available from: http://www.qgis.org, accessed on 15 May 2023). Figure 2 was generated with Microsoft Excel 2013 (Microsoft, Redmond, WA, USA). Figure 3 was created with the software RStudio v. 4.2.0 [34].

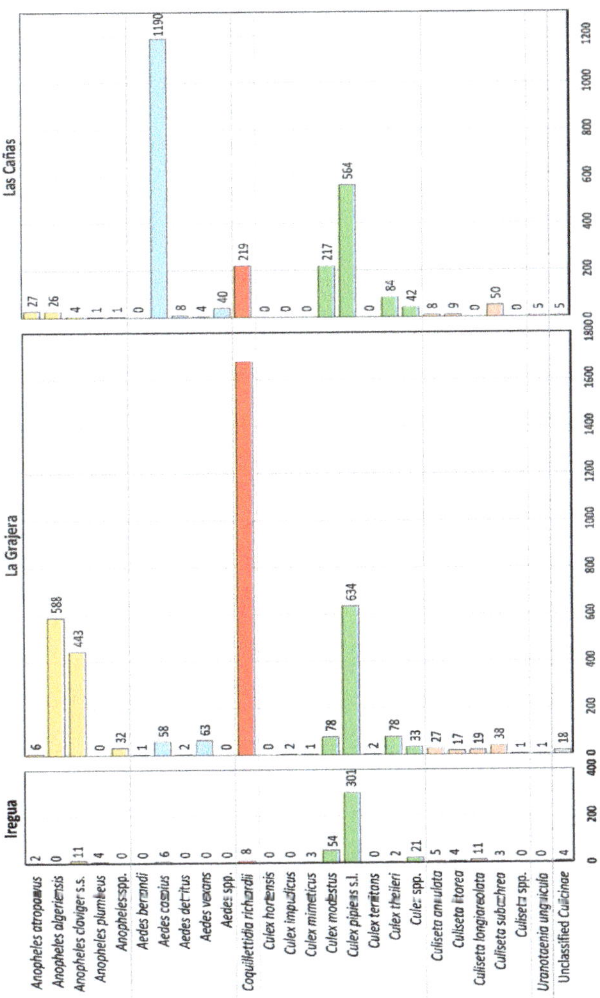

Figure 2. Total number of mosquitoes of each species caught with baited BG-Sentinel in aquatic ecosystems traps in northern Spain, 2016–2017. The specimens categorised as "spp" in each genus could not be assigned to species level because of their poor conservation status. Genera are represented with different colours.

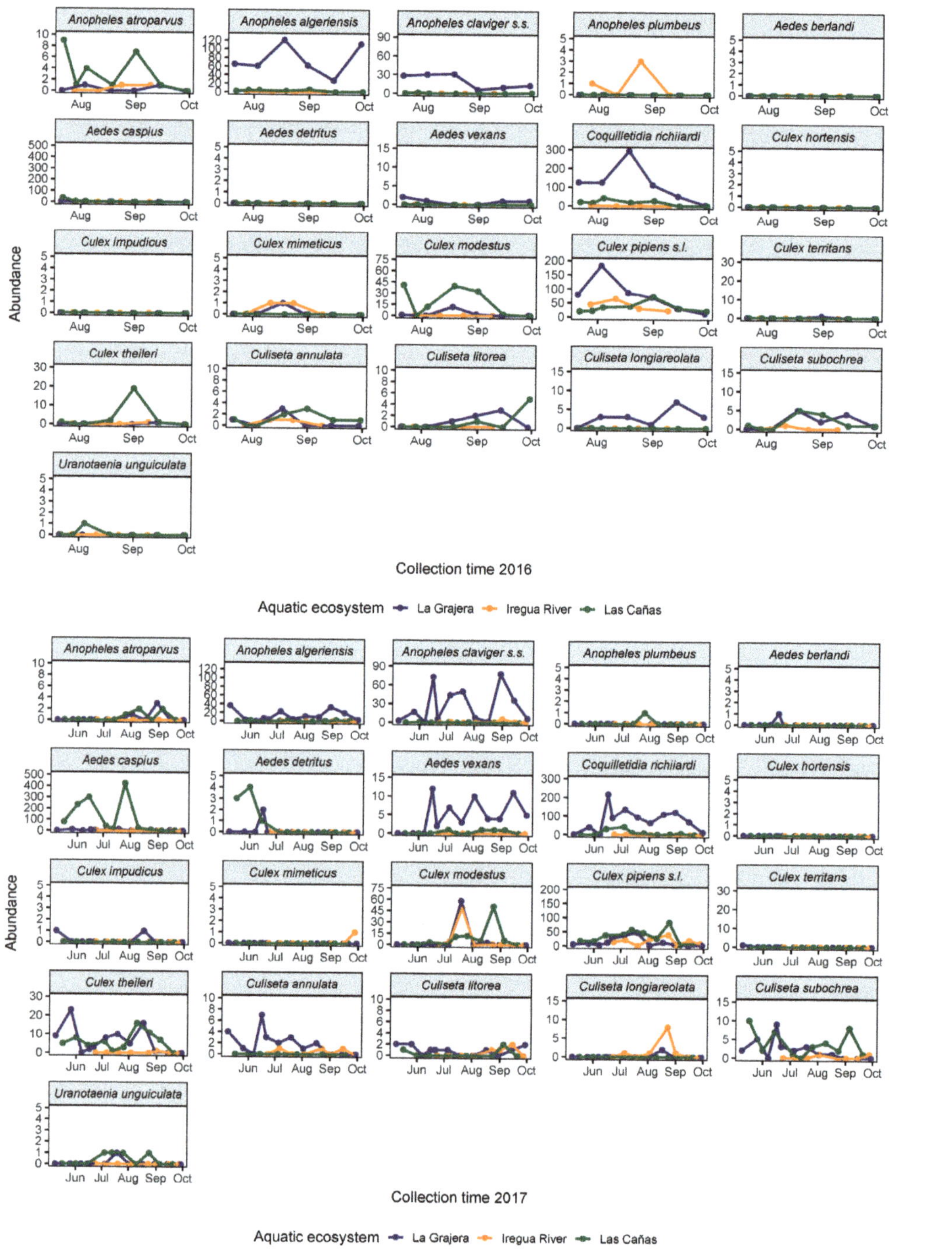

Figure 3. Population dynamic of each species caught by aquatic ecosystem with baited BG-Sentinel traps in aquatic ecosystems in northern Spain, 2016–2017 (Jun: June; Jul: July; Aug: August; Sep: September; Oct: October).

2.4. Ethics Approval and Consent to Participate

Ethical clearance for HLC was obtained from the regional ethics committee (Comité de Ética de Investigación con Medicamentos de La Rioja, Ref. CEImLAR PI-688). All procedures were in accordance with the ethical standards of the research committee and with the 1964 Helsinki Declaration and its later amendments.

3. Results

The morphological characteristics and genetic approach to the Claviger complex identified at least 24 taxa of mosquitoes arranged in six genera: *Aedes* (7 species), *Anopheles* (4 species), *Coquillettidia* (1 species), *Culex* (7 species), *Culiseta* (4 species), and *Uranotaenia* (1 species). The species and the number of specimens found in each aquatic ecosystem are displayed in Figure 2.

3.1. Molecular Identification of the Claviger Complex

An analysis of 17 *An. claviger* s.l. yielded 10 positive identifications for the species *An. claviger* s.s. as all bands were below 300 bp. The subsequent sequencing and alignment with the sequences published by Kampen et al. [33] reconfirmed that identification. Our sequences of *An. claviger* s.s. (amplicon length sequences between 235 and 238 bp) showed an identity of 96.93–98.92% with respect to the *Anopheles claviger* s.l. from Italy (MK625346). In turn, our sequences had a 95.61–96.89% similarity to the sequence of *An. claviger* s.s. (AY129232) deposited by Kampen et al. [33] and 86.67–87.56% similarity to the unique sequence of *An. petragnani* (AY129233.1) deposited in GenBank [33]. The detailed specimen records and sequence information of *An. claviger* s.s. were submitted to the GenBank public database under the following accession numbers: OQ955587–OQ955589.

3.2. Species Composition and Abundance

A total of 6735 mosquitoes were captured in the BG-Sentinel traps, of which 6479 were females and 256 males. The species composition and abundance varied depending on the aquatic ecosystem. La Grajera was the environment with the highest total catches and also with the highest relative abundances in relation to the sampling effort ($n = 3801$; 42.23 specimens/trap), followed by Las Cañas ($n = 2499$; 30.11 specimens/trap) and the Iregua River ($n = 435$; 8.70 specimens/trap) (Table 1). The most abundant genus was *Culex*, followed by *Coquillettidia richiardii* (Ficalbi, 1889), *Aedes* sp., and *Anopheles* sp.

Table 1. Total number of mosquitoes of each genus caught with baited BG-Sentinel traps in aquatic ecosystems in northern Spain, 2016–2017.

Genus	Iregua River ($n = 50$)	La Grajera ($n = 90$)	Las Cañas ($n = 83$)	Total ($n = 223$)
Anopheles	17	1069	59	**1145**
Aedes	6	124	1242	**1372**
Coquillettidia	8	1677	219	**1904**
Culex	381	828	907	**2116**
Culiseta	23	102	67	**192**
Uranotaenia	0	1	5	**6**
Unidentified	4	18	5	**27**
Total	**435**	**3801**	**2499**	**6735**

n indicates the sampling effort represented by the total number of traps used per aquatic ecosystem.

The abundance of the different genera varied according to the freshwater ecosystem. The genera *Anopheles* and *Coquillettidia* were dominant in La Grajera ($p < 0.001$), while the genera *Aedes*, *Culex*, and *Uranotaenia* dominated in Las Cañas ($p < 0.001$, $p = 0.007$, and $p = 0.047$, respectively) (Table 2). Catches of the genus *Culiseta* did not differ significantly by aquatic ecosystem. The average number of captures of the genera *Anopheles*, *Aedes*, and *Coquillettidia* showed large differences between the area with the highest number of captures and the rest of the areas. This was not the case for the genus *Culex*, where the averages were more homogeneous between the different zones.

Table 2. Mean ± standard deviation of catches of each mosquito genus with BG-Sentinel traps in aquatic ecosystems in northern Spain, 2016–2017.

Genus	Iregua River (n = 50)	La Grajera (n = 90)	Las Cañas (n = 83)	p-Value
Anopheles	0.34 ± 0.72 ^{a}	11.88 ± 16.26 ^{b}	0.71 ± 1.31 ^{a}	<0.001
Aedes	0.12 ± 0.33 ^{a}	1.38 ± 2.14 ^{a}	14.96 ± 36.70 ^{b}	<0.001
Coquillettidia	0.16 ± 0.47 ^{a}	18.63 ± 25.43 ^{b}	2.64 ± 3.90 ^{a}	<0.001
Culex	7.62 ± 8.73 ^{a}	9.20 ± 12.93 ^{ab}	10.93 ± 10.73 ^{b}	0.007
Culiseta	0.46 ± 1.16	1.13 ± 1.86	0.81 ± 1.57	0.080
Uranotaenia	0.00 ± 0.00 ^{a}	0.01 ± 0.11 ^{ab}	0.06 ± 0.24 ^{b}	0.047
All	8.70 ± 8.99 a	42.23 ± 42.41 c	30.11 ± 38.26 ^{b}	<0.001

n indicates the sampling effort represented by the total number of traps with mosquito catches used per aquatic ecosystem. Different superscripts for each gender indicate significant differences according to Duncan's *post hoc* test ($p < 0.05$).

Table 3 shows the percentage distribution of mosquitoes in each zone to see which was dominant relative to the total number of mosquitoes captured in each zone. For instance, in the Iregua River, the dominant genus was *Culex* sp. (87.6%) compared to the rest of the genera ($p < 0.001$). Although there were more *Culex* specimens in the other two aquatic ecosystems in absolute terms (Table 2), in relative terms they were less frequent (21.8% in La Grajera and 36.3% in Las Cañas) (Table 3). According to this analysis, the genera *Anopheles* and *Coquillettidia* remained dominant in relative terms in La Grajera ($p < 0.001$) and the genus *Aedes* in Las Cañas ($p < 0.001$). *Coquillettidia* was the genus with the highest relative captures in La Grajera (44.1%) and *Aedes* sp. in Las Cañas (49.7%).

Table 3. Mean ± standard deviation of the percentage of catches of each genus over the total catches per area (only if catches were present and weighted by the number of catches per trap) with baited BG-Sentinel traps in aquatic ecosystems in northern Spain, 2016–2017.

Genus	Iregua River (n = 47)	La Grajera (n = 88)	Las Cañas (n = 82)	p-Value
Anopheles	3.9 ± 8.8 ^{a}	28.1 ± 21.3 ^{b}	2.4 ± 5.2 ^{a}	<0.001
Aedes	1.4 ± 5.3 ^{a}	3.3 ± 5.0 ^{a}	49.7 ± 39.4 ^{b}	<0.001
Coquillettidia	1.8 ± 5.8 ^{a}	44.1 ± 24.1 ^{b}	8.8 ± 13.3 ^{a}	<0.001
Culex	87.6 ± 16.2 ^{c}	21.8 ± 20.2 ^{a}	36.3 ± 34.3 ^{b}	<0.001
Culiseta	5.3 ± 11.3	2.7 ± 4.0	2.7 ± 6.6	0.187
Uranotaenia	0.0 ± 0.0	0.0 ± 0.1	0.2 ± 1.0	0.226

n indicates the sampling effort represented by the total number of traps used per aquatic ecosystem. Different superscripts for each gender indicate significant differences according to Duncan's *post hoc* test ($p < 0.05$).

Table 4 shows the most captured species with respect to the captures of other species of the same genus in each sampling point. For instance, *Aedes caspius* (Pallas, 1771) represents all the captures of this genus in the Iregua River and 95.8% in Las Cañas ($p < 0.001$), while the Claviger complex represents 64.7% of the captures of the genus *Anopheles* in the Iregua River and 41.4% in La Grajera ($p < 0.001$). *Culex modestus* Ficalbi, 1890 accounted for 23.9% of the captures of *Culex* sp. in Las Cañas ($p = 0.007$). *Culex pipiens* s.l. Linnaeus, 1758 was the most abundant species of this genus in the three ecosystems, but it did not show significant differences according to the freshwater environment (p = 0.129). The same occurred with *Anopheles algeriensis* Theobald, 1903, being the most abundant *Anopheles* species in La Grajera and Las Cañas with 55% and 44.1%, respectively.

Table 4. Mean ± standard deviation of the percentage of catches of the most representative mosquito species with respect to catches of its genus (only if there were catches of that genus and weighted by the no. of catches of the corresponding genus per trap) with baited BG-Sentinel traps in aquatic ecosystems in northern Spain, 2016–2017.

	Iregua River		**La Grajera**		**Las Cañas**		
Species	n	**Avg ± SD**	n	**Avg ± SD**	n	**Avg ± SD**	***p*-Value**
Anopheles algeriensis	12	0.0 ± 0.0	77	55.0 ± 34.0	30	44.1 ± 42.0	0.123 [1]
Anopheles claviger s.s.	12	64.7 ± 49.3^{b}	77	41.4 ± 34.9^{b}	30	6.8 ± 21.0^{a}	0.032
Aedes caspius	6	100.0 ± 0.0^{b}	46	46.8 ± 39.5^{a}	51	95.8 ± 5.8^{b}	<0.001
Culex pipiens s.l.	47	79.0 ± 33.3	73	76.6 ± 31.4	79	62.2 ± 26.5	0.129
Culex modestus	47	14.2 ± 33.2^{a}	73	9.4 ± 20.1^{a}	79	23.9 ± 25.8^{b}	0.007

[1] The Iregua River was excluded from statistical analysis. n indicates the sampling effort represented by the total number of traps with catches of each genus per aquatic ecosystem. Different superscripts for each species indicate significant differences according to Duncan's *post hoc* test ($p < 0.05$).

3.3. Population Dynamics in the Peak Abundance Period

Different patterns in the population dynamics of culicids were observed depending on the species and aquatic ecosystem (Figure 3).

Two of the most abundant species in La Grajera, *An. algeriensis* and *An. claviger* s.s., had a constant presence during the study period, even presenting similar dynamics with three peaks of abundance in July, August, and October 2016 and another three peaks of abundance in May–June, July, and September–October 2017. Within the Maculipennis complex, *An. atroparvus* was the only species previously identified in the study area [15]. This species was the only anopheline species with a very limited temporal distribution, between mid-July and September in La Grajera and Las Cañas. The abundance of *Anopheles plumbeus* Stephens, 1828 was so low that no clear patterns in its population dynamic could be observed, except that it was present during July and August.

Aedines did not have a constant presence in the aquatic ecosystems studied, but rather had peaks of abundance depending on the time of year. For example, there were hardly any captures of *Ae. caspius* in Las Cañas in 2016, but there were two notable peaks of abundance between June and August 2017. *Aedes detritus* (Haliday, 1833) was a species that concentrated its presence in the late spring and early summer in 2017 in the two swamps where it was identified. The presence of *Aedes vexans* (Meigen, 1830) seemed to be constant from June onwards in La Grajera.

The presence of the most abundant species at La Grajera, *Cq. richiardii*, was constant during the two years of this study. However, its abundance peaks differed between the two years. The main peak in 2016 was in August, while in 2017 we found four peaks from May to September. Despite its lower presence in Las Cañas, two peaks were observed during August and September 2016 and a single peak in June–July 2017.

Culex pipiens s.l. was the only species present throughout the study period in all three habitats. A similar pattern was observed in all three studied ecosystems, with a peak in July and a peak in August. In Las Cañas, a third peak was also observed during September. Adults of *Cx. modestus* appeared in June, with two peaks of abundance in July and August–September. *Culex theileri* Theobald, 1903, also showed several peaks of abundance depending on the aquatic ecosystem, with a clear peak in August–September in all habitats.

Adults of *Culiseta* species were also active throughout the study period, except for the species *Culiseta longiareolata* (Macquart, 1838). The other species, either because they were secondary species or because of their own population patterns, had a temporal distribution that was not constant over time.

Uranotaenia unguiculata Edwards, 1913 appeared mainly from July to September.

3.4. Human Landing Collection

Six species were collected in the HLC: *An. plumbeus* (*n* = 1, Iregua River); *Ae. caspius* (*n* = 26, Las Cañas; *n* = 2, La Grajera); *Ae. detritus* (*n* = 1, Las Cañas; *n* = 19, La Grajera); *Ae. vexans* (*n* = 1, La Grajera); *Cx. modestus* (*n* = 2, Iregua River); and *Cq. richiardii* (*n* = 2, Las Cañas; *n* = 4, La Grajera). Specimens of *Ae. detritus* were identified biting humans in La Grajera on 24 May 2016 during the first aquatic ecosystem awareness visit.

4. Discussion

Understanding the risk of transmission of mosquito-borne diseases requires knowledge of the culicid fauna associated with environments frequented by humans, such as urban and peri-urban areas. This knowledge serves as a fundamental tool and cornerstone of the "One Health" approach. Therefore, this study highlights the large mosquito fauna that inhabit peri-urban environments in northern Spain.

In Spain, a total of 65 mosquito species from seven different genera have been described, including *Aedes*, *Anopheles*, and *Culex* [9,35], the main vectors of pathogens in Europe. In La Rioja, the culicid fauna identified include 25 species in six genera: *Aedes*, *Anopheles*, *Coquillettidia*, *Culex*, *Culiseta*, and *Uranotaenia* [15,16,36–38]. In Navarra, 21 species belonging to the same six genera have been recorded [14,35].

4.1. Molecular Identification of the Claviger Complex

PCR has been employed to specifically identify members of the Claviger complex, marking a novel application in Spain. We also provide sequences of *An. claviger* s.s. for the country. Moreover, the identification of this anopheline implies its incorporation into the list of human malaria vectors in La Rioja, together with *An. atroparvus* and *An. plumbeus*. Although *An. claviger* s.s. is regarded as a secondary malaria vector in Europe [39], with a preference for biting humans and large mammals outdoors after dusk [40], its vector competence should not be overlooked. In the case of *An. petragnani*, although we did not capture any adults in our study, it has previously been described in the larval stage in La Rioja [37].

4.2. Species Composition and Abundance

The composition and abundance of mosquito fauna in these three closely spaced habitats varied according to the type of aquatic ecosystem and the hydrological management practices employed. La Grajera, a naturalised reservoir with minimal human intervention, had the highest number of mosquito captures. In Las Cañas, naturalised conservation of the swamp is carried out by managing the reeds and promoting the conservation of native fish species and amphibians, which exert some degree of control over the pre-imaginal populations of mosquitoes. The Iregua River, on the other hand, lacks standing water bodies, so its mosquitoes depend mainly on the floods that allow parts of its groves to be flooded. The genera *Anopheles* and *Coquillettidia* dominated in La Grajera due to its stable water level during the summer season. *Coquillettidia richiardii* is also a dominant species in other wetlands of northern Spain [18]. In contrast, *Aedes* was the predominant species in Las Cañas due to its high-water fluctuation, which results in a considerable drop in water levels during summer. *Aedes* species like *Aedes rusticus* (Rossi, 1790) are also dominant in other wetlands in northern Spain [18]. The most abundant genus, *Culex*, was found to be well represented in the two reservoirs, despite being dominant only in the Iregua River. This may be attributed to the fact that *Cx. pipiens* s.l. is a species adapted to breeding in different types of natural breeding sites, including riverbanks, extremely eutrophic and polluted waters, habitats with abundant invasive hydrophytes [41], and even artificial reservoirs in urban areas [42].

Anopheles atroparvus was the least commonly trapped *Anopheles* species; nevertheless, multiple resting captures at various locations in La Grajera suggest that it may be more prevalent (Ruiz-Arrondo, unpublished data). Regarding *An. algeriensis*, this species historically inhabits the Mediterranean coast, where it commonly breeds in coastal marshy areas and saline or oligosaline channels with abundant helophytic vegetation [43]. Although this anopheline species has also been found in the province of Teruel [35], its detection in northern Spain is of particular interest, as it is far from its known distribution area [15].

Aedes caspius appeared as the most abundant species of *Aedes* in both the Iregua River and Las Cañas, although it did not represent more than 50% of the total of aedines in La Grajera due to the presence of *Ae. vexans* and *Ae. detritus*. *Aedes caspius* is one of the most abundant species on the Spanish Mediterranean coast, and *Ae. detritus* is also frequent [44]. Regarding the invasive *Aedes* mosquitoes, these species have adapted to breed in a variety of man-made and disused containers in urban/rural areas [45]. Although natural ecosystems are not favourable for the development of certain species such as *Ae. albopictus*, proactive surveillance remains crucial in preventing potential introductions and effectively managing mosquito-related risks. In fact, in La Rioja and Navarra regions, specific entomological surveillance programmes employ ovitraps and citizen science applications for invasive *Aedes* mosquito detection [14,16].

A potential limitation of this study was the use of a single trap model, namely BG-Sentinel traps, to investigate the diversity and abundance of mosquito populations in these aquatic ecosystems. In addition to the HLC method, other approaches such as the capture of adult resting mosquitoes and sampling of immature stages should be employed to address this limitation and obtain a more comprehensive understanding of the local mosquito fauna.

4.3. Population Dynamics in the Peak Abundance Period

The two most abundant anophelines, *An. algeriensis* and *An. claviger* s.s., showed a constant abundance throughout the summer season. *Anopheles claviger* s.l. has also been observed as a multivoltine species in northern Spain [18]. Schaffner et al. [31] described *An. claviger* s.s. as bivoltine, once in the early spring and once in the late summer. According to our findings, this species would have one more generation between those two periods. The same was observed for *An. algeriensis*, where the peak population periods occur during the spring and autumn [31], and yet our observations point to another peak of abundance during the summer.

The population dynamic of certain aedine species might be influenced by the availability of suitable breeding sites, especially those that are temporarily flooded, as is the case with *Ae. caspius*, *Ae. detritus*, and *Ae. vexans*, which are commonly referred to as "open floodwater species". The alternation of flooding and drought creates favourable conditions for the high-density hatching of these mosquitoes, so environmental factors such as rainfall significantly influence their population dynamics [46]. In fact, the amount and distribution of rain have a powerful effect on mosquitoes' population dynamics in species that overwinter as drought-resistant eggs. *Aedes caspius* preferentially oviposits in moister sites with abundant vegetation, while *Ae. detritus* oviposits in more saline soils [47]. However, their larvae can also be found on the margins of permanent bodies of water such as rivers, along with in trees, reeds, and canes [48]. Although rarely, we also captured adult *Ae. caspius* in the Iregua River in our study. In the Levante region, both species are closely linked to temporary bodies of water, especially in the coastal area. For example, it can be observed that the peaks of *Ae. caspius* capture, a multivoltine species (with an annual average of 14 generations in Huelva, Spain [44]), coincide with the periods close to the equinoxes, when the highest rainfall occurs in this area of eastern Spain. In the case of *Ae. detritus*, the difference between the spring and autumn capture peaks is more pronounced [49], a phenomenon also observed in other southern European countries, suggesting a possible univoltinism of the species [31]. However, on the south-western coast of Spain, they have a multivoltine character, with an annual average of five to seven generations between the spring and autumn [44]. Despite the low capture rates of this species, our results support this hypothesis, as only one peak of activity was observed between June and July in the two lotic water bodies sampled and bites occurring on human bait by *Ae. detritus* in spring 2016 in La Grajera. The same flight range for *Ae. detritus* has been noted in northern Spain [18]. In the marshy areas typical of the Valencian Community, after prolonged periods of rainfall, there are waterlogged soils with salinity conditions that greatly favour the hatching of larvae of both species, although with a predominance of *Ae. caspius*, with an increase in the abundance of this species observed in May and September [50], similar to what was observed in 2017 in La Grajera and Las Cañas. Regarding *Ae. vexans*, the seasonal population dynamic of this floodwater-associated aedine species shows a typical multivoltine pattern of abundance in European countries [32,51] and specifically in Spain [48], which is in line with our results.

Coquillettidia richiardii has several generations per year in La Grajera, a behaviour described in southern European populations [32]. This species shows at least four peaks of abundance during the summer, although a reduction in the population is observed at the end of the summer [31]. Nevertheless, Gonzalez et al. [18] noted its univoltine character in an aquatic ecosystem in northern Spain.

Culex pipiens s.l., a multivoltine species, shows the highest population peak in August, with the number of individuals decreasing progressively on each capture date. Although *Cx. modestus* is not widely represented in the wetland, two distinct peaks of abundance are observed during the summer, in line with the observations of other authors [18,31,32]. *Culex theileri* is a polycyclic species whose larvae occur in the spring [32]. In our environment, its abundance differs from one year to another; however, in 2017, its multivoltine character was observed with three distinct maxima in the two reservoirs. Meanwhile, in other Spanish regions such as Doñana and the area of influence of the Lower Guadalquivir rice fields (south-western Spanish coast), it is by far the most frequent and abundant species in the spring months [44].

The few individuals of *Ur. unguiculata* captured were concentrated during the central part of the summer, although Becker et al. [32] explained that this species is more abundant in the late summer.

We did not evaluate the population dynamics of Culicidae during May and June in 2016; therefore, the population dynamics of the species is not completed for the warm season of this year.

4.4. Human Landing Collection: Epidemiological Implications of the Identified Species

Human landing collection is considered the current "gold standard" for estimating human-biting rates, but it is labour-intensive and implies the risk of exposure to infectious mosquito bites [52]. This could explain why this technique has been so scarcely used in Spain to date [18], this being one of the first works to incorporate it. However, this technique has been used for the study of other Diptera such as simulids in the country [53]. The chance of being bitten by a mosquito in the three aquatic ecosystems studied is not very high, except during the spring hatching of *Ae. detritus* at La Grajera and the peaks of *Ae. caspius* abundance in the summer at Las Cañas.

The species most abundantly caught using HLC were *Ae. caspius* (n = 28) and *Ae. detritus* (n = 20), two crepuscular and anthropophilic mosquito species. These aedines are known for their aggressive biting behaviour towards humans and animals, both during the day and at night [54–56]. In Spain, there are serious biting nuisances for humans in certain regions such as the Levante, the eastern region of the Iberian Peninsula, where these anthropophilic species breed in flooded lagoons and coastal marshes near touristic areas [57]. Although *Ae. caspius* exhibits a preference for mammalian hosts, it has been demonstrated that they can also feed on birds, albeit when alternative host classes are not readily available [58]. In relation to disease transmission, invasive *Aedes* are of the greatest importance, as they are responsible for autochthonous cases of arbovirus, such as dengue virus (DENV) [59]. However, the potential role of indigenous species in the transmission of other pathogens of medical and veterinary importance in Spain should not be underestimated. Previous studies revealed that *Ae. caspius* populations from the Camargue (region of southern France) possess the potential to act as vectors of chikungunya virus (CHIKV) [60] and Rift Valley fever virus (RVFV) [61]. However, *Ae. caspius* populations from the Camargue and Andalusia regions (Spain) were incapable of transmitting West Nile virus (WNV) [62] and Zika virus (ZIKV) [58,63].

Coquillettidia richiardii was the third most abundant species caught with HLC (n = 6). This member of the Mansoniini feeds opportunistically on vertebrates and can naturally transmit Batai virus (BATV), Tahyna Orthobunyavirus (TAHV), Sindbis virus (SINV), and WNV [31]. With respect to its anthropophily, *Cq. richiardii* is confirmed as a species that eventually bites humans in Spain. Its anthropophily in our study is not as marked as in the region of the Danube Delta [64] or in southern England, where this species was responsible for the highest recorded biting rate in an HLC assay in the UK, with up to 161 bites per hour [65].

The low capture rate of *Culex*, with only two specimens of *Cx. modestus* caught, may be attributed to the species' preference for nocturnal activity over crepuscular activity (which was when the captures were made). Despite its abundance in the BG traps, no *Cx. pipiens* s.l. was captured in the HLC. However, it is essential not to underestimate the anthropophilic preferences and vectorial capacity of *Culex* mosquitoes. For example, the recent WNV outbreak in south-western Spain in 2020 highlighted the central role played by this genus in the transmission of this zoonotic arboviral disease [66]. Specifically, *Culex perexiguus* Theobald, 1903 and *Cx. pipiens* s.l. played a central role in the transmission of this arbovirus [67]. The apparent ornithophilic character of this latter species in these aquatic ecosystems means that the risk of human cases of WNV by this bridge vector might be lower.

As far as the genus *Anopheles* is concerned, only one specimen of *An. plumbeus* was captured in the HLC. This species is an aggressive and persistent biter, feeding principally on mammalian hosts [52] but also on birds and reptiles. Some populations even show a strong anthropophilic preference [68]. In fact, this species is usually found in urban environments, where immatures develop in tree holes in gardens and parks, causing a serious nuisance in some countries of Central Europe [31]. In our peri-urban environment, *An. plumbeus* does not pose a great risk due to its limited abundance. However, because it was the only species of anopheline captured in the HLC, it could be the species attributed as a vector in the second case of autochthonous malaria in the country, which occurred in the study area in 2014 [14]. It should be noted that despite the relative abundance of *An. claviger* s.s. in our study, it has not been a species implicated in human disturbance as González et al. [18] noted in their study in northern Spain. This behaviour might be attributed to the possibility that they are, in fact, *An. petragnani* rather than *An. claviger* s.s.

Another *Anopheles* species that in principle does not pose an obvious risk to the human population in the study area is *An. maculipennis* s.l., more especially, *An. atroparvus* as the only member of this complex identified molecularly in this peri-urban area [15]. This species played an important role in human malaria transmission in Europe. In fact, it was the species responsible for the first autochthonous case of malaria that occurred in northern Spain in early 2010 [69].

Public reports of mosquito bites in the urban area are rare and, in all cases, have been due to localised *Aedes* breeding sites. This means that a priori the mosquito populations in the ecosystems studied do not cause nuisance in the urban area. The reasons could be several: (1) the large population of mammals and birds in the two reservoirs could provide a food source for the mosquitoes without them having to look for hosts outside the area of influence of the aquatic ecosystems; (2) the low productivity of its breeding sites in the Iregua River prevents the massive emergence of species that could cause problems in the urban area; (3) in the case of Las Cañas, there is an industrial area between the reservoir and the urban environment which may be acting as a barrier to the natural dispersal of the mosquito population; and (4) not all species detected in the HLC have enough flight range to reach the urban area, with the exception of some aedine species.

These results on mosquito species composition and population dynamics during the peak abundance period contribute to a better understanding of the transmission risk of pathogenic agents of medical and veterinary importance. This study demonstrates the importance of monitoring the mosquito fauna in naturalised environments in areas very close to the city, such as peri-urban environments. The large influx of people enjoying leisure time in these environments and the constant presence of several species with vectorial potential in the summer season favours encounters between vectors and their hosts, highlighting the importance of including peri-urban areas in arthropod vector monitoring programmes.

5. Conclusions

This study highlights the large culicid fauna that inhabit peri-urban environments in northern Spain. Among the 24 taxa identified in the three aquatic environments, at least 5 are of medicalveterinary importance dueto their aggressiveness, distribution, and abundance. The different hydrological management practices could play a key role in determining the abundance of mosquito genera. Our study can provide valuable information to guide public health policies and strategies aimed at controlling mosquito populations and preventing disease outbreaks in the future. Moreover, peri-urban areas are believed to pose a higher risk of zoonotic disease transmission to humans due to their location on the fringes of urban areas and their proximity to farmland, freshwater ecosystems, and other wildlife habitats, as observed in our study. This underscores the significance of such investigations for public health.

Author Contributions: I.R.-A. and J.A.O. designed the study. I.R.-A. conducted the field work. I.R.-A. identified the females and performed the molecular identification. L.B.-S. and S.D.-E. identified the males. I.d.B. and I.R.-A. performed the statistical analyses. L.B.-S. developed Figures 1 and 2. I.R.-A. and P.M.A.-E. compiled the main information and wrote the first draft of the manuscript. All authors have read and agreed to the published version of the manuscript.

Funding: This research was funded by the European Regional Development Fund (ERDF). Partially funded by the Instituto de Estudios Riojanos in its 2019–2020 call for proposals.

Institutional Review Board Statement: The study was conducted in accordance with the Declaration of Helsinki and approved by the Ethics Committee (Comité de Ética de Investigación con medicamentos de La Rioja) (Ref. CEImLAR PI-688).

Informed Consent Statement: Informed consent was obtained from all subjects involved in the study.

Data Availability Statement: Not applicable.

Acknowledgments: We are grateful to Mikel A. González for his comments about the manuscript. We appreciate the authorities of Logroño and Gobierno de Navarra for providing the permits to conduct the fieldwork in La Grajera and Iregua River and Las Cañas reservoirs respectively.

Conflicts of Interest: The authors declare no conflict of interest.

References

1. Benelli, G.; Senthil-Nathan, S. Together in the Fight against Arthropod-Borne Diseases: A One Health Perspective. *Int. J. Environ. Res. Public Health* **2019**, *16*, 4876. [CrossRef] [PubMed]
2. Portillo, A.; Ruiz-Arrondo, I.; Oteo, J.A. Artrópodos vectores en España y sus enfermedades transmisibles. *Med. Clin.* **2018**, *151*, 450–459. [CrossRef] [PubMed]
3. Lucientes, J.; Alarcón-Elbal, P.M. *Culicoides* biting midges in Spain: A brief overview. *Small Rum. Res.* **2016**, *142*, 69–71. [CrossRef]
4. Bravo-Barriga, D.; Ruiz-Arrondo, I.; Peña, R.E.; Lucientes, J.; Delacour-Estrella, S. Phlebotomine sand flies (Diptera, Psychodidae) from Spain: An updated checklist and extended distributions. *ZooKeys* **2022**, *1106*, 81–99. [CrossRef] [PubMed]

5. Ruiz-Arrondo, I.; Alarcón-Elbal, P.M.; Figueras, L.; Delacour-Estrella, S.; Muñoz, A.; Kotter, H.; Pinal, R.; Lucientes, J. Expansión de los simúlidos (Diptera: Simuliidae) en España: Un nuevo reto para la salud pública y la sanidad animal. *Bol. SEA* **2014**, *54*, 193–200. Available online: http://sea-entomologia.org/PDF/Boletin54/193200BSEA54ExpansionsimulidosenEspa%C3%B1a.pdf (accessed on 15 May 2023).
6. Kerkow, A.; Wieland, R.; Früh, L.; Hölker, F.; Jeschke, J.M.; Werner, D.; Kampen, H. Can data from native mosquitoes support determining invasive species habitats? Modelling the climatic niche of *Aedes japonicus japonicus* (Diptera, Culicidae) in Germany. *Parasitol. Res.* **2020**, *119*, 31–42. [CrossRef]
7. Collantes, F.; Delgado, J.A.; Alarcón-Elbal, P.M.; Delacour, S.; Lucientes, J. First confirmed outdoor winter reproductive activity of Asian tiger mosquito (*Aedes albopictus*) in Europe. *An. Biol.* **2014**, *36*, 71–76. [CrossRef]
8. Delacour, S.; Barandika, J.F.; García-Pérez, A.L.; Collantes, F.; Ruiz Arrondo, I.; Alarcón-Elbal, P.M.; Bengoa, M.; Delgado, J.A.; Juste, R.A.; Molina, R.; et al. Detección temprana del mosquito tigre (*Aedes albopictus* Skuse, 1894) en el País Vasco (España). *An. Biol.* **2015**, *37*, 25–30. [CrossRef]
9. Eritja, R.; Ruiz-Arrondo, I.; Delacour-Estrella, S.; Schaffner, F.; Álvarez-Chachero, J.; Bengoa, M.; Puig, M.Á.; Melero-Alcíbar, R.; Oltra, A.; Bartumeus, F. First detection of *Aedes japonicus* in Spain: An unexpected finding triggered by citizen science. *Parasites Vectors* **2019**, *12*, 53. [CrossRef]
10. Eritja, R.; Delacour-Estrella, S.; Ruiz-Arrondo, I.; González, M.A.; Barceló, C.; García-Pérez, A.L.; Lucientes, J.; Miranda, M.Á.; Bartumeus, F. At the tip of an iceberg: Citizen science and active surveillance collaborating to broaden the known distribution of *Aedes japonicus* in Spain. *Parasites Vectors* **2021**, *14*, 375. [CrossRef]
11. Martínez-Barciela, Y.; González, A.P.; Rial, D.G.; González, J.G. First records of five species of mosquitoes (Diptera: Culicidae) in Galicia, including the first evidence of the genus *Coquillettidia* in northwestern Spain. *J. Vector Ecol.* **2021**, *46*, 96–102. [CrossRef] [PubMed]
12. González, M.A.; Cevidanes, A.; Goiri, F.; Barandika, J.F.; García-Pérez, A.L. Diversity and distribution of larval habitats of mosquitoes (Diptera: Culicidae) in northern Spain: From urban to natural areas. *J. Vector Ecol.* **2021**, *46*, 173–185. [CrossRef]
13. González, M.A.; Delacour-Estrella, S.; Bengoa, M.; Barceló, C.; Bueno-Marí, R.; Eritja, R.; Ruiz-Arrondo, I. A Survey on Native and Invasive Mosquitoes and Other Biting Dipterans in Northern Spain. *Acta Parasitol.* **2022**, *67*, 867–877. [CrossRef] [PubMed]
14. Ruiz-Arrondo, I.; McMahon, B.J.; Hernández-Triana, L.M.; Santibañez, P.; Portillo, A.; Oteo, J.A. Surveillance of Mosquitoes (Diptera, Culicidae) in a Northern Central Region of Spain: Implications for the Medical Community. *Front. Vet. Sci.* **2019**, *6*, 86. [CrossRef]
15. Ruiz-Arrondo, I.; Hernández-Triana, L.M.; Nikolova, N.I.; Fooks, A.R.; Oteo, J.A. Integrated Approaches in Support of Taxonomic Identification of Mosquitoes (Diptera: Culicidae) in Vector Surveillance in Spain. *Vector-Borne Zoonotic Dis.* **2020**, *20*, 831–842. [CrossRef] [PubMed]
16. Ruiz-Arrondo, I.; Delacour-Estrella, S.; Santibáñez, P.; Oteo, J.A. Primera detección del mosquito tigre, *Aedes albopictus* (Diptera: Culicidae), en La Rioja: Implicaciones en salud pública. *An. Biol.* **2021**, *43*, 117–122. [CrossRef]
17. Heym, E.C.; Kampen, H.; Walther, D. Mosquito species composition and phenology (Diptera, Culicidae) in two German zoological gardens imply different risks of mosquito-borne pathogen transmission. *J. Vector Ecol.* **2018**, *43*, 80–88. [CrossRef] [PubMed]
18. González, M.A.; Goiri, F.; Cevidanes, A.; Hernández-Triana, L.M.; Barandika, J.F.; García-Pérez, A.L. Mosquito community composition in two major stopover aquatic ecosystems used by migratory birds in northern Spain. *Med. Vet. Entomol.* **2023**, *37*, 616–629. [CrossRef]
19. González, M.; López, S.; Alarcón-Elbal, P.M. Blood-feeding Diptera (Culicidae and Ceratopogonidae) in an urban park of the city of Vitoria-Gasteiz (Basque Country, Spain). *J. Eur. Mosq. Control Assoc.* **2015**, *33*, 10–14. Available online: https://e-m-b.myspecies.info/sites/e-m-b.org/files/JEMCA%2833%2910-14.pdf (accessed on 10 April 2023).
20. Ferraguti, M.; Martínez-de la Puente, J.; Roiz, D.; Ruiz, S.; Soriguer, R.C.; Figuerola, J. Effects of landscape anthropization on mosquito community composition and abundance. *Sci. Rep.* **2016**, *6*, 29002. [CrossRef]
21. Rueda, J.; Hernández, R.; Alarcón-Elbal, P.M. Nuevas aportaciones al conocimiento de dos especies de culícidos agrestes de España: *Aedes (Rusticoidus) refiki* Medschid, 1928 y *Aedes (Ochlerotatus) pullatus* (Coquillett, 1904) (Diptera, Culicidae). *An. Biol.* **2017**, *39*, 191–197. [CrossRef]
22. Honnen, A.C.; Monaghan, M.T. City-Dwellers and Country Folks: Lack of Population Differentiation Along an Urban–Rural Gradient in the Mosquito *Culex pipiens* (Diptera: Culicidae). *J. Insect Sci.* **2017**, *17*, 107. [CrossRef]
23. Medlock, J.M.; Vaux, A.G.C. Seasonal dynamics and habitat specificity of mosquitoes in an English wetland: Implications for UK wetland management and restoration. *J. Vector Ecol.* **2015**, *40*, 90–106. [CrossRef] [PubMed]
24. Bravo-Barriga, D.; Gomes, B.; Almeida, A.P.G.; Serrano-Aguilera, F.J.; Pérez-Martín, J.E.; Calero-Bernal, R.; Reina, D.; Frontera, E.; Pinto, J. The mosquito fauna of the western region of Spain with emphasis on ecological factors and the characterization of *Culex pipiens* forms. *J. Vector Ecol.* **2017**, *42*, 136–147. [CrossRef] [PubMed]
25. González, M.A.; Prosser, S.W.; Hernández-Triana, L.M.; Alarcón-Elbal, P.M.; López, S.; Ruiz-Arrondo, I.; Hebert, P.D.; García-Pérez, A.L. Avian feeding preferences of *Culex pipiens* and *Culiseta* spp. along an urban-to-wild gradient in northern Spain. *Front. Ecol. Evol.* **2020**, *8*, 568835. [CrossRef]
26. Casades-Martí, L.; Frías, M.; Delacour, S.; Ruiz-Fons, F. Confirmed presence of *Aedes (Rusticoidus) refiki* Medschid, 1928 in a continental dry Mediterranean peri-urban environment in south-central Spain. *BMC Zool.* **2022**, *7*, 21. [CrossRef]

27. Climate Data. Website. Available online: https://es.climate-data.org/europe/espana/la-rioja/logrono-3148/ (accessed on 10 April 2023).
28. Confederación Hidrográfica del Ebro. Río Iregua. 2023. Available online: https://www.chebro.es/en/rio-iregua (accessed on 15 May 2023).
29. La Rioja Turismo. *Parque de La Grajera*. 2023. Available online: https://lariojaturismo.com/lugar-de-interes/parque-de-la-grajera/9e375360-9c84-6db4-aa4f-0697115b3439 (accessed on 15 May 2023).
30. Navarra Espacios Naturales. *Embalse de Las Cañas*. 2018. Available online: https://espaciosnaturales.navarra.es/es/embalse-de-las-canas (accessed on 10 April 2023).
31. Schaffner, F.; Angel, G.; Geoffroy, B.; Hervy, J.O.; Rhaeim, A. *The Mosquitoes of Europe*; CD-Rom; IRD Éditions and EID Méditerranée: Montpellier, France, 2001.
32. Becker, N.; Petric, D.; Zgomba, M.; Boase, C.; Madon, M.; Dahl, C.; Kaiser, A. *Mosquitoes and Their Control*, 2nd ed.; Springer: Berlin/Heidelberg, Germany, 2010.
33. Kampen, H.; Sternberg, A.; Proft, J.; Bastian, S.; Schaffner, F.; Maier, W.A.; Seitz, H.M. Polymerase chain reaction-based differentiation of the mosquito sibling species *Anopheles claviger* S.S. and *Anopheles petragnani* (Diptera: Culicidae). *Am. J. Trop. Med. Hyg.* **2003**, *69*, 195–199. [CrossRef]
34. R Core Team. *R: A Language and Environment for Statistical Computing*; R Foundation for Statistical Computing: Vienna, Austria, 2022; Available online: https://www.R-project.org/ (accessed on 10 April 2023).
35. Bueno-Marí, R.; Bernués-Bañeres, A.; Jiménez-Peydró, R. Updated checklist and distribution maps of mosquitoes (Diptera: Culicidae) of Spain. *J. Eur. Mosq. Control Assoc.* **2012**, *30*, 91–126. Available online: https://e-m-b.myspecies.info/sites/e-m-b.org/files/EMB%2830%2991-126-REVISED.pdf (accessed on 10 April 2023).
36. Alarcón-Elbal, P.M.; Delacour-Estrella, S.; Ruiz-Arrondo, I.; Pinal, R.; Muñoz, A.; Oropeza, V.; Carmona-Salido, V.J.; Estrada, R.; Lucientes, J. Los culícidos (Diptera Culicidae) del valle medio del Ebro I: La Rioja (Norte de España). *Bol. SEA* **2012**, *50*, 359–365. Available online: http://sea-entomologia.org/Publicaciones/PDF/BOLN_50/359365BSEA50CulicidaeEbroI.pdf (accessed on 15 May 2023).
37. Bueno Marí, R. Estudio faunístico y eco-epidemiológico de los mosquitos (Diptera, Culicidae) de La Rioja. *Zubía* **2012**, *30*, 141–161. Available online: https://dialnet.unirioja.es/descarga/articulo/6868149.pdf (accessed on 15 May 2023).
38. Ruiz-Arrondo, I.; Hernández Triana, L.; Oteo Revuelta, J.A. Fauna de mosquitos (Diptera, Culicidae) presentes en el humedal de La Grajera (Logroño) y sus implicaciones en Salud Pública. *Zubía* **2017**, *35*, 123–140. Available online: https://dialnet.unirioja.es/descarga/articulo/6338493.pdf (accessed on 10 April 2023).
39. Zamburlini, R.; Cargnus, E. Anofelismo residuo nel litorale altoadriatico a 50 anni dalla scomparsa della malaria [Residual mosquitoes in the northern Adriatic seacoast 50 years after the disappearance of malaria]. *Parassitologia* **1998**, *40*, 431–437. [PubMed]
40. Walter Reed Biosystematics Unit. *Anopheles claviger* Species Page. Walter Reed Biosystematics Unit. 2023. Available online: https://wrbu.si.edu/vectorspecies/mosquitoes/claviger (accessed on 15 May 2023).
41. Alarcón-Elbal, P.M.; Sánchez Murillo, J.M.; Delacour Estrella, S.; Ruiz Arrondo, I.; Pinal Prieto, R.; Lucientes Curdi, J. Asociación de vector del VNO e hidrófito invasor: *Culex pipiens* Linnaeus, 1758 y *Ludwigia grandiflora* (Michaux) Greuter and Burdet en el marjal de Xeraco-Xeresa, Valencia. *An. Biol.* **2013**, *35*, 17–27. [CrossRef]
42. Gangoso, L.; Aragonés, D.; Martínez-de la Puente, J.; Lucientes, J.; Delacour-Estrella, S.; Estrada Peña, R.; Montalvo, T.; Bueno-Marí, R.; Bravo-Barriga, D.; Frontera, E.; et al. Determinants of the current and future distribution of the West Nile virus mosquito vector *Culex pipiens* in Spain. *Environ. Res.* **2020**, *188*, 109837. [CrossRef]
43. Bueno Marí, R.; Bernués Bañeres, A.; Chordá Olmos, F.A.; Jiménez Peydró, R. Aportes al conocimiento de la distribución y biología de *Anopheles algeriensis* Theobald, 1903 en España. *Bol. Mal. Salud. Amb.* **2011**, *51*, 93–96.
44. Observatorio de Mosquitos Del Guadalquivir. Website. Available online: https://mosquitos.ebd.csic.es/ (accessed on 15 May 2023).
45. Alarcón-Elbal, P.M.; Rodríguez-Sosa, M.A.; Ruiz-Matuk, C.; Tapia, L.; Arredondo Abreu, C.A.; Fernández González, A.A.; Rodríguez Lauzurique, R.M.; Paulino-Ramírez, R. Breeding Sites of Synanthropic Mosquitoes in Zika-Affected Areas of the Dominican Republic. *J. Am. Mosq. Control. Assoc.* **2021**, *37*, 10–19. [CrossRef]
46. Metge, G.; Hassaïne, K. Study of the environmental factors associated with oviposition by *Aedes caspius* and *Aedes detritus* along a transect in Algeria. *J. Am. Mosq. Control. Assoc.* **1998**, *14*, 283–288. Available online: https://www.biodiversitylibrary.org/content/part/JAMCA/JAMCA_V14_N3_P283-288.pdf (accessed on 10 April 2023).
47. López Sánchez, S. *Control Integral de Mosquitos en Huelva*; Junta de Andalucía, Conserjería de Salud y Servicios Sociales: Andalusia, Spain, 1989; p. 340.
48. Encinas Grandes, A. *Taxonomía y Biología de los Mosquitos del Área Salmantina (Diptera, Culicidae)*; CSIC: Madrid, Spain, 1982; p. 437.
49. Bueno Marí, R.; Corella, E.; Jiménez Peydró, R. Culícidofauna (Diptera: Culicidae) presente en los distintos enclaves hídricos de la ciudad de Valencia (España). *Rev. Colomb. Entomol.* **2010**, *36*, 235–241. [CrossRef]
50. Alarcón-Elbal, P.M. Incidencia de Ochlerotatus (Ochlerotatus) Caspius (Pallas, 1771) (Diptera: Culicidae) En El Término Municipal de Sagunto (Valencia, EspañA). Master's Thesis, Facultad de Farmacia, Universidad de Valencia, España, Valencia, Spain, 2007; p. 111.

51. Camp, J.V.; Kniha, E.; Obwaller, A.G.; Walochnik, J.; Nowotny, N. The transmission ecology of *Tahyna orthobunyavirus* in Austria as revealed by longitudinal mosquito sampling and blood meal analysis in floodplain habitats. *Parasites Vectors* **2021**, *14*, 561. [CrossRef]
52. Service, M.W. Flight periodicities and vertical distribution of *Aedes cantans* (Mg.), *Ae. geniculatus* (Ol.), *Anopheles plumbeus* Steph. and *Culex pipiens* L. (Dipt.; Culicidae) in southern England. *Bull. Entomol. Res.* **1971**, *60*, 639–651. [CrossRef]
53. Ruiz-Arrondo, I.; Garza-Hernández, J.A.; Reyes-Villanueva, F.; Lucientes-Curdi, J.; Rodríguez-Pérez, M.A. Human-landing rate, gonotrophic cycle length, survivorship, and public health importance of *Simulium erythrocephalum* in Zaragoza, northeastern Spain. *Parasites Vectors* **2017**, *10*, 175. [CrossRef] [PubMed]
54. Gutsevich, A.V.; Monchadski, A.S.; Shtakelberg, A. Fauna of the U.S.S.R., Diptera Family Culicidae. *Acad. Sci. USSR* **1974**, *III*, 408.
55. Soliman, D.E.; Farid, H.A.; Hammad, R.E.; Gad, A.M.; Bartholomay, L.C. Innate cellular immune responses in *Aedes caspius* (Diptera: Culicidae) mosquitoes. *J. Med. Entomol.* **2016**, *53*, 262–267. [CrossRef]
56. Napp, S.; Petrić, D.; Busquets, N. West Nile virus and other mosquito-borne viruses present in Eastern Europe. *Pathog. Glob. Health* **2018**, *112*, 233–248. [CrossRef]
57. Chordá Olmos, A. Biología de Mosquitos (Diptera: Culicidae) En Enclaves Representativos de la Comunidad Valenciana. Ph.D. Thesis, Facultat de Farmàcia, Universitat de València, Valencia, Spain, 2014. Available online: https://roderic.uv.es/bitstream/handle/10550/35297/TESIS%20DOCTORAL%20ALBERTO%20CHORD%C3%81%20OLMOS.pdf (accessed on 15 May 2023).
58. Núñez, A.I.; Talavera, S.; Aranda, C.; Birnberg, L.; Rivas, R.; Pujol, N.; Verdún, M.; Failloux, A.B.; Busquets, N. European *Aedes caspius* mosquitoes are experimentally unable to transmit Zika virus. *Parasites Vectors* **2019**, *12*, 363. [CrossRef] [PubMed]
59. Ministerio de Sanidad Consumo y Bienestar Social. Primeros Casos de Dengue Autóctono en España, 2018 [Updated 23 November 2018; Cited 20 September 2019]. Available online: https://www.mscbs.gob.es/profesionales/saludPublica/ccayes/alertasActual/docs/ERR_Dengue_autoctono_Espana.pdf (accessed on 15 May 2023).
60. Vazeille, M.; Jeannin, C.; Martin, E.; Schaffner, F.; Failloux, A.B. Chikungunya: A risk for Mediterranean countries? *Acta Trop.* **2008**, *105*, 200–202. [CrossRef]
61. Moutailler, S.; Krida, G.; Schaffner, F.; Vazeille, M.; Failloux, A.B. Potential Vectors of Rift Valley Fever Virus in the Mediterranean Region. *Vector-Borne Zoonotic Dis.* **2008**, *8*, 749–753. [CrossRef]
62. Balenghien, T.; Vazeille, M.; Grandadam, M.; Schaffner, F.; Zeller, H.; Reiter, P.; Sabatier, P.; Fouque, F.; Bicout, D.J. Vector competence of some French *Culex* and *Aedes* mosquitoes for West Nile virus. *Vector-Borne Zoonotic Dis.* **2008**, *8*, 589–596. [CrossRef]
63. Gutiérrez-López, R.; Bialosuknia, S.M.; Ciota, A.T.; Montalvo, T.; Martínez-de la Puente, J.; Gangoso, L.; Figuerola, J.; Kramere, L.D. Vector competence of *Aedes caspius* and *Ae. albopictus* mosquitoes for Zika Virus, Spain. *Emerg. Infect. Dis.* **2019**, *25*, 346–348. [CrossRef]
64. Reiter, P. West Nile virus in Europe: Understanding the present to gauge the future. *Eurosurveillance* **2010**, *15*, 19508. [CrossRef]
65. Brugman, V.A.; England, M.E.; Stoner, J.; Tugwell, L.; Harrup, L.E.; Wilson, A.J.; Medlock, J.M.; Logan, J.G.; Fooks, A.R.; Mertens, P.P.C.; et al. How often do mosquitoes bite humans in southern England? A standardised summer trial at four sites reveals spatial, temporal and site-related variation in biting rates. *Parasites Vectors* **2017**, *10*, 420. [CrossRef] [PubMed]
66. García San Miguel Rodríguez-Alarcón, L.; Fernández-Martínez, B.; Sierra Moros, M.J.; Vázquez, A.; Julián Pachés, P.; García Villacieros, E.; Gómez Martín, M.B.; Figuerola Borras, J.; Lorusso, N.; Ramos Aceitero, J.M.; et al. Unprecedented increase of West Nile virus neuroinvasive disease, Spain, summer 2020. *Euro Surveill.* **2021**, *26*, 2002010. [CrossRef] [PubMed]
67. Figuerola, J.; Jiménez-Clavero, M.Á.; Ruíz-López, M.J.; Llorente, F.; Ruiz, S.; Hoefer, A.; Aguilera-Sepúlveda, P.; Jiménez-Peñuela, J.; García-Ruiz, O.; Herrero, L.; et al. A One Health view of the West Nile virus outbreak in Andalusia (Spain) in 2020. *Emerg. Microbes Infect.* **2022**, *11*, 2570–2578. [CrossRef] [PubMed]
68. Petri, D. *Seasonal and Daily Activity of Mosquitoes (Diptera: Culicidae) In Vojvodina*, 1st ed; University of Novi Sad: Novi Sad, Serbia, 1989.
69. Santa-Olalla, P.; Vazquez-Torres, M.C.; Latorre-Fandos, E.; Mairal-Claver, P.; Cortina-Solano, P.; Puy-Azon, A.; Adiego Sancho, B.; Leitmeyer, K.; Lucientes-Curdi, J.; Sierra-Moros, M.J. First autochthonous malaria case due to *Plasmodium vivax* since eradication, Spain, October 2010. *Euro Surveill.* **2010**, *15*, 19684. [CrossRef]

diversity

MDPI

Article

Coexistence and Spatial Distribution of Invasive and Sylvatic Container-Breeding Mosquitoes in City–Forest Ecotone within the Brazilian Semi-arid

Hosana Rosa-Silva [1], Julia G. Cardoso [1], Ronaldo Reis-Júnior [1], Paulo H. C. Corgosinho [1], Maurício L. Faria [1], Sérvio P. Ribeiro [2], Filipe V. S. Abreu [3], Pablo Cuevas-Reyes [4] and Magno A. Z. Borges [1,*]

1 Departamento de Biologia, Centro de Biológicas e da Saúde, Universidade Estadual de Montes Claros, Montes Claros 39401-089, MG, Brazil; hosanarosa.bio@gmail.com (H.R.-S.); julia.gomescardoso@gmail.com (J.G.C.); chrysopa@gmail.com (R.R.-J.); pcorgo@gmail.com (P.H.C.C.); fariaml@gmail.com (M.L.F.)

2 Departamento de Biodiversidade, Evolução e Meio Ambiente, Instituto de Ciências Exatas e Biológicas, Universidade Federal de Ouro Preto, Ouro Preto 35400-000, MG, Brazil; serviopr@gmail.com

3 Instituto Federal do Norte de Minas Gerais, Campus Salinas, Salinas 39560-000, MG, Brazil; filipe.vieira@ifnmg.edu.br

4 Facultad de Biología, Edificio R, Universidad Michoacana de San Nicolás de Hidalgo, Ciudad Universitaria, Morelia 58030, MI, Mexico; pcragalla@gmail.com

* Correspondence: magno.borges@unimontes.com

Abstract: *Aedes* species are vectors of the most significant arboviruses in Brazil with the greatest health and economic impact in the country. However, little is known about the factors that influence the distribution of *Aedes*, as well as other container-breeding mosquitoes, especially on the border between urban and forest environments. Here, we tested the effect of three phytophysiognomies within the city–forest ecotone in the Brazilian semi-arid region on the spatial distribution of vector mosquitoes. We also investigated the effect of rainfall on the spatial distribution and abundance of mosquitoes and described the interspecific interactions between invasive and native mosquito species. Immatures were sampled during the rainy seasons of 2019–2020 (October 2019 to February 2020) and 2020–2021 (November 2020 to February 2021). In each sampled location, 30 ovitraps were installed in shaded areas. There was a predominance of the *Aedes* mosquitoes, especially *Ae. albopictus* and *Ae. aegypti*, while three species of *Haemagogus* (*Hg. spegazzinii*, *Hg. janthinomys* and *Hg. leucocelaenus*) were rarely found. The sylvatic mosquito *Ae. terrens* was abundant in areas with lower anthropic influence and during higher rainfall regimes with minimal pluviometric variation. This rainfall was also favorable for the presence of the predatory mosquito larvae *Toxorhynchites theobaldi*. The abundance of invasive *Aedes* species showed positive correlations with each other and negative correlations with the sylvatic *Ae. terrens*. Our results demonstrate that human occupation and the rainfall regime affect the interactions between invasive and sylvatic species of container mosquitoes.

Keywords: mosquito ecology; *Aedes*; *Toxorhynchites*; semi-arid; invasive species

Citation: Rosa-Silva, H.; Cardoso, J.G.; Reis-Júnior, R.; Corgosinho, P.H.C.; Faria, M.L.; Ribeiro, S.P.; Abreu, F.V.S.; Cuevas-Reyes, P.; Borges, M.A.Z. Coexistence and Spatial Distribution of Invasive and Sylvatic Container-Breeding Mosquitoes in City–Forest Ecotone within the Brazilian Semi-arid. *Diversity* **2023**, *15*, 822. https://doi.org/10.3390/d15070822

Academic Editors: Carlos Barceló, Ignacio Ruiz-Arrondo and Luc Legal

Received: 13 May 2023
Revised: 8 June 2023
Accepted: 27 June 2023
Published: 29 June 2023

1. Introduction

Globalization and urbanization are the main causes of the emergence and reemergence of significant diseases in the world, leading to a central concern to understand the role of insects as vectors in the transmission of diseases in humans and their distribution at different spatial scales [1,2]. The land use change from natural habitats to urban areas can negatively impact ecological processes (i.e., collapse of biotic interactions) as well as biodiversity, resulting in the loss of native species and the establishment and colonization of urban and peri-urban areas by opportunistic and invasive species [3–5]. For example, as a result of the urbanization process, many disease vector mosquitoes are able to increase their presence and distribution in urban and peri-urban areas, which has important implications for public health [6,7].

The distribution and occurrence of mosquitoes is affected by several factors and varies at different spatial scales [8]. Mosquito vector distribution is largely determined by climate at broad spatial scales [9]. At local and regional scales, such as in natural environments or peri-urban areas of municipalities, distribution and abundance patterns are affected by a multitude of biotic factors, including vegetation cover [10–12], host availability [13,14], and competition and predation interactions with other species [15,16]. Abiotic aspects such microclimate [17,18], availability of larval habitats [19,20], water temperature [21], conductivity and PH [22,23] have also impacted the abundance of vectors.

Invasive mosquito species of the genus *Aedes* such as *Aedes aegypti* and *Ae. albopictus* are the primary vectors of several human arboviruses, such as Dengue, Zika and Chikungunya [24]. These species are well established in Brazil and are excellent models for the study of vector dynamics in peri-urban areas [25–27]. *Ae. aegypti* are highly anthropophilic and prefer urban and suburban environments with high human population and residence density [28,29]. Comparatively, *Ae. albopictus* is found in vegetation-covered regions and areas with low-density human populations, but it has also been observed in transitional areas [30,31]. A variety of hosts has been already observed [32] and their eggs are drought resistant, remaining viable for months during dry season, and hatching when the rainy season starts [33]. Mosquitoes can occupy tree holes in natural environments or artificial containers in anthropogenic ones during the dry season, allowing them to resist drought [6].

Currently, there is great concern about the potential expansion and distribution of *Ae. Albopictus* into urban and peri-urban areas, which, in turn, would facilitate the arrival and colonization of pathogens that normally occur in natural habitats [34,35]. Although *Ae. aegypti* is considered the main vector of Dengue virus (DENV), *Ae. albopictus* can also contribute to the transmission of this disease in several regions around the world [36]. Beyond DENV, *Ae. albopictus* is also vector of the Chikungunya virus (CHIKV), which is responsible of recent outbreaks of this disease in France [37] and in countries of Central Africa [38]. During October 2019, transmission of the Zika virus (ZIKV) in France was also reported [39]. Recent studies have questioned the possible transmission of yellow fever and West Nile Virus by *Ae. albopictus* [34,40]. In Brazil, the infection of *Ae. albopictus* with DENV and ZIKV was reported for the first time during an outbreak in the rural zone of the coastal state of Espírito-Santo [41]. Hence, considering the species' ability of colonizing natural, rural, peri-urban, and urban regions [42], it is considered a potential bridge vector between sylvatic (i.e., Yellow fever virus) and urban (i.e., DENV) arboviruses cycles [43], increasing the importance of detecting and monitoring its presence in different environments in Brazil.

Here, we tested whether the vector mosquito species differ in their spatial distribution considering three phytophysiognomies of a peri-urban area in Montes Claros city, north of the Minas Gerais state, Brazil. We also tested whether the pluviosity can influence the spatial distribution and abundance of mosquitoes. We also described the interspecific interactions between the native and invasive species.

2. Materials and Methods

Sampling was carried out in the municipality of Montes Claros, north of Minas Gerais. The climate classification is Aw (tropical–semi-arid), with high temperatures (annual average of 24.1 °C) and average annual rainfall of 1085 mm [44]. The study was conducted in a peri-urban area (16°45′31″ S 43°53′40″ W) located close to residential areas and formed by different vegetation types. The three phytophysiognomies where sampling was conducted correspond to a parcel of regenerating dry forest over limestone outcrops; a degraded Cerrado surrounded by an active subsistence pasture matrix (cattle and horses); and a riparian forest of a small watercourse. All areas are close to three residential neighborhoods, separated from them by a minimum distance of 400 m and a maximum distance of 1400 m.

We collected mosquitoes during two different rainy seasons. The first (2019–2020), occurred between October 2019 and February 2020. The second collection (2020–2021) was performed between November 2020 and February 2021. To capture immature mosquitoes, oviposition traps (ovitraps) were used [45]. Ovitraps are used as a standard for epidemiological surveillance as recommended by the WHO, especially in low infestations and when larval research proves to be unproductive [46]. For epidemiological surveillance, the ovitraps should preferably stay in the field for one week until the eggs are removed. In our case, the ovitraps were used as an artificial tree hole, modified from that proposed by Yanoviak and Fincke [47]. Before the beginning of each rainy season, in each sampled phytophysiognomy (dry forest, pasture, and riparian forest), 30 traps of the ovitrap type were arbitrarily installed in shaded or partially shaded places, totaling 90 traps in each rainy season. Each trap was numbered and tied to trees at breast height with a minimum distance of 10 m from each other. The traps were installed empty and remained in the field during the rainy season, naturally receiving the input of water and resources that accumulated during the trap's exposure period. After 30 days of the first event of rain, 10 ovitraps were collected in each sampling area. Then, this procedure was repeated at intervals of 30 days until all traps were collected, totaling three mosquito collections in the 2019/2020 rainy season. In the 2020/2021 rainy season, the rain interruption allowed only two mosquito collections to be carried out. After each sampling, the traps were taken to the Laboratory of Ecology and Biological Control of Insects at the State University of Montes Claros, to rear the mosquitoes. The water collected from each trap was stored in an insectarium with controlled temperature (27 ± 2 °C) and light conditions (12 h dark/light). The original volume of the ovitraps was maintained by replacing the water with distilled water in each trap. The larvae were inspected daily until pupae were obtained, which were then transferred to 100 mL plastic cups and placed in a hatching trap to allow the adult mosquitoes to emerge. Mosquito species were identified at the species level using identification keys proposed by Forattini (2002) and Consoli and Oliveira-Filho (1994). To obtain adult mosquitoes of the *Toxorhynchites* species, predatory larvae were fed with larvae of other species collected from the study areas. Rainfall data were obtained from the online platform of the National Institute of Meteorology (INMET).

Generalized linear models (GLMs) were constructed to test the effects of biotic (interaction between mosquito species) and abiotic factors (sampled sites and rainfall regimes) (i.e., explanatory variables) on mosquito abundance (response variable). The rainfall regime was assessed based on the total rainfall during the sampling periods and the accumulated rainfall in the 14 days preceding the removal of the ovitraps from the field. These time intervals were chosen to ensure adequate time for the complete development of mosquitoes. The complete models were reduced through a stepwise procedure, and the results were subjected to analysis of variance (ANOVA), considering p-values < 0.05 as statistically significant. Contrast analysis was performed to assess the significant differences among the study sites. Residual analysis was conducted to examine the homogeneity of variance and the adequacy of error distribution using the diagnostic function of the 'RT4Bio' package. Analyses were performed using statistical software R (R Development Core Team, 2015).

3. Results

A total of 2689 adult mosquitoes were obtained in all study sites, grouped into two subfamilies (Culicinae and Toxorhynchitinae), three genera, and seven species (Table 1). There was a predominance of mosquitoes of the genus Aedes, which corresponded to 97.73% of the total number of sampled mosquitoes (Table 1).

Table 1. Total and relative abundance of mosquito species sampled in three peri-urban phytophysiognomies of Montes Claros, MG, during the rainy seasons of 2019–2020 and 2020–2021.

Species	2019–2020			2020–2021		
	Dry Forest	Pasture	Riparian Forest	Dry Forest	Pasture	Riparian Forest
Aedes albopictus (Skuse, 1984)	428 (85.3%)	533 (83.2%)	240 (70.3%)	240 (55.7%)	253 (46.9%)	69 (29.4%)
Aedes aegypti (Linneaus, 1762)	35 (6.9%)	102 (15.9%)	87 (25.5%)	65 (15.08%)	137 (25.4%)	17 (7.3%)
Aedes terrens (Walker, 1856)	28 (5.6%)	0	3 (0.87%)	117 (27.2%)	141 (26.2%)	133 (56.5%)
Toxorhynchites theobaldi (Dyar and Knab, 1906)	11 (2.2%)	6 (0.9%)	10 (2.93%)	9 (2.08%)	8 (1.5%)	11(4.7%)
Haemagogus spegazzini (Brethes, 1912)	0	0	1 (0.3%)	0	0	1(0.4%)
Haemagogus janthinomys (Dyar 1921)	0	0	0	0	0	3 (1.3%)
Haemagogus leucocelaenus (Dyar, 1925)	0	0	0	0	0	1 (0.4%)
Total	502 (33.8%)	641 (43.1%) 1484	341 (22.9%)	431	539 1205	235

3.1. *Influence of Sampled Sites and Habitat Characteristics*

The presence of *Ae. aegypti*, *Ae. albopictus*, and *Tx. theobaldi* was detected in all sampling areas. Alternatively, *Hg. leucocelaenus* and *Hg. janthinomys* were found only in the Riparian forest. There was a significant variation in the abundance of *Ae. albopictus* (deviance = 17.597; $p < 0.05$) through the phytophysiognomies studied. However, sampling sites did not influence the abundance of *Ae. aegypti* and *Ae. terrens*.

3.2. *Rainfall Regime*

The abundance of *Ae. albopictus* did not exhibit a significant relationship between rainfall variation and the accumulated rainfall 14 days before sampling. However, the abundance of *Ae. aegypti* showed a significant relationship with the interaction between rainfall variation and the accumulated rainfall 14 days before sampling (deviance = 4.7331, $p < 0.05$). The highest densities were observed in rainfall regimes of slight variation and intermediate accumulated volumes in the riparian and dry forest. In the pasture area, the highest densities were found in regimes with slight variation, but relatively small accumulated volumes in the previous 14 days (Figure 1). The abundance of *Ae. terrens* showed a significant relationship with rainfall variation × accumulated rainfall 14 days before sampling in all analyzed environments (deviance = 26.818, $p < 0.05$). The results indicate that the highest abundances of *Ae. terrens* were associated with the interaction between the low volumes of rainfall accumulated during the 14 days prior to sampling, particularly in regimes of high variance in the riparian forest and pasture areas. On the other hand, in the dry forest, the highest abundances were observed when there was an interaction between regimes with higher accumulated volumes and lower variance (Figure 2). The abundance of *Tx. theobaldi* was found to be related to the total volume of accumulated rainfall (deviance = 11.0903, $p < 0.05$) and with the variation of the rainfall regime (deviance = 8.7467, $p < 0.05$) (Figure 3), not differing in abundance between the sampled areas.

3.3. *Interaction between Species*

The abundance of *Ae. albopictus* was positively correlated with the abundance of *Ae. aegypti* (Y = e(2.31 + 031 × LogX))(deviance = 9.3901, $p < 0.05$) (Figure 4). In contrast, the abundance of *Ae. albopictus* in each trap did not vary with the abundance of *Ae. terrens*. The abundance of *Ae. aegypti* in each trap was negatively correlated with the abundance of the native mosquito *Ae. terrens* (Y = e(1.32 − 0.54logX)) (deviance = 4.6207, $p < 0.05$) (Figure 5). The abundance of *Tx. theobaldi* in each trap was unrelated to the abundance of any other mosquito in this study.

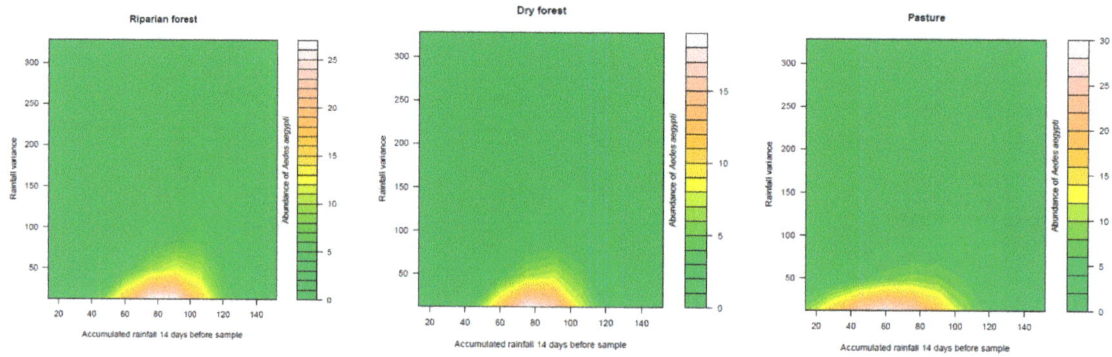

(a) (b) (c)

Figure 1. Effect of rainfall variation and accumulated precipitation on the abundance of *Aedes aegypti* in three peri-urban environments ((**a**)—riparian forest, (**b**)—dry forest, and (**c**)—pasture) in the city of Montes Claros, northern region of Minas Gerais state, Brazil.

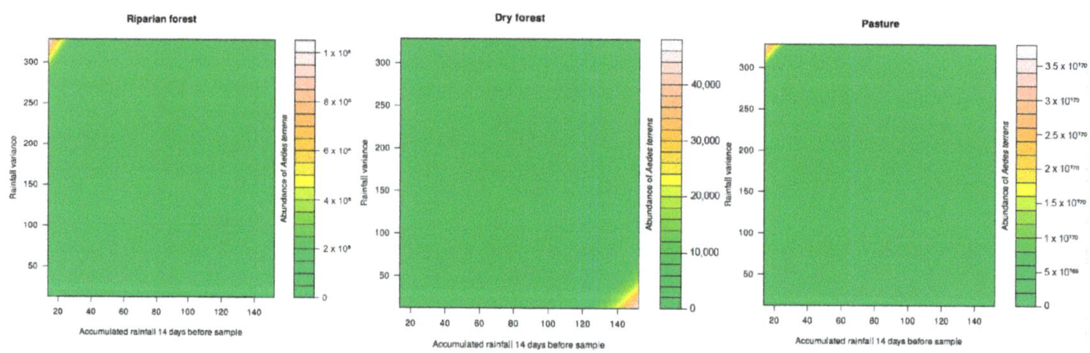

(a) (b) (c)

Figure 2. Effect of rainfall variation and accumulated precipitation on the abundance of *Aedes terrens* in three peri-urban environments ((**a**)—riparian forest, (**b**)—dry forest, and (**c**)—pasture) in the city of Montes Claros, northern region of Minas Gerais state, Brazil.

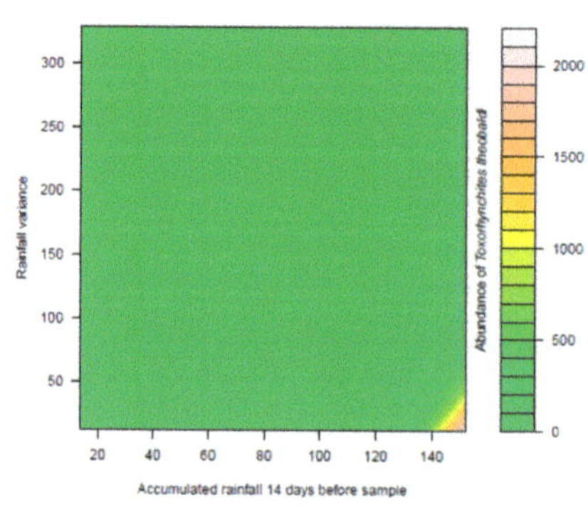

Figure 3. Effect of rainfall variation and accumulated precipitation on the abundance of *Toxorhynchites theobaldi* in peri-urban environments in the city of Montes Claros, northern region of Minas Gerais state, Brazil.

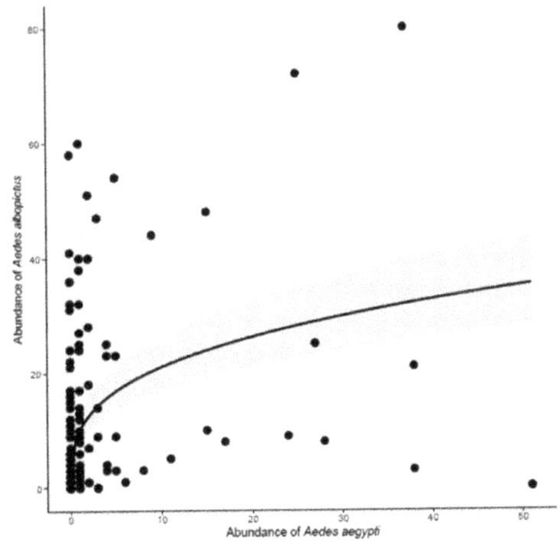

Figure 4. Effect of *Aedes albopictus* abundance on the variation in *Ae. aegypti* abundance, sampled in ovitraps distributed in three peri-urban environments in the city of Montes Claros, northern region of Minas Gerais state, Brazil. Each point represents an individual ovitrap collected in the study area.

second round.

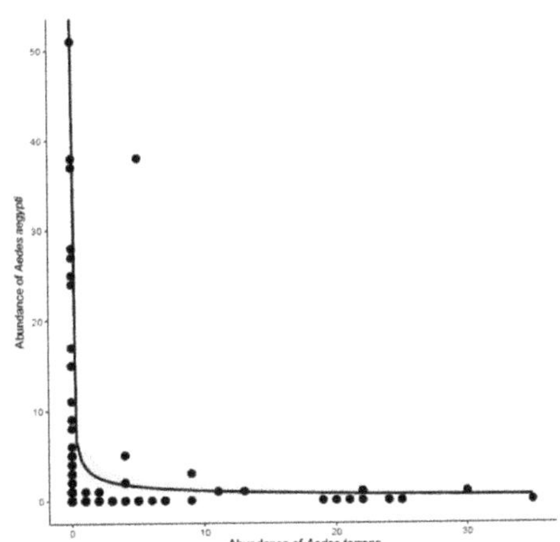

Figure 5. Effect of *Aedes aegypti* abundance on the variation in *Ae. terrens* abundance, sampled in ovitraps distributed in three peri-urban environments in the city of Montes Claros, northern region of Minas Gerais state, Brazil. Each point represents an individual ovitrap collected in the study area.

4. Discussion

In this study, invasive *Ae. albopictus* and *Ae. aegypti* were found to be the predominant species among the three phytophysiognomies tested, followed by the native mosquito *Ae. terrens* and the predatory larvae mosquito *Toxorhynchites theobaldi*. Additionally, there was also the punctual presence of *Hg. spegazzini* and the wild YFV vectors, *Hg. leucocelaenus* and *Hg. janthinomys*. The observed low level of diversity can be attributed to the urban matrix surrounding the sampled sites, which is often associated with low species richness and diversity in tropical areas [48,49]. The region of northern Minas Gerais also exhibits strong seasonality, characterized by distinct rainy and dry seasons with extended periods of drought lasting more than six months [50]. Due to the semi-arid climate and prolonged intervals between rainfall, oviposition sites may be less productive, resulting in decreased abundance and diversity among container-breeding species [51,52]. In strongly seasonal environments like these, the presence of urban areas tends to favor the establishment of invasive species, which are able to survive in artificial containers within cities, although with reduced abundance during dry seasons.

Habitat characteristics are often described as a determining factor in spatial distribution studies of *Aedes* species [53,54]. In this study, *Ae. albopictus* was found to be predominant in all sampled locations throughout both years of the study. This finding is consistent with the widely accepted view that this mosquito is more abundant in peri-urban and wild areas [28,55]. Recent studies have provided evidence that the food preference patterns of *Ae. albopictus* can be more accurately classified as anthropophagy and opportunism rather than anthropophilia [32,43]. As a result of this behavior, *Ae. albopictus* is more likely to transmit zoonotic pathogens in environments where humans are present at intermediate levels, such as peri-urban areas and transition areas [43,56]. Several studies suggest that peri-domestic mosquitoes, like *Ae. malayensis*, escape vector control measures targeting domestic vector species and may contribute to the reemergence of arbovirus [57]. Hence, the comparable abundance of *Ae. albopictus* in both pasture and dry forest areas could be attributed to the presence of hosts such as horses and cattle, as well as the availability of the traps as breeding sites.

Despite the well-known highly anthropophilic behavior of *Ae. aegypti* [28,55,58], our study revealed the presence of this mosquito species in all sampled sites within the peri-urban environment. The occurrence of *Ae. aegypti* has also been reported in forest parks within Porto Alegre, in an Atlantic Forest fragment located far from the edge of the forest [59]. This mosquito was also found 100 m inside urban parks in Manaus, including areas with dense vegetation that are inaccessible to the public [60], highlighting the ability of this species to inhabit different environments. Hence, our study demonstrates that this invasive mosquito coexists with native and transitional species in peri-urban areas. Despite the lower abundance compared to that of *Ae. albopictus*, the presence of *Ae. aegypti* raises concerns about transmission of urban arboviruses in peri-urban environments and should be monitored.

The abundance patterns of mosquitoes in these peri-urban areas may have also been influenced by rainfall in the days prior to sampling. *Aedes* species are known to colonize temporary breeding sites [61], and outbreaks of diseases transmitted by these mosquitoes regularly coincide with the rainy seasons [62,63]. The abundance of *Ae. aegypti* did not differ between the sampled sites, which suggests that during the rainy season, this mosquito increases its densities in urban environments where there is greater availability of human hosts, absence of predators and natural competitors, and the ability to reproduce in artificial containers [64]. Thus, denser populations on the edges of cities favor the penetration of individuals in peri-urban areas. The abundance of *Ae. aegypti* was higher in rainfall regimes with slight variation in the 14 days preceding the collections, as well as in rainfall regimes of small/intermediate intensity and greater amplitude in the pasture area. This species often occurs in areas with suboptimal environmental conditions, and *Ae. aegypti* females exhibit opportunistic laying behavior, distributing their eggs among several breeding sites and being undemanding regarding the biotic conditions of these breeding places [33]. Furthermore, the eggs are resistant to desiccation and most of them are deposited individually on the reservoir wall, just above the water surface [25]. However, some eggs (varying from 4% to 62%) can also be deposited directly on the water surface, which helps to maintain the mosquito population in the dry season, when there is no new water input [65,66]. Such conditions, therefore, facilitate the establishment of this species even in regions with low rainfall.

The relationship between the abundance of *Ae. terrens* and rainfall variables indicates a strong relationship with the dry forest, which is associated with high intensity of rainfall with minimal variation. In contrast, in areas closer to human activities such as the pasture and the riparian forest, the presence of *Ae. terrens* occasionally occurs under conditions of high rainfall variability, particularly during the peak of the rainfall. This species has a wide geographic distribution, recorded from the south of the Amazon basin in Brazil to the northern region of Argentina and in several countries of South America [67]. Despite the limited literature, experimental studies have described this mosquito as a possible vector of arboviruses. An experimental study found that populations of *Ae. terrens* were competent in transmitting both lineages of the CHIKV circulating in the Americas [68].

The same pattern can be observed for the predatory mosquito *Tx. theobaldi*. Our findings agree with those of other studies that suggest that wild species are typically less prone to adaptation [69] and are influenced by the volume of water retained in the containers and the stability of rainfall in these environments. Females of the genus *Toxorhynchites* lay their eggs individually on the water's surface in both natural and artificial containers [70,71]. Additionally, females of this species disperse their eggs widely to minimize the risk of cannibalism among the offspring and ensure the availability of prey resources [72]. Therefore, this oviposition behavior may be associated with the low densities of *Tx. theobaldi* observed in all sampled sites. Species belonging to this genus exhibit predatory behavior during their larval stages and have been extensively evaluated for their potential role as biological control agents [73].

The establishment of invasive species is related to the characteristics of the invader, the environment, and the resident community [74,75]. The consequences of competitive interactions between species of the genus *Aedes* depend on both the environment and on the level of urbanization in the surrounding areas [31]. The density of *Ae. albopictus* does not vary in relation to the densities of native species, *Ae. terrens* and *Tx. theobaldi*. The mosquito *Ae. albopictus* is native to Southeast Asia [76,77] and is considered one of the most widespread invasive species worldwide [78–80]. Although the competitive superiority of *Ae. albopictus* over *Ae. aegypti* has been observed in field and laboratory experiments [81], the effects of interspecific competition were not enough to displace *Ae. aegypti* in the study areas. In this scenario, our findings corroborate the idea that in peri-urban or transitional areas *Ae. aegypti* and *Ae. albopictus* coexist in large numbers [28,55]. Furthermore, the differential effects of interactions between invasive species and the native community may promote coexistence.

The abundance of the *Ae. aegypti* mosquito decreases with the presence of the native mosquito *Ae. terrens*. Therefore, despite the spillover of *Ae. aegypti* from urban to peri-urban areas in drier regimes, the conditions and adaptation of native *Ae. terrens* may have buffered the persistence of *Ae. aegypti* in these locations. A similar effect has been observed in urban neighborhoods of Baltimore, MD, USA, where competitive interactions between the resident mosquito *Culex pipiens*, the primary vector of West Nile virus, and *Ae. albopictus* depend on specific interaction conditions and the types of containers, thereby facilitating the persistence of *Cx. pipiens* despite the invasion of *Ae. albopictus* [82]. Recently, studies have demonstrated that *Ae. terrens* can be infected with the Guapiaçu virus, a newly identified virus phylogenetically similar to host-related insect-specific flavivirus (dISFV) [83]. Thus, the possibility of host switching exists, as reported with other arboviruses.

Overall, our study shows that complex interactions between abiotic and biotic factors determine distribution patterns and abundance of mosquito species in peri-urban areas. Our findings suggest that in peri-urban environments within strongly seasonal regions, the buffering effect exerted by native biota on the spillover of invasive species, typically associated with urban areas, is relatively weaker. This effect is primarily attributed to the drastic reduction in wild mosquito densities during the dry season, which, in turn, facilitates the local colonization by typically urban species, at least at the beginning of the rainy season. Consequently, in seasonal environments, invasive mosquito species coexist with native species. This overlapping of distributions can facilitate the flow of diseases and pathogens from wild to urban environments and vice versa. Therefore, our research demonstrates the importance of monitoring ecotonal areas as a central aspect of health surveillance measures.

Author Contributions: Conceptualization, H.R.-S., R.R.-J., P.H.C.C., M.L.F. and M.A.Z.B.; methodology, H.R.-S., R.R.-J., P.H.C.C., M.L.F. and M.A.Z.B.; material collection, H.R.-S., J.G.C., P.H.C.C., M.L.F. and M.A.Z.B.; formal analysis, H.R.-S., R.R.-J., M.L.F., P.C.-R. and M.A.Z.B.; investigation, H.R.-S., J.G.C., R.R.-J., P.H.C.C., M.L.F., S.P.R., F.V.S.A., P.C.-R. and M.A.Z.B.; resources, H.R.-S. and M.A.Z.B.; writing—original draft preparation, H.R.-S., M.L.F., S.P.R., F.V.S.A., P.C.-R. and M.A.Z.B.; writing—review and editing, H.R.-S., J.G.C., R.R.-J., P.H.C.C., M.L.F., S.P.R., F.V.S.A., P.C.-R. and M.A.Z.B.; visualization, H.R.-S., J.G.C., R.R.-J., P.H.C.C., M.L.F., S.P.R., F.V.S.A., P.C.-R. and M.A.Z.B.; supervision, H.R.-S., P.H.C.C., M.L.F., S.P.R., F.V.S.A., P.C.-R. and M.A.Z.B.; project administration, H.R.-S., F.V.S.A. and M.A.Z.B.; funding acquisition, H.R-S. and M.A.Z.B. All authors have read and agreed to the published version of the manuscript.

Funding: This research was funded by COORDENAÇÃO DE APRIMORAMENTO DE PESSOAL DE NÍVEL SUPERIOR (CAPES), grant number 88882.436098/2019-01.

Data Availability Statement: Not applicable.

Acknowledgments: The authors express their gratitude to all institutions and technicians who contributed to and supported the fieldwork. Special thanks are extended to Alice Meiry Silva Dias and Luana dos Reis Fiúza for their invaluable contributions during the fieldwork and mosquito identification. Additionally, we would like to thank the residents of the collection sites for their willingness to assist the researchers. We are also grateful to Coordenação de Aperfeiçoamento de Pessoal de Nível Superior (CAPES) for awarding the scholarship.

Conflicts of Interest: The authors declare no conflict of interest. The funders had no role in the design of the study; in the collection, analyses, or interpretation of data; in the writing of the manuscript; or in the decision to publish the results.

References

1. Weaver, S.C.; Charlier, C.; Vasilakis, N.; Lecuit, M. Zika, Chikungunya, and Other Emerging Vector-Borne Viral Diseases. *Annu. Rev. Med.* **2018**, *69*, 395–408. [CrossRef] [PubMed]
2. Wilke, A.B.B.; Beier, J.C.; Benelli, G. Complexity of the Relationship between Global Warming and Urbanization—An Obscure Future for Predicting Increases in Vector-Borne Infectious Diseases. *Curr. Opin. Insect Sci.* **2019**, *35*, 1–9. [CrossRef] [PubMed]
3. Johnson, M.T.J.; Munshi-South, J. Evolution of Life in Urban Environments. *Science* **2017**, *358*, eaam8327. [CrossRef] [PubMed]
4. Knop, E. Biotic Homogenization of Three Insect Groups due to Urbanization. *Glob. Chang. Biol.* **2016**, *22*, 228–236. [CrossRef]
5. McKinney, M.L. Urbanization as a Major Cause of Biotic Homogenization. *Biol. Conserv.* **2006**, *127*, 247–260. [CrossRef]
6. Rose, N.H.; Sylla, M.; Badolo, A.; Lutomiah, J.; Ayala, D.; Aribodor, O.B.; Ibe, N.; Akorli, J.; Otoo, S.; Mutebi, J.-P.; et al. Climate and Urbanization Drive Mosquito Preference for Humans. *Curr. Biol.* **2020**, *30*, 3570–3579.e6. [CrossRef]
7. Kraemer, M.U.G.; Sinka, M.E.; Duda, K.A.; Mylne, A.Q.N.; Shearer, F.M.; Barker, C.M.; Moore, C.G.; Carvalho, R.G.; Coelho, G.E.; Van Bortel, W.; et al. The Global Distribution of the Arbovirus Vectors Aedes Aegypti and Ae. Albopictus. *Elife* **2015**, *4*, e08347. [CrossRef]
8. Chandrasegaran, K.; Lahondère, C.; Escobar, L.E.; Vinauger, C. Linking Mosquito Ecology, Traits, Behavior, and Disease Transmission. *Trends Parasitol.* **2020**, *36*, 393–403. [CrossRef]
9. Laporta, G.Z.; Potter, A.M.; Oliveira, J.F.A.; Bourke, B.P.; Pecor, D.B.; Linton, Y.-M. Global Distribution of Aedes Aegypti and Aedes Albopictus in a Climate Change Scenario of Regional Rivalry. *Insects* **2023**, *14*, 49. [CrossRef]
10. Gardner, A.M.; Anderson, T.K.; Hamer, G.L.; Johnson, D.E.; Varela, K.E.; Walker, E.D.; Ruiz, M.O. Terrestrial Vegetation and Aquatic Chemistry Influence Larval Mosquito Abundance in Catch Basins, Chicago, USA. *Parasites Vectors* **2013**, *6*, 9. [CrossRef]
11. Wimberly, M.C.; Davis, J.K.; Evans, M.V.; Hess, A.; Newberry, P.M.; Solano-Asamoah, N.; Murdock, C.C. Land Cover Affects Microclimate and Temperature Suitability for Arbovirus Transmission in an Urban Landscape. *PLoS Negl. Trop. Dis.* **2020**, *14*, e0008614. [CrossRef] [PubMed]
12. Yang, L.; Turo, K.J.; Riley, C.B.; Inocente, E.A.; Tian, J.; Hoekstra, N.C.; Piermarini, P.M.; Gardiner, M.M. Can Urban Greening Increase Vector Abundance in Cities? The Impact of Mowing, Local Vegetation, and Landscape Composition on Adult Mosquito Populations. *Urban Ecosyst.* **2019**, *22*, 827–839. [CrossRef]
13. Burkett-Cadena, N.D.; McClure, C.J.W.; Estep, L.K.; Eubanks, M.D. Hosts or Habitats: What Drives the Spatial Distribution of Mosquitoes? *Ecosphere* **2013**, *4*, 30. [CrossRef]
14. Faraji, A.; Egizi, A.; Fonseca, D.M.; Unlu, I.; Crepeau, T.; Healy, S.P.; Gaugler, R. Comparative Host Feeding Patterns of the Asian Tiger Mosquito, Aedes Albopictus, in Urban and Suburban Northeastern USA and Implications for Disease Transmission. *PLoS Negl. Trop. Dis.* **2014**, *8*, e3037. [CrossRef]
15. Fader, J.E. The Importance of Interspecific Interactions on the Present Range of the Invasive Mosquito Aedes Albopictus (Diptera: Culicidae) and Persistence of Resident Container Species in the United States. *J. Med. Entomol.* **2016**, *53*, 992–1001. [CrossRef] [PubMed]

16. Juliano, S.A.; Westby, K.M.; Ower, G.D. Know Your Enemy: Effects of a Predator on Native and Invasive Container Mosquitoes. *J. Med. Entomol.* **2019**, *56*, 320–328. [CrossRef]
17. Lounibos, L.P.; O'Meara, G.F.; Juliano, S.A.; Nishimura, N.; Escher, R.L.; Reiskind, M.H.; Cutwa, M.; Greene, K. Differential Survivorship of Invasive Mosquito Species in South Florida Cemeteries: Do Site-Specific Microclimates Explain Patterns of Coexistence and Exclusion? *Ann. Entomol. Soc. Am.* **2010**, *103*, 757–770. [CrossRef]
18. Murdock, C.C.; Evans, M.V.; McClanahan, T.D.; Miazgowicz, K.L.; Tesla, B. Fine-Scale Variation in Microclimate across an Urban Landscape Shapes Variation in Mosquito Population Dynamics and the Potential of Aedes Albopictus to Transmit Arboviral Disease. *PLoS Negl. Trop. Dis.* **2017**, *11*, e0005640. [CrossRef]
19. Evans, M.V.; Hintz, C.W.; Jones, L.; Shiau, J.; Solano, N.; Drake, J.M.; Murdock, C.C. Microclimate and Larval Habitat Density Predict Adult Aedes Albopictus Abundance in Urban Areas. *Am. J. Trop. Med. Hyg.* **2019**, *101*, 362–370. [CrossRef]
20. Wilke, A.B.B.; Vasquez, C.; Carvajal, A.; Medina, J.; Chase, C.; Cardenas, G.; Mutebi, J.-P.; Petrie, W.D.; Beier, J.C. Proliferation of Aedes Aegypti in Urban Environments Mediated by the Availability of Key Aquatic Habitats. *Sci. Rep.* **2020**, *10*, 12925. [CrossRef]
21. Krol, L.; Gorsich, E.E.; Hunting, E.R.; Govender, D.; van Bodegom, P.M.; Schrama, M. Eutrophication Governs Predator-Prey Interactions and Temperature Effects in Aedes Aegypti Populations. *Parasites Vectors* **2019**, *12*, 179. [CrossRef] [PubMed]
22. Medeiros-Sousa, A.R.; de Oliveira-Christe, R.; Camargo, A.A.; Scinachi, C.A.; Milani, G.M.; Urbinatti, P.R.; Natal, D.; Ceretti-Junior, W.; Marrelli, M.T. Influence of Water's Physical and Chemical Parameters on Mosquito (Diptera: Culicidae) Assemblages in Larval Habitats in Urban Parks of São Paulo, Brazil. *Acta Trop.* **2020**, *205*, 105394. [CrossRef] [PubMed]
23. Kinga, H.; Kengne-Ouafo, J.A.; King, S.A.; Egyirifa, R.K.; Aboagye-Antwi, F.; Akorli, J. Water Physicochemical Parameters and Microbial Composition Distinguish Anopheles and Culex Mosquito Breeding Sites: Potential as Ecological Markers for Larval Source Surveillance. *J. Med. Entomol.* **2022**, *59*, 1817–1826. [CrossRef]
24. Jones, R.; Kulkarni, M.A.; Davidson, T.M.V.; RADAM-LAC Research Team; Talbot, B. Arbovirus Vectors of Epidemiological Concern in the Americas: A Scoping Review of Entomological Studies on Zika, Dengue and Chikungunya Virus Vectors. *PLoS ONE* **2020**, *15*, e0220753. [CrossRef]
25. Carvalho, F.D.; Moreira, L.A. Why Is Aedes Aegypti Linnaeus so Successful as a Species? *Neotrop. Entomol.* **2017**, *46*, 243–255. [CrossRef] [PubMed]
26. Kamal, M.; Kenawy, M.A.; Rady, M.H.; Khaled, A.S.; Samy, A.M. Mapping the Global Potential Distributions of Two Arboviral Vectors Aedes Aegypti and Ae. Albopictus under Changing Climate. *PLoS ONE* **2018**, *13*, e0210122. [CrossRef] [PubMed]
27. Kotsakiozi, P.; Gloria-Soria, A.; Caccone, A.; Evans, B.; Schama, R.; Martins, A.J.; Powell, J.R. Tracking the Return of Aedes Aegypti to Brazil, the Major Vector of the Dengue, Chikungunya and Zika Viruses. *PLoS Negl. Trop. Dis.* **2017**, *11*, e0005653. [CrossRef]
28. Braks, M.A.H.; Honório, N.A.; Lourençqo-De-Oliveira, R.; Juliano, S.A.; Lounibos, L.P. Convergent Habitat Segregation of Aedes Aegypti and Aedes Albopictus (Diptera: Culicidae) in Southeastern Brazil and Florida. *J. Med. Entomol.* **2003**, *40*, 785–794. [CrossRef]
29. Benitez, E.M.; Ludueña-Almeida, F.; Frías-Céspedes, M.; Almirón, W.R.; Estallo, E.L. Could Land Cover Influence Aedes Aegypti Mosquito Populations? *Med. Vet. Entomol.* **2019**, *34*, 138–144. [CrossRef]
30. Leisnham, P.T.; LaDeau, S.L.; Juliano, S.A. Spatial and Temporal Habitat Segregation of Mosquitoes in Urban Florida. *PLoS ONE* **2014**, *9*, e91655. [CrossRef]
31. Hopperstad, K.A.; Sallam, M.F.; Reiskind, M.H. Estimations of Fine-Scale Species Distributions of Aedes Aegypti and Aedes Albopictus (Diptera: Culicidae) in Eastern Florida. *J. Med. Entomol.* **2021**, *58*, 699–707. [CrossRef] [PubMed]
32. Fikrig, K.; Harrington, L.C. Understanding and Interpreting Mosquito Blood Feeding Studies: The Case of Aedes Albopictus. *Trends Parasitol.* **2021**, *37*, 959–975. [CrossRef]
33. Brady, O.J.; Hay, S.I. The Global Expansion of Dengue: How Aedes Aegypti Mosquitoes Enabled the First Pandemic Arbovirus. *Annu. Rev. Entomol.* **2020**, *65*, 191–208. [CrossRef] [PubMed]
34. Couto-Lima, D.; Madec, Y.; Bersot, M.I.; Campos, S.S.; Motta, M.d.A.; Santos, F.B.D.; Vazeille, M.; Vasconcelos, P.F.d.C.; Lourenço-de-Oliveira, R.; Failloux, A.-B. Potential Risk of Re-Emergence of Urban Transmission of Yellow Fever Virus in Brazil Facilitated by Competent Aedes Populations. *Sci. Rep.* **2017**, *7*, 4848. [CrossRef] [PubMed]
35. Ferreira-de-Lima, V.H.; Câmara, D.C.P.; Honório, N.A.; Lima-Camara, T.N. The Asian Tiger Mosquito in Brazil: Observations on Biology and Ecological Interactions since Its First Detection in 1986. *Acta Trop.* **2020**, *205*, 105386. [CrossRef]
36. Paupy, C.; Delatte, H.; Bagny, L.; Corbel, V.; Fontenille, D. Aedes Albopictus, an Arbovirus Vector: From the Darkness to the Light. *Microbes Infect.* **2009**, *11*, 1177–1185. [CrossRef]
37. Delisle, E.; Rousseau, C.; Broche, B.; Leparc-Goffart, I.; L'Ambert, G.; Cochet, A.; Prat, C.; Foulongne, V.; Ferre, J.B.; Catelinois, O.; et al. Chikungunya Outbreak in Montpellier, France, September to October 2014. *Eurosurveillance* **2015**, *20*, 21108. [CrossRef]
38. Paupy, C.; Kassa Kassa, F.; Caron, M.; Nkoghé, D.; Leroy, E.M. A Chikungunya Outbreak Associated with the Vector Aedes Albopictus in Remote Villages of Gabon. *Vector-Borne Zoonotic Dis.* **2012**, *12*, 167–169. [CrossRef]
39. Brady, O.J.; Hay, S.I. The First Local Cases of Zika Virus in Europe. *Lancet* **2019**, *394*, 1991–1992. [CrossRef]
40. Gould, E.; Pettersson, J.; Higgs, S.; Charrel, R.; de Lamballerie, X. Emerging Arboviruses: Why Today? *One Health* **2017**, *4*, 1–13. [CrossRef]

41. Ricas Rezende, H.; Malta Romano, C.; Morales Claro, I.; Santos Caleiro, G.; Cerdeira Sabino, E.; Felix, A.C.; Bissoli, J.; Hill, S.; Rodrigues Faria, N.; Cardoso da Silva, T.C.; et al. First Report of Aedes Albopictus Infected by Dengue and Zika Virus in a Rural Outbreak in Brazil. *PLoS ONE* **2020**, *15*, e0229847. [CrossRef] [PubMed]
42. Pereira-Dos-Santos, T.; Roiz, D.; Lourenço-de-Oliveira, R.; Paupy, C. A Systematic Review: Is Aedes Albopictus an Efficient Bridge Vector for Zoonotic Arboviruses? *Pathogens* **2020**, *9*, 266. [CrossRef] [PubMed]
43. Pereira-Dos-Santos, T.; Roiz, D.; Abreu, F.V.S.; Luz, S.L.B.; Santalucia, M.; Jiolle, D.; Santos Neves, M.S.A.; Simard, F.; Lourenço-de-Oliveira, R.; Paupy, C. Potential of *Aedes albopictus* as a bridge vector for enzootic pathogens at the urban-forest interface in Brazil. *Emerg. Microbes Infect.* **2018**, *7*, 191. [CrossRef]
44. Alvares, C.A.; Stape, J.L.; Sentelhas, P.C.; Gonçalves, J.L.M.; Sparovek, G. Köppen's Climate Classification Map for Brazil. *Meteorol. Z.* **2014**, *22*, 711–728. [CrossRef]
45. Eliason, D.A. A Preferred Oviposition Site as a Surveillance Method for Aedes Aegypti. *Mosq. News* **1966**, *26*, 531–535.
46. World Health Organization. *Dengue Guidelines for Diagnosis, Treatment, Prevention and Control: New Edition*; World Health Organization: Geneva, Switzerland, 2009.
47. Yanoviak, S.P.; Fincke, O.M. Sampling Methods for Water-Filled Tree Holes and Their Artificial Analogues. In *Insect Sampling in Forest Ecosystems*; Leather, S.R., Ed.; Blackwell Science Ltd.: Oxford, UK, 2005; pp. 168–185. ISBN 978047075051348.
48. Perfecto, I.; Vandermeer, J. Biodiversity Conservation in Tropical Agroecosystems: A New Conservation Paradigm. *Ann. N. Y. Acad. Sci.* **2008**, *1134*, 173–200. [CrossRef]
49. Wilke, A.B.B.; Benelli, G.; Beier, J.C. Anthropogenic Changes and Associated Impacts on Vector-Borne Diseases. *Trends Parasitol.* **2021**, *37*, 1027–1030. [CrossRef] [PubMed]
50. de Oliveira Souza, D.; Oliveira, F.G.; de Castro, I.L.; de Souza Soares, J.B.; Reis, M.M.; de Figueiredo, F.P. Frequência de ocorrência de precipitação pluviométrica em Montes Claros-MG. *Agrarian* **2018**, *11*, 337–342. [CrossRef]
51. Petermann, J.S.; Gossner, M.M. Aquatic Islands in the Sky: 100 Years of Research on Water-Filled Tree Holes. *Ecol. Evol.* **2022**, *12*, e9206. [CrossRef]
52. Ptatscheck, C.; Traunspurger, W. Meio- and Macrofaunal Communities in Artificial Water-Filled Tree Holes: Effects of Seasonality, Physical and Chemical Parameters, and Availability of Food Resources. *PLoS ONE* **2015**, *10*, e0133447. [CrossRef]
53. Bennett, K.L.; McMillan, W.O.; Enríquez, V.; Barraza, E.; Díaz, M.; Baca, B.; Whiteman, A.; Cerro Medina, J.; Ducasa, M.; Gómez Martínez, C.; et al. The Role of Heterogenous Environmental Conditions in Shaping the Spatiotemporal Distribution of Competing Aedes Mosquitoes in Panama: Implications for the Landscape of Arboviral Disease Transmission. *Biol. Invasions* **2021**, *23*, 1933–1948. [CrossRef] [PubMed]
54. Chaves, L.F.; Friberg, M.D. Aedes Albopictus and Aedes Flavopictus (Diptera: Culicidae) Pre-Imaginal Abundance Patterns Are Associated with Different Environmental Factors along an Altitudinal Gradient. *Curr. Res. Insect Sci.* **2021**, *1*, 100001. [CrossRef]
55. Honório, N.A.; Castro, M.G.; Barros, F.S.M.d.; Magalhães, M.d.A.F.M.; Sabroza, P.C. The Spatial Distribution of Aedes Aegypti and Aedes Albopictus in a Transition Zone, Rio de Janeiro, Brazil. *Cad. Saúde Pública* **2009**, *25*, 1203–1214. [CrossRef] [PubMed]
56. Fikrig, K.; Rose, N.; Burkett-Cadena, N.; Kamgang, B.; Leisnham, P.T.; Mangan, J.; Ponlawat, A.; Rothman, S.E.; Stenn, T.; McBride, C.S.; et al. Aedes Albopictus Host Odor Preference Does Not Drive Observed Variation in Feeding Patterns across Field Populations. *Sci. Rep.* **2023**, *13*, 130. [CrossRef]
57. Mendenhall, I.H.; Manuel, M.; Moorthy, M.; Lee, T.T.M.; Low, D.H.W.; Missé, D.; Gubler, D.J.; Ellis, B.R.; Ooi, E.E.; Pompon, J. Peridomestic Aedes Malayensis and Aedes Albopictus Are Capable Vectors of Arboviruses in Cities. *PLoS Negl. Trop. Dis.* **2017**, *11*, e0005667. [CrossRef] [PubMed]
58. Lounibos, L.P.; Kramer, L.D. Invasiveness of Aedes Aegypti and Aedes Albopictus and Vectorial Capacity for Chikungunya Virus. *J. Infect. Dis.* **2016**, *214*, S453–S458. [CrossRef]
59. Montagner, F.R.G.; Silva, O.S.; Jahnke, S.M. Mosquito Species Occurrence in Association with Landscape Composition in Green Urban Areas. *Braz. J. Biol.* **2018**, *78*, 233–239. [CrossRef]
60. Hendy, A.; Hernandez-Acosta, E.; Chaves, B.A.; Fé, N.F.; Valério, D.; Mendonça, C.; Lacerda, M.V.G.d.; Buenemann, M.; Vasilakis, N.; Hanley, K.A. Into the Woods: Changes in Mosquito Community Composition and Presence of Key Vectors at Increasing Distances from the Urban Edge in Urban Forest Parks in Manaus, Brazil. *Acta Trop.* **2020**, *206*, 105441. [CrossRef]
61. Soghigian, J.; Andreadis, T.G.; Livdahl, T.P. From Ground Pools to Treeholes: Convergent Evolution of Habitat and Phenotype in Aedes Mosquitoes. *BMC Evol. Biol.* **2017**, *17*, 262. [CrossRef]
62. Viana, D.V.; Ignotti, E. The Ocurrence of Dengue and Weather Changes in Brazil: A Systematic Review. *Rev. Bras. Epidemiol.* **2013**, *16*, 240–256. [CrossRef]
63. Valdez, L.D.; Sibona, G.J.; Condat, C.A. Impact of Rainfall on Aedes Aegypti Populations. *Ecol. Model.* **2018**, *385*, 96–105. [CrossRef]
64. Talaga, S.; Dejean, A.; Azémar, F.; Dumont, Y.; Leroy, C. Impacts of Biotic and Abiotic Parameters on Immature Populations of Aedes Aegypti. *J. Pest Sci.* **2020**, *93*, 941–952. [CrossRef]
65. Soares, F.A.; Silva, J.C.; Oliveira, J.B.B.S.; Abreu, F.V.S. Study of oviposition behavior of *Aedes aegypti* in two neighborhoods under the influence of semi-arid climate in the municipality of Salinas, State of Minas Gerais, Brazil. *Rev. Patol. Trop.* **2015**, *44*, 77–88.
66. Madeira, N.G.; Macharelli, C.A.; Carvalho, L.R. Variation of the oviposition preferences of Aedes aegypti in function of substratum and humidity. *Memórias Inst. Oswaldo Cruz* **2002**, *97*, 415–420. [CrossRef] [PubMed]

67. Schick, R.X. Mosquito Studies (Diptera, Culicidae) XX. The Terrens Group of Aedes (Finlaya). *Contrib. Am. Entomol. Inst.* **1970**, *5*, 1–158.
68. Lourenço-de-Oliveira, R.; Failloux, A.-B. High Risk for Chikungunya Virus to Initiate an Enzootic Sylvatic Cycle in the Tropical Americas. *PLoS Negl. Trop. Dis.* **2017**, *11*, e0005698. [CrossRef]
69. Silva, S.O.F.; de Mello, C.F.; Julião, G.R.; Dias, R.; Alencar, J. Sexual Proportion and Egg Hatching of Vector Mosquitos in an Atlantic Forest Fragment in Rio de Janeiro, Brazil. *Life* **2022**, *13*, 13. [CrossRef]
70. Trimble, R.M. Laboratory Observations on Oviposition by the Predaceous Tree-Hole Mosquito, Toxorhynchites Rutilus Septentrionalis (Diptera: Culicidae). *Can. J. Zool.* **1979**, *57*, 1104–1108. [CrossRef]
71. Donald, C.L.; Siriyasatien, P.; Kohl, A. Toxorhynchites Species: A Review of Current Knowledge. *Insects* **2020**, *11*, 747. [CrossRef]
72. Focks, D.A.; Sackett, S.R.; Dame, D.A.; Bailey, D.L. Ability of Toxorhynchites Amboinensis (Doleschall) (Diptera: Culicidae) to Locate and Oviposit in Artificial Containers in an Urban Environment. *Environ. Entomol.* **1983**, *12*, 1073–1077. [CrossRef]
73. Focks, D.A. Toxorhynchites as Biocontrol Agents. *J. Am. Mosq. Control. Assoc.* **2007**, *23*, 118–127. [CrossRef]
74. Lounibos, L.P. Invasions by Insect Vectors of Human Disease. *Annu. Rev. Entomol.* **2002**, *47*, 233–266. [CrossRef]
75. Giunti, G.; Becker, N.; Benelli, G. Invasive Mosquito Vectors in Europe: From Bioecology to Surveillance and Management. *Acta Trop.* **2023**, *239*, 106832. [CrossRef]
76. Bonizzoni, M.; Gasperi, G.; Chen, X.; James, A.A. The Invasive Mosquito Species Aedes Albopictus: Current Knowledge and Future Perspectives. *Trends Parasitol.* **2013**, *29*, 460–468. [CrossRef] [PubMed]
77. Battaglia, V.; Agostini, V.; Moroni, E.; Colombo, G.; Lombardo, G.; Rambaldi Migliore, N.; Gabrieli, P.; Garofalo, M.; Gagliardi, S.; Gomulski, L.M.; et al. The Worldwide Spread of Aedes Albopictus: New Insights from Mitogenomes. *Front. Genet.* **2022**, *13*, 931163. [CrossRef] [PubMed]
78. Benedict, M.Q.; Levine, R.S.; Hawley, W.A.; Lounibos, L.P. Spread of the Tiger: Global Risk of Invasion by the Mosquito Aedes Albopictus. *Vector-Borne Zoonotic Dis.* **2007**, *7*, 76–85. [CrossRef] [PubMed]
79. Lwande, O.W.; Obanda, V.; Lindström, A.; Ahlm, C.; Evander, M.; Näslund, J.; Bucht, G. Globe-Trotting Aedes Aegypti and Aedes Albopictus: Risk Factors for Arbovirus Pandemics. *Vector-Borne Zoonotic Dis.* **2020**, *20*, 71–81. [CrossRef] [PubMed]
80. Carvalho, R.G.; Lourenço-de-Oliveira, R.; Braga, I.A. Updating the Geographical Distribution and Frequency of Aedes Albopictus in Brazil with Remarks Regarding Its Range in the Americas. *Memórias Inst. Oswaldo Cruz* **2014**, *109*, 787–796. [CrossRef]
81. Lounibos, L.P.; Juliano, S.A. Where Vectors Collide: The Importance of Mechanisms Shaping the Realized Niche for Modeling Ranges of Invasive Aedes Mosquitoes. *Biol. Invasions* **2018**, *20*, 1913–1929. [CrossRef]
82. Leisnham, P.T.; LaDeau, S.L.; Saunders, M.E.M.; Villena, O.C. Condition-Specific Competitive Effects of the Invasive Mosquito Aedes Albopictus on the Resident Culex Pipiens among Different Urban Container Habitats May Explain Their Coexistence in the Field. *Insects* **2021**, *12*, 993. [CrossRef]
83. de Oliveira Ribeiro, G.; da Costa, A.C.; Gill, D.E.; Ribeiro, E.S.D.; da S. Rego, M.O.; Monteiro, F.J.C.; Villanova, F.; Nogueira, J.S.; Maeda, A.Y.; de Souza, R.P.; et al. Guapiaçu Virus, a New Insect-Specific Flavivirus Isolated from Two Species of Aedes Mosquitoes from Brazil. *Sci. Rep.* **2021**, *11*, 4674. [CrossRef] [PubMed]

diversity

MDPI

Article

The Mosquitoes of Querétaro, Mexico: Distribution, Ecology, and Discovery of *Shannoniana huasteca* n. sp. (Diptera: Culicidae)

Aldo I. Ortega-Morales [1], Luis M. Hernández-Triana [2,*] and Quetzaly K. Siller-Rodríguez [3,*]

[1] Departamento de Parasitología, Universidad Autónoma Agraria Antonio Narro, Unidad Laguna, Torreón 27054, Coahuila, Mexico; agrortega@hotmail.com
[2] Animal and Plant Health Agency, Virology Department, Vector Borne Diseases Research Group, Surrey KT153NB, UK
[3] Facultad de Ciencias Biológicas, Universidad Juárez del Estado de Durango, Gómez Palacio 35010, Durango, Mexico
* Correspondence: luis.hernandez-triana@apha.gov.uk (L.M.H.-T.); qksr@hotmail.com (Q.K.S.-R.)

Abstract: In order to document the diversity and distribution of mosquitoes inhabiting the Querétaro State of México, collection trips were conducted in all physiographic regions and sub-regions of the state (Sierra Madre Oriental, Central Plateau, and Neo-Volcanic Axis). In addition, mosquito specimens collected in Querétaro and deposited in the Collection of Arthropods of Medical Importance (CAIM) were re-examined. A total of 2718 specimens (570 larvae, 384 larval exuviae, 537 pupal exuviae, 30 pupae, 807 females, 368 males, and 22 male genitalia) were analyzed. In total, 2 subfamilies, namely Anophelinae and Culicinae, 5 tribes, 12 genera, 20 subgenera, and 50 species were found. Of these, 3 tribes, 8 genera, 11 subgenera, and 33 species are new records for the mosquito fauna of Querétaro. Two undescribed species were found, and one of them, *Shannoniana huasteca* Ortega n. sp., is described here using morphology and Cytochrome oxidase subunit 1 (*COI*) DNA barcoding. Taxonomic notes, new distribution limits, comments about the medical importance of species, and a key to identify adult females of *Shannoniana* species are provided.

Keywords: mosquitoes; Querétaro; new records; *Shannoniana huasteca*; new species

Citation: Ortega-Morales, A.I.; Hernández-Triana, L.M.; Siller-Rodríguez, Q.K. The Mosquitoes of Querétaro, Mexico: Distribution, Ecology, and Discovery of *Shannoniana huasteca* n. sp. (Diptera: Culicidae). *Diversity* **2023**, *15*, 697. https://doi.org/10.3390/d15060697

Academic Editors: Carlos Barceló, Ignacio Ruiz-Arrondo and Luc Legal

Received: 7 April 2023
Revised: 10 May 2023
Accepted: 15 May 2023
Published: 23 May 2023

1. Introduction

Mexico is divided into 32 political states of which only nine have been systematically studied in terms of the taxonomy, ecology, and distribution of mosquito species: Tlaxcala (26 spp.) [1], Quintana Roo (88 spp.) [2–9], Veracruz (141 spp.) [10–12], Tamaulipas (82 spp.) [13], Hidalgo (57 spp.) [14,15], Nuevo León (67 spp.) [16–18], Tabasco (107 spp.) [19–22], Mexico City (28 spp.) [23,24], and Mexico State (51 spp.) [25]. In Querétaro state, 17 mosquito species had been previously reported. However, most records are based on collections made in urban and sub-urban regions, not on collections from conserved forest and jungle regions, primarily in the north of the state. In this study, all physiographical regions of Querétaro (Sierra Madre Oriental, Central Plateau, and Neo-Volcanic Axis) were sampled with special emphasis on conserved forest regions and other sylvan regions of the state during the dry and rainy seasons.

A current checklist of the mosquito species that inhabit Querétaro state is provided in this study. Moreover, biological notes and medical importance are provided for the newly reported species and for species that reach their distributional limits within the state. Two undescribed species were found during the field collections, one of which is described here: *Shannoniana huasteca* Ortega n. sp. using morphological characters of adult stages and the analysis of DNA-barcodes. The second species, belonging to the genus *Culiseta*, subgenus *Culiseta*, is left undescribed until more material becomes available. Querétaro is the tenth state of Mexico to have the list of mosquito species updated. At present, 50 species are currently known (Table 1). Specimens collected and examined during this study were deposited in the Culicidae Collection of the Parasitology Department of the Autonomous Agrarian University Antonio Narro, Laguna unit, Torreón, Coahuila, Mexico.

Table 1. Checklist of the mosquito species that occur in Querétaro state. VM-P: [26]. V: [27]. D-NV: [28]. IBMC: [29]. STEA: [30]. WEEA [31] OMEA [a]: [32]. OMEA [b]: [33]. F.R.: First Record. NSR: New State Record (in bold). NS: New Species (in bold). [1] Reported as *Aedes atropalpus* (Coquillett). [2] Reported as *Culex peus* Speiser.

Taxa	F.R.	Taxa	F.R.
Anopheles (*Anopheles*)		26. ***Cx. coronator*** **Dyar and Knab**	**NSR**
1. ***An. apicimacula*** **Dyar and Knab**	**NSR**	27. ***Cx. declarator*** **Dyar and Knab**	**NSR**
2. ***An. eiseni*** **Coquillett**	**NSR**	28. ***Cx. erythrothorax*** **Dyar**	**NSR**
3. ***An. franciscanus*** **McCracken**	**NSR**	29. *Cx. quinquefasciatus* Say	D-NV
4. *An. pseudopunctipennis* Theobald	VM-P	30. ***Cx. restuans*** **Theobald**	**NSR**
5. *An. punctipennis* (Say)	VM-P	31. ***Cx. salinarius*** **Coquillett**	**NSR**
Anopheles (*Nyssorhynchus*)		32. *Cx. stigmatosoma* [2] Dyar	D-NV
6. *An. albimanus* Wiedemann	IBMC	33. ***Cx. tarsalis*** **Coquillett**	**NSR**
Aedes (*Aedimorphus*)		34. ***Cx. thriambus*** **Dyar**	**NSR**
7. *Ae. vexans* (Meigen)	V	***Culex*** **(*Melanoconion*)**	
Aedes (*Georgecriagius*)		35. ***Cx. erraticus*** **(Dyar and Knab)**	**NSR**
8. *Ae. epactius* [1] Dyar and Knab	D-NV	35. ***Cx. peccator*** **Dyar and Knab**	**NSR**
Aedes (*Howardina*)		***Culex*** **(*Neoculex*)**	
9. *Ae. allotecnon* Kumm, Komp, and Ruiz	V	36. ***Cx. apicalis*** **Adams**	**NSR**
10. ***Ae. quadrivittatus*** **(Coquillett)**	**NSR**	37. ***Cx. arizonensis*** **Bohart**	**NSR**
Aedes (*Ochlerotatus*)		***Culex*** **(*Phenacomyia*)**	
11. ***Ae. angustivittatus*** **Dyar and Knab**	**NSR**	38. ***Cx. lactator*** **Dyar and Knab**	**NSR**
12. ***Ae. euplocamus*** **Dyar and Knab**	**NSR**	***Lutzia*** **(*Lutzia*)**	
13. *Ae. scapularis* (Rondani)	D-NV	40. ***Lt. bigoti*** **(Bellardi)**	**NSR**
14. *Ae. shannoni* Vargas and Downs	V	***Culiseta*** **(*Culiseta*)**	
15. *Ae. trivittatus* (Coquillett)	V	41. ***Cs. inornata*** **(Williston)**	**NSR**
Aedes (*Protomacleaya*)		42. ***Cs. particeps*** **(Adams)**	**NSR**
16. ***Ae. amabilis*** **Schick**	**NSR**	43. ***Cs.*** **n. sp.**	**NS**
17. *Ae. brelandi* Zavortink	STEA	***Limatus***	
18. *Ae. podographicus* Dyar and Knab	OMEA [a]	44. ***Li. durhamii*** **Theobald**	**NSR**
19. ***Ae. schicki*** **Zavortink**	**NSR**	***Sabethes*** **(*Sabethoides*)**	
20. *Ae. triseriatus* (Say)	STEA	45. ***Sa. chloropterus*** **(von Humboldt)**	**NSR**
Aedes (*Stegomyia*)		***Shannoniana***	
21. *Ae. aegypti* (Linneaus)	IBMC	46. ***Sh. huasteca*** **n. sp.**	**NS**
22. *Ae. albopictus* (Skuse)	OMEA [b]	***Wyeomyia*** **(*Triamyia*)**	
Haemagogus (*Haemagogus*)		47. ***Wy. aporonoma*** **Dyar and Knab**	**NSR**
23. *Hg. equinus* Theonald	WEEA	***Wyeomyia*** **(*Wyeomyia*)**	
Psorophora **(*Grabhamia*)**		48. ***Wy. adelpha/guatemala***	**NSR**
24. ***Ps. signipennis*** **(Coquillett)**	**NSR**	49. ***Wy. mitchellii*** **(Theobald)**	**NSR**
Culex **(*Anoedioporpa*)**		***Toxorhynchites*** **(*Lynchiella*)**	
25. ***Cx. restrictor*** **Dyar and Knab**	**NSR**	50. ***Tx. moctezuma*** **(Dyar and Knab)**	**NSR**
Culex (*Culex*)			

2. Materials and Methods

2.1. Study Area

Querétaro state is located in the north-central part of Mexico, between 21°40′12″ and 20°00′54″ north latitude and the meridians 99°02′35″ and 100°35′48″ west longitude. The state has an area of 11,699 km^2. It is bordered to the north by the state of San Luis Potosí; to the west by the state of Guanajuato; to the east by the state of Hidalgo; to the southeast by the state of Mexico; and to the southwest by the state of Michoacán. The state is divided into three physiographic regions and four subregions (Figure 1): Sierra Madre Oriental (Carso Huasteco); Central Plateau (Mountains and Plains of Northern Guanajuato); and Neo-Volcanic Axis (Plains and Mountains of Querétaro and Hidalgo, Thousand Peaks, and

Lakes and Volcanoes of Anáhuac) [34]. A description of the regions and subregions of Querétaro and a list of the municipalities sampled are given in Table 2.

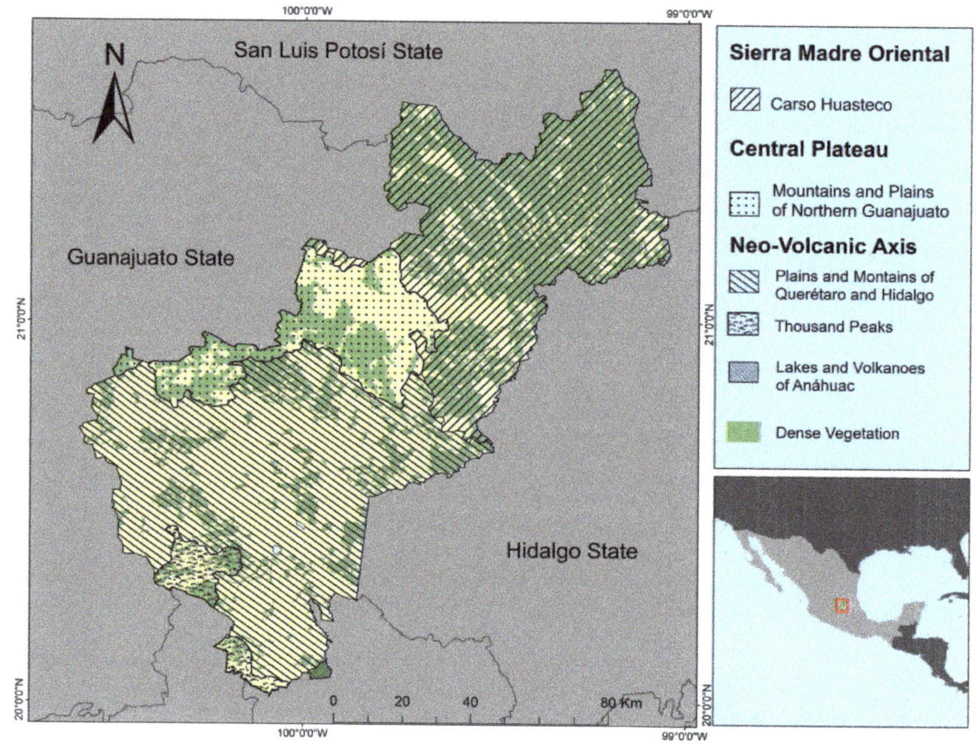

Figure 1. Physiography of Querétaro state. Red bow shows the study area.

Table 2. Description of the physiography of Querétaro state and list of municipalities sampled.

Region (Subregion)	Municipalities Sampled	Description of Subregions
Sierra Madre Oriental (Carso Huasteco)	Arroyo Seco, Jalpan de Serra, Landa de Matamoros, Peñamiller, Pinal de Amoles, San Joaquín,	Located in the south of the Sierra Madre Oriental, extends from San Luis Potosí and Guanajuato states to Querétaro, Hidalgo, Veracruz, and Puebla states. This subregion includes cloud and tropical forests, with oak and pine predominate.
Central Plateau (Mountains and Plains of Northern Guanajuato)	Cadereyta de Montes	This region is characterized by being an elevated region made up of wide plains interrupted by scattered mountain ranges, covered for the most part by Cenozoic volcanic rocks. Its average altitude is from 1700 to 2300 m above sea level.
Neo-Volcanic Axis (Plains and Mountains of Querétaro and Hidalgo)	Amealco de Bonfil, Cadereyta de Montes, Colón, Corregidora, El Marqués, Ezequiel Montes, Huimilpan, Pedro Escobedo, San Juan del Río, Santiago de Querétaro, Tequisquiapan, Tolimán	This subregion is an area of rough terrain where rounded hills predominate, some reaching elevations of 2400 m; the weather is dry and semi-warm, with warm temperatures during the summer, and cold during the winter. Includes grasslands and shrublands.

Table 2. *Cont.*

Region (Subregion)	Municipalities Sampled	Description of Subregions
Neo-Volcanic Axis (Thousand Peaks)	Huimilpan	This small subregion has mountainous relief that is made up of volcanic mountains, staggered lava plateaus, basaltic hills and the wide valley of the Lerma River. The climate is temperate and warm sub-humid, the rains occurring between June and October.
Neo-Volcanic Axis (Lakes and Volcanoes of Anáhuac)	Amealco de Bonfil	This subregion has a relief with hills and mountains with elevations above 3000 m; the climate is temperate humid, cool in summer. There are extensive regions of pine and oak forest, as well shrublands and grasslands.

2.2. Mosquito Collection

Immature stages and adult mosquitoes were collected in specific locations in the four physiographic regions of Querétaro (Table 2). The collections were conducted in both the dry and rainy seasons from 2012 to 2021. Immature stages were collected from all bodies of water found in the study area. Larvae and pupae were placed alive in cups with water from the aquatic habitat and transported to the Laboratorio de Biología Molecular of the Parasitology Department of the Universidad Autónoma Agraria Antonio Narro unidad laguna (LBM-UAAAN-UL). A portion of fourth-instar larvae from each collection was mounted on microscope slides using Euparal, whereas the rest of the live larvae were placed in individual emergence tubes to obtain adults with associated larval and pupal exuviae. Male genitalia was dissected to assist identification when required. Adults were collected in the field using CDC light traps, Shannon traps, and/or biting/landing on humans, and they were killed using triethylamine vapors and later mounted on insect pins. Mosquitoes mounted on insect pins were identified using a stereomicroscope Zeiss Discovery V8, while immature stages and exuviae were identified using a microscope Zeiss Primostar. The morphological terminology proposed by Harbach and Knight [35] for mosquito taxonomy was followed in this study.

2.3. Review of Entomological Collections

The Collection of Arthropods of Medical Importance (CAIM) deposited in the Diagnostic and Epidemiologic Reference Institute (InDRE) was reviewed for additional records of mosquitoes of Querétaro. The species found in the CAIM collection that were not collected by us are *Anopheles eiseni* Coquillett, *An. albimanus* Wiedemann, *Haemagogus equinus* Theobald, and *Culex salinarius* Coquillett. The traditional classification of Culicidae [36] was followed in large part, except that we consider only two subfamilies, incorporating *Toxorhynchites* from the tribe Toxorhynchitini into Culicinae, and we followed [37] the arrangement of Aedini taxa that was incorporated into the online classification of the Walter Reed Biosystematics Unit (WRBU) [38]. Generic and subgeneric abbreviations of Culicidae names also followed the WRBU [38].

2.4. DNA Extraction and COI Amplification

For DNA extraction, a modified Hotshot technique [39–41] was employed. Two legs were placed directly into 50 µL of alkaline lysis buffer in micro vials, which were then sonicated in a water bath for 20 min. Micro vials were subsequently incubated in a thermocycler for 30 min at 94 °C and cooled for 5 min. at 4 °C, after which 50 µL of the neutralizing buffer was added to each vial. PCR amplification of the full-length *COI* barcode region [42,43] was performed using Folmer primers (LCO1490 and HCO2198) and a Qiagen PCR system with the following reaction mix with a final volume of 50 µL: 2 µL of the DNA template, 25 µL of H_2O, 5 µL of NH_4, 5 µL of dNTPs (2 mM/µL), 2.5 µL of $MgCl_2$ (25 mM/µL), 0.1 µL of Bioline Taq Polymerase (Bioline Reagents Ltd., London, UK), 5 µL of each primer (each at 10 pmol/µL), and 0.38 µL of bovine serum albumin

(20 mg/mL) [39–41]. The thermal profile consisted of the following: An initial denaturation step at 94 °C for 1 min., 5 cycles of preamplification of 94 °C for 1 min., 45 °C for 1.5 min., 72 °C for 1.5 min., followed by 35 cycles of amplification of 94 °C for 1 min., 57 °C for 1.5 min., and 72 °C for 1 min., followed by a final elongation step of 72 °C for 5 min. All PCR products were visualized with a 1.5% agarose gel, and samples showing bands of the correct size were bidirectionally sequenced using the ABI PRISM® BigDye® Terminator v3.1 Cycle Sequencing Kit (Applied Biosystems, Waltham, MA, USA) at the Sequencing Unit, APHA.

2.5. Sequence Analysis

DNA sequences generated in both directions were edited manually using BioEdit sequence alignment Editor version 7.0.5.3 [44], and a consensus sequence was generated using ClustalW [45]. Full details for each specimen and sequence information can be found in the Barcode of Life Database (BOLD) within the "Human Pathogens and Zoonoses Initiative", Working Group 1.4. The Digital Object Identifier (DOI) for the publicly available project in BOLD is dx.doi.org/requested. Accession numbers for the sequences of *Shannoniana huasteca* n. sp. were obtained from NCBI (accession numbers requested). For certain species, we used public *COI* barcode sequences publicly available in BOLD: *Shannoniana fluviatilis* (Grench Guiana) (FGMOS1099-16, FGMOS816-16), *Sh. shcedocyclia* (French Guiana) (FGMOS817-16, FGMOS946-16, FGMOS947-16, FGMOS1126-16, FGMOS1134-16). We also compared published sequences of *Sh. moralesi* (Mexico) (MOSQV056-18, MQCCHP015-16, MQCHP016-16, MQCHP017-16, MQCHP018-16, MQCHP019-16, XNSLC054-18, XNSLC055-18), and *Trichoprosopon digitatum* (Mexico) (MQCHP064-16, MQCHP080-16, MQCHP082-16). The dataset was analyzed in MEGA v.6 [41]. The Maximum Likelihood (ML) analysis was performed using the Kimura 2-Parameter distance metric to determine their distribution pattern, and the tree was rooted to *Tr. digitatum*. The tree robustness was measured by the bootstrap approach using 1000 pseudoreplicates [46].

2.6. Nomenclatural Acts

The electronic edition of this article conforms to the requirements of the amended International Code of the Zoological Nomenclature (ICZN), and hence the new name contained herein is available under that Code from the electronic edition of this article. This published work and the nomenclatural acts it contains have been registered in ZooBank, the online registration system for the ICZN. The ZooBank LSIDs (Life Science Identifiers) can be resolved and the associated information viewed through any standard web browser by appending the LSID to the prefix http://zoobank.org/ (access date on 15 January 2023). The LSID for this publication is 4DD7EB32-56DD-41B6-AACC-877947FE26D4. The electronic edition of this work was published in a journal with an ISSN and has been archived and is available from the MDPI digital repositories.

3. Results

3.1. Mosquito Identification

A total of 2718 specimens from 203 collections were studied. Among the specimens were 570 fourth-instar larvae, 384 larval exuviae, 537 pupal exuviae, 30 pupae, 807 adult females, 368 adult males, and 22 dissected male genitalia. The mosquito fauna of Querétaro state consists of 50 species representing the subfamilies Anophelinae and Culicinae, 5 tribes of the subfamily Culicinae, 12 genera, and 20 subgenera (Table 1). Three tribes (Culisetini, Sabethini, and Toxorhynchitini), eight genera (*Psorophora*, *Lutzia*, *Culiseta*, *Limatus*, *Sabethes*, *Shannoniana*, *Wyeomyia*, and *Toxorhynchites*), 11 subgenera (*Grabhamia*, *Anoedioporpa*, *Melanoconion*, *Neoculex*, *Phenacomyia*, *Lutzia*, *Culiseta*, *Sabethoides*, *Triamyia*, *Wyeomyia*, and *Lynchiella*), and 33 species (*Anopheles apicimacula*, *An. eiseni*, *An. franciscanus*, *Aedes quadrivittatus*, *Ae. angustivittatus*, *Ae. euplocamus*, *Ae. amabilis*, *Ae. schicki*, *Psorophora signipennis*, *Culex restrictor*, *Cx. coronator*, *Cx. declarator*, *Cx. erythrothorax*, *Cx. restuans*, *Cx. salinarius*, *Cx. tarsalis*, *Cx. thriambus*, *Cx. erraticus*, *Cx. peccator*, *Cx. apicalis*, *Cx. arizonensis*, *Cx. lactator*,

Lutzia bigoti, *Culiseta inornata*, *Cs. particeps*, *Cs.* n. sp., *Limatus durhamii*, *Sabethes chloropterus*, *Shannoniana huasteca*, *Wyeomyia aporonoma*, *Wy. adelpha/guatemala*, *Wy. mitchellii*, and *Toxorhynchites moctezuma*) are new records for the mosquito fauna of Querétaro. Finally, two new species were discovered, one of which (*Sh. huasteca*) is described herein. The species accumulation curve of 46 of the 50 mosquito species collected is shown in Figure 2.

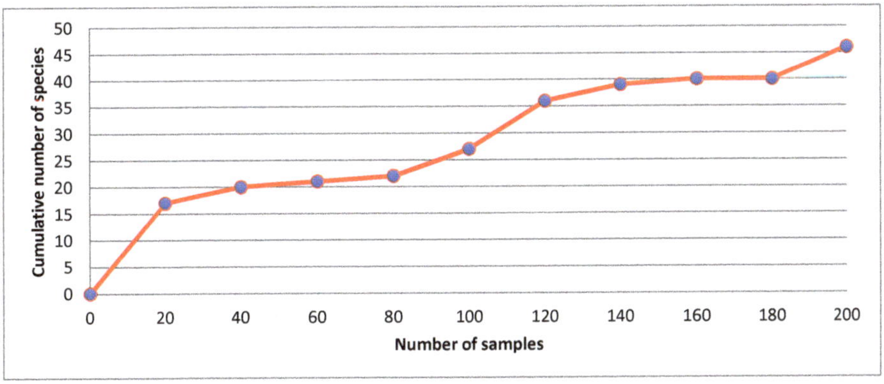

Figure 2. Species accumulation curve for 46 of the 50 mosquito species (203 collections) collected in Querétaro during 2012–2021.

3.2. *Biological and Ecological Notes for New State Records*

Biological and ecological notes for each species group are reported here. Specific notes including the collection site, date of collection, larval habitat, aquatic parameters, and associated species are shown in Table 3.

3.2.1. Genus *Anopheles*

Both the subgenera *Anopheles* and *Nyssorhynchus* had been previously reported in Querétaro. However, three species within the subgenus *Anopheles* are first recorded in the state: *Anopheles apicimacula*, *An. eiseni*, and *An. franciscanus*. One female of *An. apicimacula* was collected, while immature stages of *An. franciscanus* were collected from swamps, ponds, and stream margins in several locations. One record of *An. eiseni* was obtained from the CAIM collection; this species was not found during our collection trips.

3.2.2. Genus *Aedes*

Of the 16 species of *Aedes* known from Querétaro, five are reported for the first time in the state: *Aedes quadrivittatus*, *Ae. angustivittatus*, *Ae. euplocamus*, *Ae. amabilis*, and *Ae. schicki*. *Aedes quadrivittatus* was one of the most common species within the genus in the forested and conserved regions of the northern part of state; females of this species were collected approaching humans, while immature stages were collected from water in bromeliad axils. Adult females of *Ae. angustivittatus* were collected approaching humans at a single location while immature stages of *Ae. euplocamus* were collected from aquatic habitats at ground level in two sites. *Aedes amabilis* was very common in the conserved regions of the state, and this species frequently approached humans. Only one larvae of *Ae. schicki* was collected from a tree hole, making it the rarest species within the genus in Querétaro.

3.2.3. Genus *Psorophora*

In Mexico, the genus *Psorophora* is a common group of mosquitoes during the rainy season; however, this is the first record of this genus, subgenus *Grabhamia*, and *Ps. signipennis* in Querétaro. Immature stages of this species were collected from ponds during the rainy season. *Psorophora signipennis* is the only species within the genus *Psorophora* known for the state.

Table 3. Collection records of the newly reported mosquito species found in Querétaro, Mexico. The position (geographic coordinates in scale of degrees, minutes, and seconds); elevation (meters above sea level); type of aquatic habitat of the immature stages or environmental condition where adults were collected; environmental parameters of the aquatic habitat (pH, temperature, and dissolved salts in scale of parts per million); and associated species for each collection.

Taxa	Collection Date	Collection Site			Habitat	Aquatic Parameters			Associated Species
		Latitude N	Longitude W	Elev.		pH	Temp	Salts	
An. apicimacula	17 July 2015	20°4′41.2″	100°6′2.0″	1949	Resting on peri-domiciliary	—	—	—	—
An. franciscanus	14 November 2012	20°8′53.3″	99°58′38.5″	2329	Stream margin with clear water, green algae and total shade	6.40	14.5	66	*Ae. epactius*, *Cx. apicalis*
	14 November 2012	20°9′10.7″	100°1′9.6″	2463	Pond with colored water, brown algae and partial shade	6.55	17.2	100	*Cx. tarsalis*
	15 November 2012	20°6′41.9″	100°6′46.9″	2556	Swamp with colored water, green algae and absent shade	7.48	15.3	90	—
	16 November 2012	20°44′38.5″	99°56′19.7″	2058	Spring with clear water, floating vegetation and total shade	7.18	14.1	720	—
	16 November 2012	20°53′30.1″	99°41′44.1″	1539	Pond with clear water, green algae and absent shade	7.19	25.8	701	—
	17 November 2012	21°1′52″	99°36′2.3″	562	Pond with clear water, green algae and partial shade	7.22	24.8	277	*An. pseudopunctipennis*, *Cx. coronator*, *Cx. restuans*
	17 July 2015	20°24′41.2″	100°6′2″	1949	Pond with clear water, aquatic vegetation, green brown and partial shade	8	21	535	*Cx. stigmatosoma*
Ae. quadrivittatus	27 September 2012	21°13′7″	99°5′42″	368	Bromeliad axil with colored water and abundant leaf at bottom, total shade	6.60	26.4	22	*Ae. aegypti*, *Wy. mitchellii*
	27 September 2012	21°13′7″	99°5′42″	368	Bromeliad axil with colored water and abundant leaf at bottom	6.99	24.7	12	*Ae. allotecnon*, *Wy. mitchellii*
	27 September 2012	21°16′14.4″	99°3′20.1″	897	Human biting/landing at day with total shade	—	—	—	*Ae. allotecnon*, *Ae. podographicus*, *Sa. chloropterus*, *Sh. huasteca*, *Wy. mitchellii*
	19 November 2012	21°34′18.8″	99°14′33.3″	—	Human biting/landing at day with total shade	—	—	—	*Ae. trivittatus*, *Ae. amabilis*
	20 November 2012	21°34′53.6″	99°14′6.8″	686	Human biting/landing at day with total shade	—	—	—	*Ae. trivittatus*, *Ae. amabilis*, *Ae. brelandi*, *Sa. chloropterus*, *Wy. mitchellii*
	20 November 2012	21°34′18.8″	99°14′33.3″	766	Human biting/landing at day with total shade	—	—	—	*Ae. triseriatus*, *Ae. amabilis*
	24 September 2018	21°15′29.3″	99°411.1″	1010	Human biting/landing at day with total shade	—	—	—	*Ae. allotecnon*, *Sh. huasteca*, *Wy. mitchellii*
Ae. angustivittatus	28 September 2012	21°30′59″	99°17′13.3″	1033	Human biting/landing at day with total shade	—	—	—	*Ae. trivittatus*, *Ae. allotecnon*, *Ae. amabilis*, *Ae. brelandi*
Ae. euplocamus	28 September 2012	21°18′48″	99°16′21.7″	965	Rain gutter with colored water, leaves at bottom and total shade	9.70	27.6	58	*Ae. epactius*, *Ae. shannoni*
	20 November 2012	21°34′27.7″	99°12′39.8″	715	Rock hole with clear water, leaves at bottom and total shade	7.63	24	162	*Ae. epactius*, *Cx. coronator*

Table 3. *Cont.*

Taxa	Collection Date	Collection Site			Habitat	Aquatic Parameters			Associated Species
		Latitude N	Longitude W	Elev.		pH	Temp	Salts	
Ae. amabilis	28 September 2012	21°30′32.7″	99°17′22.2″	965	Human biting/landing at day with total shade	—	—	—	—
	28 September 2012	21°30′59″	99°17′13.3″	1033	Human biting/landing at day with total shade	—	—	—	*Ae. angustivittatus*, *Ae. trivittatus*, *Ae. allotecnon*, *Ae. brelandi*
	19 November 2012	21°30′59″	99°17′13.3″	686	Shannon trap baited with human at day				*Ae. trivittatus*, *Ae. triseriatus*
	19 November 2012	21°34′18.8″	99°14′33.3″	—	Human biting/landing at day with total shade	—	—	—	*Ae. quadrivittatus*, *Ae. trivittatus*
	20 November 2012	21°34′53.6″	99°14′6.8″	686	Human biting/landing at day with total shade	—	—	—	*Ae. quadrivittatus*, *Ae. trivittatus*, *Ae. brelandi*, *Sa. chloropterus*, *Wy. mitchellii*
	20 November 2012	21°34′53.6″	99°14′6.8″	686	Human biting/landing at day with total shade	—	—	—	*Ae. quadrivittatus*, *Ae. triseriatus*
	24 September 2018	21°15′29.3″	99°411.1″	1010	Tree hole with colored water and leaves at bottom	6.73	23	180	—
Ae. schicki	15 November 2012	20°55′51.6″	99°33′16.3″	2381	Tree hole with colored water with abundant leaves at bottom	7.25	12.8	243	—
Ps. signipennis	6 July 2012	20°26′55.9″	100°15′52.1″	2184	Pond with clear water and partial shade	7.3	24	102	—
	6 July 2012	20°26′45.4″	100°16′4.5″	2187	Pond with clear water and partial shade	7.57	25	101	*Ae. trivittatus*
	6 July 2012	20°26′45.4″	100°16′4.5″	2187	Pond with colored water, emerging vegetation and partial shade	7.57	24.7	213	*Ae. trivittatus*
Cx. restrictor	12 September 2012	21°13′13.7″	99°5′42″	368	Tree hole with colored water, leaves at bottom with total shade	8.26	24.6	1132	—
	27 September 2012	21°13′11.2″	99°5′56.9″	140	Discarded tire with clear water, total shade	7.55	22.7	42	*Cx. thriambus*, *Tx. moctezuma*
Cx. coronator	2 April 2012	21°32′8.6″	99°40′59.9″	960	Stream margin with clear water and total shade	8.64	27.9	792	—
	3 April 2012	21°26′45.7″	99°37′58.1″	567	Pond with clear water, aquatic vegetation and partial shade	7.8	26.3	510	*An. punctipennis*, *Cx. restuans*
	3 April 2012	21°23′29.3″	99°34′57.9″	534	Pond with clear water, emerging vegetation and total shade	7.9	30	206	—
	3 April 2012	21°12′35.3″	99°31′59.5″	1095	Irrigation gutter with colored water and total shade	6.8	25.3	482	*Cx. tarsalis*
	4 April 2012	21°02′23.6″	99°46′06.6″	1292	Pond with clear water and partial shade	7.7	34.2	1429	—
	17 November 2012	21°1′52″	99°36′2.3″	562	Pond with clear water, green algae and partial shade	7.22	24.8	277	*An. franciscanus*, *An. pseudopunctipennis*, *Cx. restuans*
	20 November 2012	21°34′27.7″	99°12′39.8″	715	Rock hole with clear water, leaves at bottom and total shade	7.63	24	162	*Ae. epactius*, *Ae. euplocamus*
Cx. declarator	2 April 2012	21°33′45.8″	99°42′27.9″	—	Artificial container with clear water and partial shade	8.2	31.3	328	*Ae. epactius*
Cx. erythrothorax	2 April 2012	21°26′45.7″	99°38′1.7″	575	Human biting/landing at night	—	—	—	—
	3 April 2012	21°26′44.6″	99°38′6.0″	548	Resting intra-domiciliary	—	—	—	*Ae. aegypti*, *Cx. quinquefasciatus*
	8 July 2012	20°32′13.4″	99°52′59.8″	1898	Human biting/landing at night	—	—	—	*Cx. tarsalis*
	8 July 2012	20°32′13.4″	99°52′59.8″	1898	Resting in vegetation at night	—	—	—	*Cx. quinquefasciatus*, *Cx. stigmatosoma*
	14 November 2012	20°8′19.1″	99°57′8.3″	2403	Resting on caves at day, total shade	—	—	—	—
	15 November 2012	20°2′31.7″	100°4′27.4″	2393	Irrigation gutter with clear water, aquatic vegetation and partial shade	7.02	15.9	170	*Cx. stigmatosoma*, *Cx. tarsalis*, *Cs. inornata*

Table 3. *Cont.*

Taxa	Collection Date	Collection Site			Habitat	Aquatic Parameters			Associated Species
		Latitude N	Longitude W	Elev.		pH	Temp	Salts	
Cx. restuans	3 April 2012	21°26′45.7″	99°37′58.1″	567	Pond with clear water, aquatic vegetation and partial shade	7.8	26.3	510	*An. punctipennis*, *Cx. coronator*
	14 November 2012	20°8′53.3″	99°58′38.5″	2329	Stream margin with clear water and total shade	6.39	17.4	223	*Cx. quinquefasciatus*, *Cx. stigmatosoma*, *Cx. arizonensis*, *Cs. particeps*
	20 November 2012	21°33′50.2″	99°13′11.4″	—	Rock hole with clear water, leaves at bottom and total shade	7.22	17.2	101	—
Cx. tarsalis	3 April 2012	21°18′50.0″	99°30′21.9″	549	Pond with clear water, with floating vegetation and total shade	8.20	30.2	171	*An. punctipennis*, *Cx. stigmatosoma*
	3 April 2012	21°18′50.0″	99°30′21.9″	549	Pond with clear water, with floating vegetation and total shade	8.4	31.7	175	*An. punctipennis*, *Cx. tarsalis*, *Cx. peccator*
	3 April 2012	21°12′35.3″	99°31′59.5″	1095	Irrigation gutter with colored water and total shade	6.8	25.3	482	*Cx. coronator*
	4 April 2012	21°09′32.4″	99°34′37.9″	1785	Spring with clear water and total shade	8.3	13	225	*Cx. stigmatosoma*, *Cx. arizonensis*, *Lt. bigoti*, *Cs. particeps*
	6 July 2012	20°22′34.3″	100°16′28.8″	—	Rock hole with clear water and partial shade	9.1	20.5	156	*Ae. trivittatus*, *Cs. particeps*
	7 July 2012	20°25′08.9″	100°17′40.7″	2151	Pond with clear water, emerging vegetation and absent shade	7.2	19.5	111	—
	7 July 2012	20°22′22.8″	100°17′0.85″	—	Artificial container with clear water and partial shade	8.3	19.7	223	*Cx. stigmatosoma*
	7 July 2012	20°22′27.3″	100°16′30.1″	2305	Pond with clear water, emerging vegetation and partial shade	7.4	20.3	90	*Cx. quinquefasciatus*, *Cx. stigmatosoma*, *Cx. thriambus*
	7 July 2012	20°11′15.5″	100°8′57″	2639	Flower vase with clear water and absent shade	8.4	23.2	298	—
	7 July 2012	20°23′2.7″	100°0′35.8″	1915	Resting in vegetation at day with partial shade	—	—	—	*Ae. trivittatus*, *Cx. quinquefasciatus*, *Cx. stigmatosoma*
	8 July 2012	20°46′34.7″	100°03′10.4″	1941	Resting in vegetation at day with total shade	—	—	—	*Cx. stigmatosoma*
	8 July 2012	20°54′13.9″	99°55′53.9″	1756	Flower vase with colored water, leaves at bottom and partial shade	7.7	26.4	631	*Cx. stigmatosoma*
	8 July 2012	20°39′46.2″	99°53′29.1″	1996	Pond with clear water, emerging vegetation and absent shade	9.4	25.1	212	*Cx. stigmatosoma*
	8 July 2012	20°32′13.4″	99°52′59.8″	1898	Human biting/landing at night	—	—	—	*Cx. erythrothorax*
	14 November 2012	20°8′53.3″	99°58′38.5″	2329	Rock hole with clear water, leaves at bottom, partial shade	6.59	22.7	279	*Cx. quinquefasciatus*, *Cx. stigmatosoma*, *Cs. particeps*
	14 November 2012	20°9′10.7″	100°1′9.6″	2463	Pond with dark water, aquatic vegetation and total shade	7.20	15.3	777	*An. franciscanus*
	15 November 2012	20°2′31.7″	100°4′27.4″	2393	Irrigation gutter with clear water, aquatic vegetation and partial shade	7.02	15.9	170	*Cx. stigmatosoma*
	16 November 2012	20°46′12.6″	99°55′4″	1932	Irrigation gutter with clear water, aquatic vegetation and partial shade	7.11	23.5	29	*Cx. apicalis*

Table 3. *Cont.*

Taxa	Collection Date	Collection Site			Habitat	Aquatic Parameters			Associated Species
		Latitude N	Longitude W	Elev.		pH	Temp	Salts	
	7 July 2012	20°9′10.7″	100°1′9.6″	2463	Pond with clear water, emerging vegetation and partial shade	7.4	20.3	90	*Cx. quinquefasciatus*, *Cx. stigmatosoma*, *Cx. tarsalis*
	12 September 2012	21°13′7″	99°5′42″	368	Resting on caves with total shade	—	—	—	*Ae. albopictus*
	12 September 2012	21°13′11.2″	99°5′56.9″	140	Discarded tire with clear water, total shade	7.55	22.7	42	*Cx. restrictor*, *Tx. moctezuma*
	29 September 2012	21°8′6.1″	99°37′27.6″	1790	Flower vase with clear water, leaves at bottom, green algae and total shade	7.94	19.9	484	—
	29 September 2012	21°8′6.1″	99°37′27.6″	1790	Discarded tire with clear water, leaves at bottom and total shade	8.26	19.4	389	*Cx. stigmatosoma*
	16 November 2012	20°43′59.3″	99°56′16.7″	2080	Artificial container with colored water, leaves at bottom and absent shade	7.21	15.2	555	*Cx. quinquefasciatus*
	16 November 2012	20°53′23.7″	99°34′55.1″	2399	Discarded tire with clear water and partial shade	7.29	22.5	177	—
Cx. thriambus	16 November 2012	20°56′4.4″	99°33′35.3″	2416	Discarded tire with colored water, leaves at bottom and total shade	7.24	12	780	—
	16 November 2012	20°56′4.4″	99°33′35.3″	2416	Cattle drinker with clear water, leaves at bottom and partial shade	7.24	13.4	158	*Ae. trivittatus*, *Cs. particeps*
	17 November 2012	20°59′45.9″	99°34′40.3″	1884	Spring with clear water, aquatic vegetation, brown algae and total shade	7.10	15	289	*An. pseudopunctipennis*, *Cx. stigmatosoma*
	17 November 2012	20°59′45.9″	99°34′40.3″	1684	Artificial container with clear water, leaves at bottom and total shade	7.10	16	328	*Cx. quinquefasciatus*, *Cx. stigmatosoma*, *Cs. particeps*
	24 September 2018	21°8′38″	99°37′15.1″	2292	Discarded tire with clear water, leaves at bottom and total shade	7.15	20	190	*Cs.* n. sp., *Tx. moctezuma*
	22 September 2021	21°15′29.3″	99°4′11.1″	1010	Tree hole with colored water, leaves at bottom with total shade	7.50	20	1095	*Cx. lactator*
Cx. erraticus	3 April 2012	21°18′50.0″	99°30′21.9″	549	Pond with clear water with floating vegetation and total shade	8.4	31.7	175	*An. punctipennis*, *Cx. tarsalis*, *Cx. peccator*
Cx. peccator	3 April 2012	21°18′50.0″	99°30′21.9″	549	Pond with clear water with floating vegetation and total shade	8.4	31.7	175	*An. punctipennis*, *Cx. tarsalis*, *Cx. erraticus*
	3 April 2012	21°18′50.0″	99°30′21.9″	549	Resting in vegetation at day with total shade	—	—	—	—

Table 3. *Cont.*

Taxa	Collection Date	Collection Site			Habitat	Aquatic Parameters			Associated Species
		Latitude N	Longitude W	Elev.		pH	Temp	Salts	
Cx. apicalis	14 November 2012	20°8′53.3″	99°58′38.5″	2329	Stream margin with clear water, green algae and total shade	6.40	14.5	66	*An. franciscanus, Ae. epactius*
	15 November 2012	20°16′59.9″	100°10′5.7″	2296	Stream margin with clear water, emerging vegetation, green algae and partial shade	7.49	22	65	*An. punctipennis*
	15 November 2012	20°20′22.3″	100°14′53.7″	—	Pond with colored water, aquatic vegetation and absent shade	7.07	25	38	—
	15 November 2012	20°20′19.9″	100°19′50.4″	2418	Rock hole with dark water, emerging vegetation and partial shade	7.11	23.5	29	Cx. arizonensis
	16 November 2012	20°34′43″	100°19′21.7″	2037	Discarded tire with colored water, leaves at bottom and total shade	7.43	17.11	682	Ae. epactius
	16 November 2012	20°53′23.7″	99°34′55.1″	2399	Discarded tire with clear water and partial shade	7.28	20.4	110	—
Cx. arizonensis	4 April 2012	21°09′33.7″	99°34′35.9″	1813	Waterhole with clear water and total shade	7.56	16.8	269	—
	4 April 2012	21°09′32.4″	99°34′37.9″	1785	Spring with clear water and total shade	8.3	13	225	*Cx. stigmatosoma, Cx. tarsalis, Lt. bigoti, Cs. particeps*
	29 September 2012	21°8′6.1″	99°37′27.6″	1790	Flower vase with clear water, leaves at bottom, green algae and total shade	9.76	19.6	277	*Ae. albopictus, Cx. stigmatosoma, Cs. particeps*
	14 November 2012	20°8′1.1″	99°56′52.6″	2369	Resting on caves at day, total shade	—	—	—	*Cx. stigmatosoma, Cs. particeps*
	14 November 2012	20°8′53.3″	99°58′38.5″	2329	Stream margin with clear water and total shade	6.39	17.4	223	*Cx. quinquefasciatus, Cx. restuans, Cx. stigmatosoma, Cs. particeps*
	15 November 2012	20°20′19.9″	100°19′50.4″	2418	Rock hole with dark water, emerging vegetation and partial shade	7.11	23.5	29	*Cx. apicalis*
Cx. lactator	22 September 2021	21°15′29.3″	99°4′11.1″	1010	Human biting/landing at day	—	—	—	*Ae. allotecnon, Sa. chloropterus, Sh. huasteca, Wy. aporonoma, Wy. mitchellii*
	22 September 2021	21°15′29.3″	99°4′11.1″	1010	Tree hole with colored water, leaves at bottom with total shade	7.50	20	1095	*Cx. thriambus*
Lt. bigoti	4 April 2012	21°09′32.4″	99°34′37.9″	1785	Spring with clear water and total shade	8.3	13	225	*Cx. stigmatosoma, Cx. tarsalis, Cx. arizonensis, Cs. particeps*
	28 September 2017	21°16′46.9″	99°8′29.2″	1828	Artificial container with clear water, leaves at bottom and total shade	6	24.6	23	*Ae. albopictus*
Cs. inornata	15 November 2012	20°2′31.7″	100°4′27.4″	2393	Irrigation gutter with clear water, aquatic vegetation and partial shade	7.02	15.9	170	*Cx. erythrothorax, Cx. stigmatosoma, Cx. tarsalis*

Table 3. *Cont.*

Taxa	Collection Date	Collection Site			Habitat	Aquatic Parameters			Associated Species
		Latitude N	Longitude W	Elev.		pH	Temp	Salts	
Cs. particeps	4 April 2012	21°09′32.4″	99°34′37.9″	1785	Spring with clear water and total shade	8.3	13	225	*Cx. stigmatosoma*, *Cx. tarsalis*, *Cx. arizonensis*, *Lt. bigoti*
	4 April 2012	21°09′32.4″	99°34′37.9″	1811	Resting in vegetation at day with total shade	—	—	—	—
	7 July 2012	20°11′15.5″	100°8′57″	2639	Flower vase with clear water, leaves at bottom and partial shade	9.9	21.9	197	*Cx. stigmatosoma*
	29 September 2012	21°8′6.1″	99°37′27.6″	1790	Flower vase with clear water, leaves at bottom, green algae and total shade	9.76	19.6	277	*Ae. albopictus*, *Cx. stigmatosoma*, *Cx. arizonensis*
	14 November 2012	20°8′1.1″	99°56′52.6″	2369	Resting on caves at day, total shade	—	—	—	*Cx. stigmatosoma*, *Cx. arizonensis*
	14 November 2012	20°8′53.3″	99°58′38.5″	2329	Stream margin with clear water and total shade	6.39	17.4	223	*Cx. quinquefasciatus*, *Cx. restuans*, *Cx. stigmatosoma*, *Cx. arizonensis*
	16 November 2012	20°56′4.4″	99°33′35.3″	2416	Cattle drinker with clear water, leaves at bottom and partial shade	7.24	13.4	158	*Ae. trivittatus*, *Cx. thriambus*
	17 November 2012	20°59′45.9″	99°34′40.3″	1684	Artificial container with clear water, leaves at bottom and total shade	7.10	16	328	*Cx. quinquefasciatus*, *Cx. stigmatosoma*, *Cx. thriambus*
Cs. n. sp.	24 September 2018	21°8′38″	99°37′15.1″	2292	Discarded tire with clear water, leaves at bottom and total shade	7.15	20	190	*Cx. thriambus*, *Tx. moctezuma*
Li. durhamii	28 September 2017	21°16′46.9″	99°8′29.2″	1824	Artificial container with clear water, leaves at bottom and total shade	6.2	25	53	—
Sa. chloropterus	27 September 2012	21°16′14.4″	99°3′20.1″	897	Human biting/landing at day with total shade	—	—	—	*Ae. allotecnon*, *Ae. quadrivittatus*, *Ae. podographicus*, *Sh. huasteca*, *Wy. mitchellii*
	20 November 2012	21°34′53.6″	99°14′6.8″	686	Human biting/landing at day with total shade	—	—	—	*Ae. quadrivittatus*, *Ae. trivittatus*, *Ae. amabilis*, *Ae. brelandi*, *Wy. mitchellii*
	28 September 2017	21°16′46.9″	99°8′29.2″	1828	Human biting landing at day with total shade	—	—	—	*Ae. allotecnon*, *Ae. podographicus*, *Sh. huasteca*, *Wy. adelpha/guatemala*
Sh. huasteca	27 September 2012	21°16′14.4″	99°3′20.1″	897	Human biting/landing at day with total shade	—	—	—	*Ae. allotecnon*, *Ae. quadrivittatus*, *Ae. podographicus*, *Sa. chloropterus*, *Wy. mitchellii*
	28 September 2017	21°16′46.9″	99°8′29.2″	1828	Human biting/landing at day with total shade	—	—	—	*Ae. allotecnon*, *Ae. podographicus*, *Sa. chloropterus*, *Wy. adelpha/guatemala*
	24 September 2018	21°15′29.3″	99°411.1″	1010	Human biting/landing at day with total shade	—	—	—	*Ae. allotecnon*, *Ae. quadrivittatus*, *Wy. mitchellii*
	22 September 2021	21°15′29.3″	99°4′11.1″	1010	Human biting/landing at day	—	—	—	*Ae. allotecnon*, *Cx. lactator*, *Sa. chloropterus*, *Wy. aporonoma*, *Wy. mitchellii*

Table 3. *Cont.*

Taxa	Collection Date	Collection Site			Habitat	Aquatic Parameters			Associated Species
		Latitude N	Longitude W	Elev.		pH	Temp	Salts	
Wy. aporonoma	22 September 2021	21°15′29.3″	99°4′11.1″	1010	Human biting/landing at day	—	—	—	*Ae. allotecnon*, *Cx. lactator*, *Sa. chloropterus*, *Sh. huasteca*, *Wy. mitchellii*
Wy. adelpha/guatemala	28 September 2017	21°16′46.9″	99°8′29.2″	1828	Human biting/landing at day with total shade	—	—	—	*Ae. allotecnon*, *Ae. podographicus*, *Sa. chloropterus*, *Sh. huasteca*,
Wy. mitchellii	27 September 2012	21°13′7″	99°5′42″	368	Bromeliad axil with colored water and abundant leaf at bottom, total shade	6.60	26.4	22	*Ae. quadrivittatus*, *Ae. aegypti*
	27 September 2012	21°13′7″	99°5′42″	368	Bromeliad axil with colored water and abundant leaf at bottom	6.99	24.7	12	*Ae. allotecnon*, *Ae. quadrivittatus*
	27 September 2012	21°16′14.4″	99°3′20.1″	897	Human biting/landing at day with total shade	—	—	—	*Ae. allotecnon*, *Ae. quadrivittatus*, *Ae. podographicus*, *Sa. chloropterus*, *Sh. huasteca*
	18 November 2012	21°33′27.4″	99°9′20.4″	1024	Human biting/landing at day with total shade	—	—	—	*Ae. allotecnon*
	19 November 2012	21°34′39″	99°13′41.9″	806	Human biting/landing at day with total shade	—	—	—	*Ae. trivittatus*, *Ae. podographicus*
	20 November 2012	21°35′7.1″	99°14′15.7″	739	Human biting/landing at day with partial shade	—	—	—	*Ae. triseriatus*
	20 November 2012	21°34′53.6″	99°14′6.8″	686	Human biting/landing at day with total shade	—	—	—	*Ae. quadrivittatus*, *Ae. trivittatus*, *Ae. amabilis*, *Ae. brelandi*, *Sa. chloropterus*
	20 November 2012	21°35′7.1″	99°14′15.7″	739	Bromeliad axil with clear water, leaves at bottom and partial shade	7.61	20	44	—
	20 November 2012	21°34′27.7″	99°12′39.8″	715	Human biting/landing at day with total shade	—	—	—	—
	24 September 2018	21°15′29.3″	99°411.1″	1010	Human biting/landing at day with total shade	—	—	—	*Ae. allotecnon*, *Ae. quadrivittatus*, *Sh. huasteca*
	22 September 2021	21°15′29.3″	99°4′11.1″	1010	Human biting/landing at day	—	—	—	*Ae. allotecnon*, *Cx. lactator*, *Sa. chloropterus*, *Sh. huasteca*, *Wy. aporonoma*
Tx. moctezuma	27 September 2012	21°13′11.2″	99°5′56.9″	140	Discarded tire with clear water, total shade	7.55	22.7	42	*Cx. restrictor*, *Cx. thriambus*
	27 September 2012	21°13′11.2″	99°5′56.9″	140	Discarded tire with clear water, total shade	7.48	22.3	77	—
	24 September 2018	21°8′38″	99°37′15.1″	2292	Discarded tire with clear water, leaves at bottom and total shade	7.15	20	190	*Cx. thriambus*, *Cs.* n. sp.
	24 September 2018	21°15′29.3″	99°4′11.1″	1010	Tree hole with colored water, leaves at bottom and total shade	7.39	23.3	270	*Ae.* (*Protomacleaya*) sp.

3.2.4. Genus *Culex*

The genus *Culex* is the most diverse group of mosquitoes in Querétaro, including fifteen species in the state, of which two had previously been reported. The subgenera *Anoedioporpa*, *Melanoconion*, *Neoculex*, and *Phenacomyia* are recorded for the first time in Querétaro. Immature stages of *Cx. restrictor* were collected from two locations, in a tree hole and a discarded tire; this type of larval habitat is common for this species in Mexico. The new records within the subgenus *Culex* are *Cx. coronator*, *Cx. declarator*, *Cx. erythrothorax*, *Cx. restuans*, *Cx. salinarius*, *Cx. tarsalis*, and *Cx. thriambus*. The Coronator complex includes five recognized species: *Cx. camposi* Dyar, *Cx. coronator*, *Cx. ousqua* Dyar, *Cx. ousquatissimus* Dyar, and *Cx. usquatus* Dyar [47]. Based on male genitalia morphology, *Cx. coronator s.s.* was identified from material collected in Querétaro; this is a common species whose immature stages were collected from a variety of aquatic habitats, mostly natural habitats at ground level. Although *Cx. declarator* is a common species in the central-northern region of Mexico, immature stages of this species were collected from an artificial container in only a single location. *Culex erythrothorax*, *Cx. restuans*, *Cx. tarsalis*, and *Cx. thriambus* are common species that were found frequently and collected from a variety of aquatic habitats such as natural and artificial containers. The record of *Cx. salinarius* was obtained from the CAIM collection. The subgenus *Melanoconion* in Querétaro includes *Cx. erraticus* and *Cx. peccator*; immature stages of both species were collected from natural habitats at ground level in two locations. Within the subgenus *Neoculex*, the species found in Querétaro are *Cx. apicalis* and *Cx. arizonensis*. The subgenus *Phenacomyia* and *Cx. lactator* are recorded for the first time in Querétaro. Adult females of this species were collected approaching humans during the day at a single location.

3.2.5. Genus *Lutzia*

The genus *Lutzia*, subgenus *Lutzia*, and *Lt. bigoti* are recorded for the first time in Querétaro. Immature stages of this species were collected from one spring and an artificial container with clear water, predating larvae of *Ae. albopictus* and *Cx.* spp., respectively.

3.2.6. Genus *Culiseta*

The tribe Culisetini, genus *Culiseta*, and subgenera *Culiseta*, *Culiseta inornata*, and *Cs. particeps* are recorded for the first time in Querétaro. Immature stages of *Cs. inornata* were collected from an irrigation gutter in one location, while immature stages of *Cs. particeps* were collected from a variety of aquatic habitats, such as natural ponds and swamps, artificial containers, and discarded tires. Immature stages of one undescribed species within this genus were discovered in discarded tires. These specimens are in a poor condition; hence, this species will be formally described when more specimens are obtained.

3.2.7. Genus *Limatus*

The tribe Sabethini, genera *Limatus*, *Sabethes*, *Shannoniana*, and *Wyeomyia*; subgenera *Sabethoides*, *Triamyia*, and *Wyeomyia*; species *Li. durhammi*, *Sa. chloropterus*, *Sh. huasteca*, *Wy. apronoma*, *Wy. adelpha/guatemala*, and Wy. mitchellii are recorded for the first time in Querétaro. Immature stages of *Li. durhamii* were collected from an artificial container with clear water in one location with no associated species.

3.2.8. Genus *Sabethes*

In Mexico, the genus *Sabethes* is divided into two subgenera: *Sabethes* and *Sabethoides*. The latter is reported for the first time in Querétaro and is represented by *Sa. chloropterus*. Adult females of this species were collected approaching humans during the day.

3.2.9. Genus *Shannoniana*

In Mexico, three species of the genus *Shannoniana* had been previously recorded: *Sh. fluviatilis* (Theobald), *Sh. moralesi* (Dyar and Knab), and *Sh. schedocyclia* (Dyar and Knab).

In the present study, we discovered a fourth species within the genus. Adult females of *Sh. huasteca* n. sp. were collected approaching humans and males were collected resting in vegetation and approaching humans together with the females. Adults of both sexes are described herein.

3.2.10. Genus *Wyeomyia*

Three species of the genus *Wyeomyia* are reported for the first time in Querétaro: Adult females of *Wy. aporonoma* were collected approaching humans in one location in association with *Ae. allotecnon, Cx. lactator, Sa. chloropterus, Sh. huasteca*, and *Wy. mitchellii*. Since *Wy. guatemala* is possibly a synonymy of *Wy. adelpha* [48], and both species are treated here as a single taxon. Adult females of *Wy. adelpha/guatemala* were collected approaching humans during the day in one location, while immature stages of *Wy. mitchelli* were collected from bromeliad axils and adult females were collected approaching humans in several locations of Querétaro.

3.2.11. Genus *Toxorhynchites*

The tribe Toxorhynchitini, genus *Toxothynchites*, and subgenera *Lynchiella* and *Tx. moctezuma* are recorded for the first time in Querétaro. Immature stages of *Tx. moctezuma* were collected from discarded tires and one tree hole, always with clear water and predating on larvae of *Ae.* sp. and *Cx. thriambus* in tropical and conserved regions of the state.

3.3. *Molecular Analysis*

In total, we analyzed 25 DNA barcodes for five species within the genus *Shannoniana* (four taxa) and *Trichoprosopon* (one taxa) (Table 4). In general, all specimens of the same species clustered together (Figure 3), although there was a deep split in *Sh. fluviatilis* (BOLD:ACZ4319, BOLD:ACZ4320) and *Sh. schedocyclia* (BOLD:ACZ3895, BOLD:ACZ:3896), where two BINs were found in each taxon. All specimens identified as *Sh. huasteca* n. sp. were grouped closely with *Sh. moralesi*, although both groups are well separated with high support bootstrap values (Figure 3). The average genetic divergence was 0.08%; the intra-specific genetic divergence varied from 0.04% in *Sh. huasteca* n. sp., *Sh. moralesi* (0.55%), *Sh. schedocyclia* (1.31%), and *Tr. digitatum* (0.20%). In *Sh. fluviatilis*, the genetic divergence was above 2% (3.96%). Interspecific genetic divergence varied from 4.70% to 13.13%; the pair *Sh. moralesi/Sh. fluviatilis* were the more divergent species (13.13%), while the pair *Sh. huasteca* n. sp./*Sh. moralesi* were less divergent (4.70%).

Table 4. Percentage of interspecific (between groups) pairwise K2P genetic divergence of unique DNA barcodes (658 bp), representing five species of Sabethini.

	Sh. fluviatilis	***Sh. shcedocyclia***	***Sh. moralesi***	***Tr. digitatum***
Sh. fluviatilis	—	—	—	—
Sh. schedocyclia	7.43%	—	—	—
Sh. moralesi	13.13%	12.78%	—	—
Tr. digitatum	11.34%	9.93%	12.67%	—
Sh. huasteca n. sp.	12.29%	12.60%	4.70%	12.72%

3.4. *Description of New Species*

Shannoniana huasteca Ortega n. sp. 4DD7EB32-56DD-41B6-AACC-877947FE26D4. Type specimens: *Holotype*: adult female (A♀) without associated larval and pupal exuviae [CC-UL, 04240918-CN], Camino a Neblinas, Landa de Matamoros, Querétaro, Mexico (21°15′29.3″ N–99°4′11.12″ W) (Figure 4), elevation 1010 m, 24 Sep 2018, 17:00–18:00, human biting at day, tropical cloud forest with oaks and conserved vegetation (Figure 5), col. A.I. Ortega-Morales. *Paratypes*: 10A♀, (same data as *holotype*); [CAIM]. *Allotypes*: 3♂with dissected genitalia, (same data as holotype); [CC-UL] (Table 3).

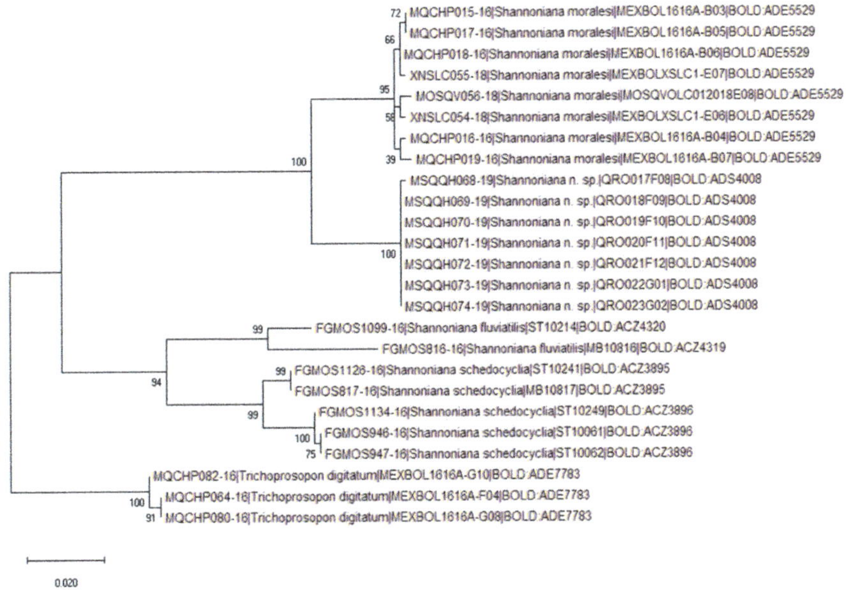

Figure 3. Maximum likelihood tree base on the Kimura 2-parameter of the *COI* DNA barcodes (658 bp) for species of *Shannoniana* ($n = 4$) and *Trichoprosopon* ($n = 1$). A divergence of >2% may be indicative of separate operational taxonomic units.

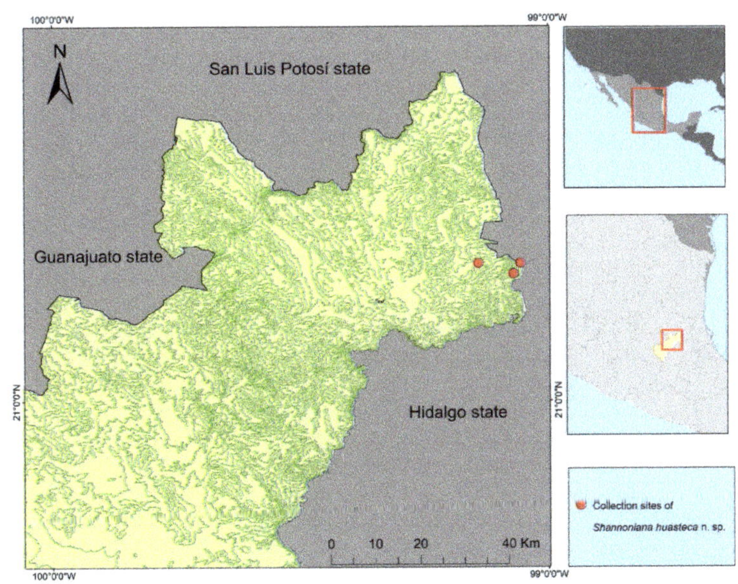

Figure 4. Distribution of *Sh. huasteca*.

Figure 5. Collection site of *Shannoniana huasteca* n. sp., showing the tropical and conserved types of vegetation. The site is near Neblinas road, Landa de Matamoros, Querétaro, Mexico.

Female. Head: Occiput and vertex covered with flat decumbent blue, green and silver scales, with purple and greenish reflections, with a row of erect scales (Figure 6A), interocular setae large, pedicel bare, yellow-brown. Antennae approximately 0.50–0.75 forefemur length. Clypeus bare, dark-brown. Maxillary palpus approximately 0.25 proboscis length, three-segmented, third palpomere longer than the first two, dark-scaled with purplish reflections. Proboscis as long as forefemur, sometimes slightly longer forefemur length of 1.10–1.20, with dark scales with purplish reflections. Thorax: Integument of scutum golden, covered with pale golden-brown decumbent narrow scales without iridescent reflections, acrostichal and dorsocentral setae absent (Figure 6B). Scutellum trilobed, with 7–10 setae on lateral lobes and 5–7 setae on mid lobe, all lobes covered with flat dark-blue scales with purplish reflections. Row of erect dark setae above the paratergite and the wing. Postpronotum covered with flat yellow-golden scales with golden reflections, without setae. Antepronotum lobe with silvery scales, with 4–5 setae. Integument of mesokatepisterum and mesanepimeron dark-golden, mostly covered with large patch of silvery flat scales (Figure 7A), mesanepimeron with 10–12 dark-brown setae. Wing: Approximately 1.30–1.50 mm, scales on veins flat and light-brown (Figure 7B). Halter: Dark-brown with blue scales. All trochanters with patches of silvery scales. All femora dark-scaled, with some iridescent scales bluish-greenish on dorsal line, with a small knee spot of pale scales, fore and midfemora predominantly dark-scaled, with bluish reflections, hindfemur with dorsal dark scaled line and ventral line withe-scaled. Hindtibia dark-scaled, with a complete ring of white-yellow scales apically. Foretarsus covered predominantly with dark scales, mid and hindtarsus with tarsomeres 1–4 dark-scaled, tarsomere 5 with dark scales on dorsal line and white scales on ventral line. Abdomen: All terga covered with dark scales with bluish reflections, apical corners of dark scales on terga extending into 0.50 of sternal segments, sterna covered with white scales.

Figure 6. Adult female of *Shannoniana huasteca*, holotype. (**A**) Occiput showing the decumbent green, blue, and silvery scales, arrow showing the row of erect scales; (**B**) scutum covered with pale golden-brown narrow scales.

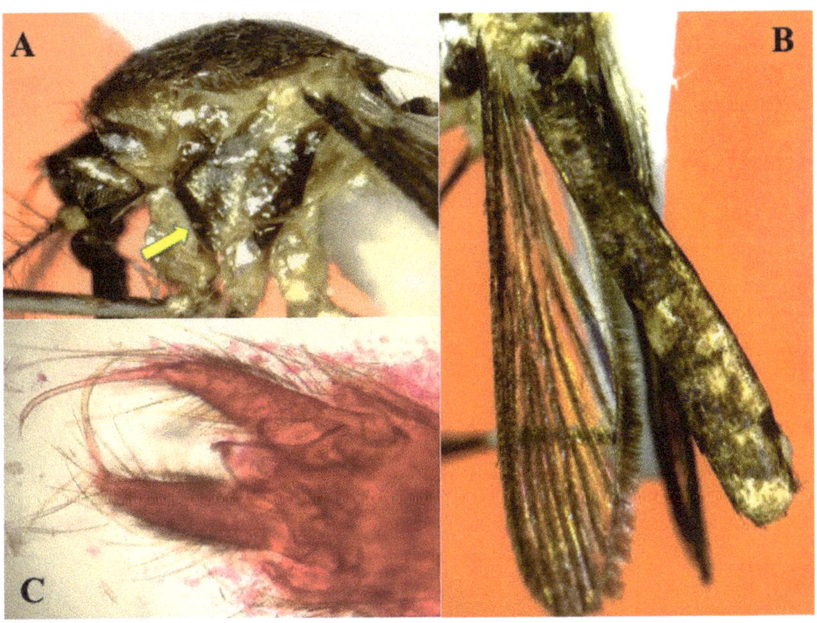

Figure 7. Adult of *Shannoniana huasteca* n. sp. (**A**) General aspect of lateral view of thorax showing the integument golden, mesokaterpisternum, and mesanepimeron dark-golden (holotype), arrow shows the posprocoxal area; (**B**) wing covered with light-brown scales and dark scales on abdominal terga (holotype); (**C**) general aspect of male genitalia (allotype).

Male. In general, as in females except for sexual characteristics and plumose antennae.

Male genitalia (Figure 7C). Segment IX: Tergum with deep emargination between tergal lobes, the lobes slightly longer than broad, with 6–8 long, curved setae. Gonocoxite: Length approximately three times median width, basal tergomesal lobe well developed, with 5–7 long and strong setae. Gonostylus: Narrow, simple, and long, slightly curved apically. Aedeagus strongly sclerotized, simple and ovate.

Larva. Unknown.

Pupa. Unknown.

Systematics. Females of *Sh. huasteca* n. sp. are distinguished from all three previously described species within the genera *Shannoniana* (*Sh. fluviatilis* (Theobald), *Sh. moralesi* (Dyar and Knab), and *Sh. schedocyclia* (Dyar and Knab)) by having all legs covered with dark scales, except for tarsomere five of hind leg, which has white scales in a ventral line; silvery, decumbent scales on vertex not extending to the ocular line, but interrupted by a patch of dark scales with bluish and greenish reflections; and the absence of a patch of silvery scales on postprocoxal area. The males are readily distinguished by the structure of the male genitalia (Figure 7C), especially the narrow and long gonostyle, slightly curved apically.

Bionomics. Although the type locality was visited on numerous occasions to search for immature stages of *Sh. huasteca* n. sp., these were not found. Immature stages of other species were collected in different aquatic habitats such as containers and phytotelmata (e.g., axils of bracts of *Xanthosoma* spp. and bromeliads). In addition, ovitraps were displayed at different elevations from ground level, but all those collections failed to find immature stages of *Sh. huasteca* n. sp. Adult females were collected approaching humans probing to bite during the day in association with *Aedes allotecnon*, *Ae. quadrivittatus*, *Culex lactator*, *Sabethes chloropterus*, *Wyeomyia aporonoma*, and *Wy. mitchellii*. The medical importance of *Sh. huasteca* n. sp. is unknown, but since females can be persistent biters of humans, the species could be involved in the transmission of pathogens.

Distribution. *Shannoniana huasteca* n. sp. has been collected in the northern region of the state of Querétaro (Neblinas road, location of Landa de Matamoros County). Locations in which the species was collected belong to Huasteco Carso of the Sierra Madre Oriental. *Shannoniana huasteca* n. sp. may occur in the forested regions of the states adjacent to Querétaro such as the southeastern San Luis Potosí state and northwestern Hidalgo state, with both states sharing physiographical conditions belonging to the Carso Huasteco of the Sierra Madre Oriental.

Etymology. This species is named *huasteca* because of the type locality in the Carso Huasteco sub-region. "Huasteco" is a word derived from the huasteco language, which means someone from an Amerindian tribe of the Mayan family that lives in the Mexican states of Tamaulipas, San Luis Potosí, Querétaro, and Veracruz.

3.5. Keys to Species of Adult Female of Shannoniana

ADULT FEMALE (Figure 8) (Modified from Lane and Cerqueira [49] and Clark-Gil and Darsie [48]).

1. Hind tarsi with basal rings of white scales on segments I–IV (Figure 9A) . . . . *Sh. schedocyclia* Distr.: Bolivia, Brazil, French Guiana, Guatemala, Mexico, Nicaragua, Panama, Venezuela [50] (Distr. Mex.: Chiapas [27,28,51–53], Oaxaca [27,28], Veracruz [10,28,52]).

- Hind tarsi with segments I–IV covered only with dark scales (Figure 9B). 2

2(1) Tarsomere V of hind leg covered completely with dark scales (Figure 9C) . . . .*Sh. fluviatilis*

Distr.: Argentina, Bolivia, Brazil, Colombia, Ecuador, French Guiana, Guatemala, Guyana, Mexico, Nicaragua, Panama [50] (Distr. Mex.: Chiapas [28,52,54], Veracruz [10,28,52,55], Oaxaca [28], Quintana Roo [2,7]).

- Tarsomere V of hind leg with white scales (Figure 9D) 3

3(2) Silver scales on occiput extending to the ocular line and reaching the inner corner of the eye, mostly with silver reflections (Figure 10); postprocoxal area with a patch of silvery scales . . . .*Sh. moralesi*

Distr.: Belize, Guatemala, Mexico, Panama [50] (Distr. Mex.: Chiapas [27,28,51–53,55,56], Veracruz [10,12,27,28,52], Oaxaca [27,28], Tabasco [20]).

Silver scales on occiput are restricted to the occiput and do not extend to the eyes, the rest of scales on ocular line are dark, with purplish and greenish reflections (Figure 6A); postprocoxal area without a patch of silvery scales (Figure 7A). *Sh. huasteca* n. sp. Ortega

Distr.: Mexico (Distr. Mex.: Querétaro).

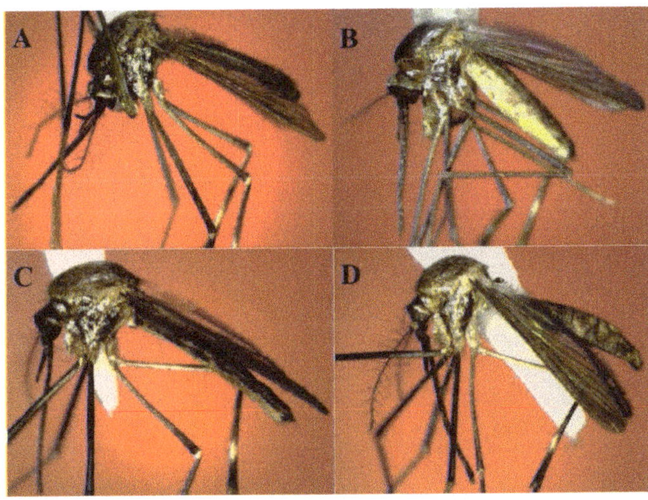

Figure 8. General aspect of adult female of *Shannoniana* spp. (**A**) *Sh. schedocyclia* (Chiapas, Mexico); (**B**) *Sh. fluviatilis* (Quintana Roo, Mexico, MX-QROO-19); (**C**) *Sh. moralesi* (Chiapas, Mexico, 01010818-EU); (**D**) *Sh. huasteca* n. sp. (holotype).

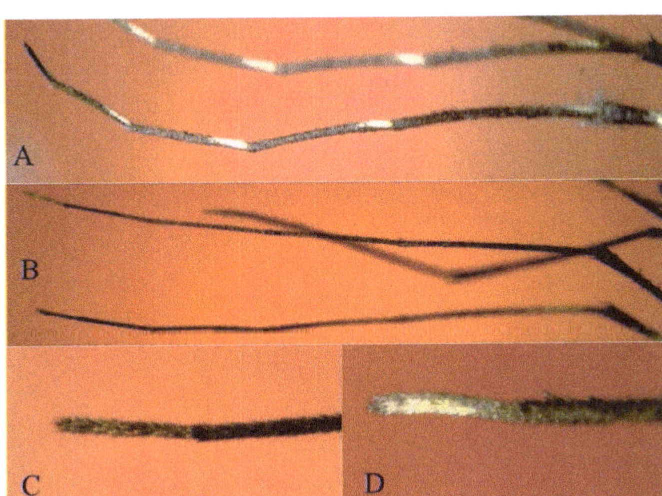

Figure 9. Hind tarsi of adult female of *Shannoniana* spp. (**A**) *Sh. schedocyclia* (Chiapas, Mexico); (**B**,**C**) *Sh. fluviatilis* (Quintana Roo, Mexico, MX-QROO-19); (**D**) *Sh. huasteca* n. sp.

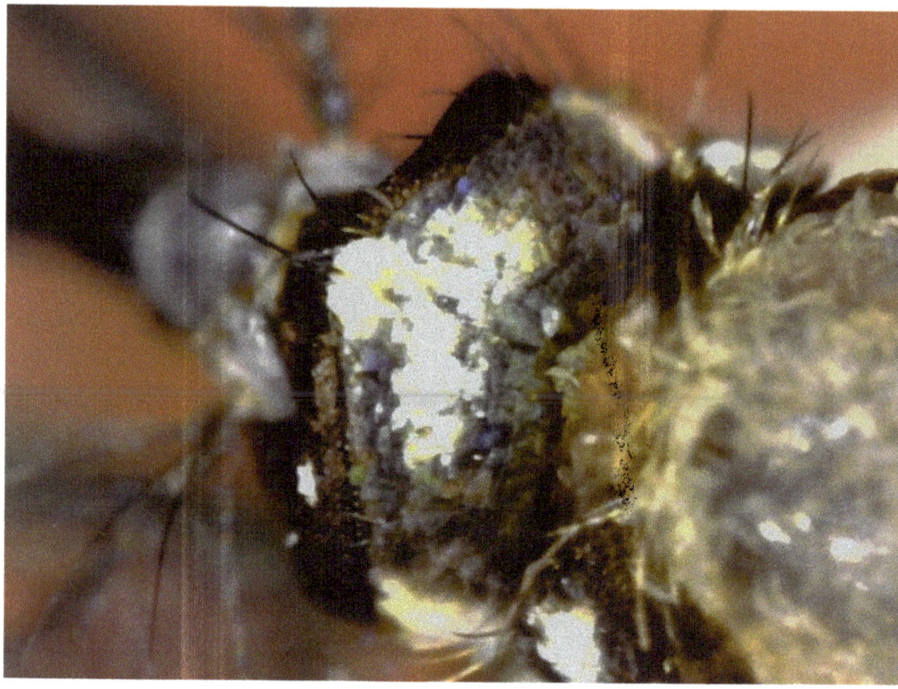

Figure 10. Occuput of *Sh. moralesi* (Chiapas, Mexico, 01010818-EU) showing the decumbent scales mostly with silver reflections.

4. Discussion and Conclusions

4.1. Ecology and Distributional Groups of Species

Based on our collection records and the known distributions of the mosquito species collected in Querétaro state, three groups of species are recognized. The species of each group have similar geographical distributions, reaching their southern or northern distributional limits across the state. The immature stages of groups 1 and 2 develop in phytotelmata such as tree holes and bromeliad axils, while immature stages of group 3 develop in ponds and swamps (Table 3).

4.1.1. Group 1

Species that occur in the Nearctic Region and extend into northern Mexico where they reach their southern limit of distribution in Querétaro include *Aedes brelandi*, *Ae. triseriatus* and the endemic species *Ae. schicki*. Immature stages of those species develop in tree holes filled with rainwater; the presence of species in this group is restricted to the forested and conserved areas of the northern part of the state, extending from the Nearctic Region into the Huasteco Carso of the Sierra Madre Oriental in Querétaro.

4.1.2. Group 2

Species of this group occur in tropical forests in the Neotropical Region and extend into Querétaro where they reach their northern limit. Only one species is reported in this group, *Wyeomyia apronoma*, whose immature stages occur in bamboo internodes, tree holes, and coconut shells. This species has been previously reported in several states of southeastern Mexico, but the distribution is restricted to the north region by the Huasteco Carso of the Sierra Madre Oriental in Querétaro state.

4.1.3. Group 3

Ground pool inhabiting species that extend from the Neotropical Region into Middle Mexico in Querétaro state but no farther north are *Aedes euplocamus*, which has been previously reported in tropical regions of southeastern Mexico where it is a common species during the rainy season, and the Mexican endemic species *Ae. shannoni*, which has been previously reported in the states of Michoacán, Morelos, Querétaro, and Mexico State, but reaches its northernmost distributional limit in Querétaro.

4.2. Species from Adjacent Regions That May Occur in Querétaro

Some species of mosquitoes that have not yet been reported from Querétaro occur in adjacent states and may occur within the state. Included among these are 20 species that have been previously recorded in the state of Hidalgo [14]: *Anopheles aztecus* Hoffmann, *An. crucians* Wiedemann, *An. parapunctipennis* Martini, *An. punctimacula* Dyar and Knab, *An. argyritarsis* Robineau-Desvoidy, *Aedeomyia sqamipennis* Lynch-Arribálzaga, *Aedes muelleri* Dyar, *Psorophora ferox* (von Humboldt), *Culex bidens* Dyar, *Cx. interrogator* Dyar and Knab, *Cx. nigripalpus* Theobald, *Cx. pinarocampa* Dyar and Knab, *Cx. pseudostigmatosoma* Strickman, *Cx. stenolepis* Dyar and Knab, *Cx. rejector* Dyar and Knab, *Cx. territans* Walker, *Cx. corniger* Theobald, *Sabethes gymnothorax* Harbach and Petersen, *Uranotaenia coatzacoalcos* Dyar and Knab, and *Ur. sapphirina* (Osten Sacken); and 12 species that have been previously recorded in the state of México [25]: *Ae. ramirezi* Vargas and Downs, *Ae. guerrero* Berlin, *Ae. lorraineae* Berlin, *Ae. chionotum* Zavortink, *Ae. gabriel* Schick, *Ae. idanus* Schick, *Ae. kompi* Vargas and Downs, *Ae. vargasi* Schick, *Ae. zoosophus* Dyar and Knab, *Haemagogus mesodentatus* Komp and Kumm, *Culiseta incidens* (Thomson), and *Ur. geometrica* Theobald.

4.3. Medical Importance of Mosquitoes of Querétaro

Some of the species reported in Querétaro are of medical and veterinary importance because they are vectors of pathogens causing diseases. In Table 5 the most important public health species that occur in Querétaro are listed.

4.4. Molecular Analysis

The DNA barcode sequences of specimens belong to five Sabethini species of mosquitoes we analyzed in this study grouped together, although a discrepancy in BINs were found in *Sh. fluviatilis* and *Sh. schedocyclia*. This agrees with Talaga [57] in their analysis of the Culicidae DNA barcodes from French Guiana. As we have not been able to examine the voucher specimens from where the DNA barcodes sequences were obtained, we cannot make further comments with regards to the taxonomic status of these two BINs. The specimen we identified as *Sh. huasteca* n. sp. separate with strong support values from those identified as *Sh. moralesi*, which supports our hypothesis that they represent a new species. The latter finding is also supported by the different morphological traits found in the adult female general coloration and the male genitalia.

4.5. Mosquitoes Diversity in Querétaro and Mexico

With the addition of the new mosquito records found in Querétaro reported here, there are currently 50 species known in the state. The state ranks eighth in species richness of the ten Mexican states that have been systematically inventoried for mosquito species. With the addition of *Shannoniana huasteca* n. sp. to the list of mosquito species in Mexico, there are currently 247 known species in the country.

Table 5. Medical importance and pathogens of mosquito vector species collected in Querétaro state, Mexico. Mal: Malaria. DI: *Dirofilaria immitis*. DENV: Dengue virus. ZIKV: Zika virus. CHIKV: Chikungunya virus. YF: Yellow fever virus. SLE: St. Louis encephalitis virus. WNV: West Nile virus. VEEV: Venezuelan equine encephalitis virus. EEEV: Eastern equine encephalitis virus. WEEV: Western equine encephalitis virus. LCV: La Crosse virus.

Taxa	Mal	DI	DENV	ZIKV	CHKV	YF	SLE	WNV	VEEV	EEEV	WEEV	LCV
Anopheles pseudopunctipennis	√											
An. punctipennis	√											
An. albimanus	√											
Aedes vexans		√		√			√	√	√	√	√	
Ae. angustivittatus									√			
Ae. scapularis		√				√	√		√			
Ae. trivittatus		√						√		√	√	√
Ae. triseriatus								√		√		√
Ae. aegypti			√	√	√	√		√	√	√		√
Ae. albopictus		√	√	√	√	√		√	√	√	√	√
Haemagogus equinus						√						
Culex quinquefasciatus					√		√	√	√	√		
Cx. restuans							√	√		√		
Cx. salinarius							√	√		√		
Cx. tarsalis								√	√		√	
Cx. erraticus								√				
Culiseta inornata								√			√	
Sabethes chloropterus						√	√					

Author Contributions: A.I.O.-M. and L.M.H.-T.—study conceptualization; A.I.O.-M. and L.M.H.-T.—methodology; A.I.O.-M.—sampling; A.I.O.-M. and L.M.H.-T.—data curation; A.I.O.-M.—mosquito identification; L.M.H.-T. and Q.K.S.-R.—molecular assays; A.I.O.-M., L.M.H.-T., and Q.K.S.-R.—analyses; A.I.O.-M. and Q.K.S.-R.—figures conceptualization; A.I.O.-M.—writing—original draft preparation; L.M.H.-T. and Q.K.S.-R.—writing—review and editing; A.I.O.-M., L.M.H.-T. and Q.K.S.-R.—visualization and supervision; A.I.O.-M. and L.M.H.-T.—project administration and funding acquisition. All authors have read and agreed to the published version of the manuscript.

Funding: Aldo I. Ortega-Morales was the recipient of a grant supported by the Consejo Nacional de Ciencia y Tecnología-Convocatoria Ciencia Básica (National Council for Science and Technology-Basic Science Call) CONACyT (Project: Catálogo Nomenclatural de los mosquitos (Diptera: Culicidae) presentes en el Eje Neovolcánico de algunos estados de México) [Nomenclatural Catalogue of the Mosquitoes (Diptera: Culicidae) from Neo-Volcanic Axis of some states of Mexico] (Grant 182216/QH1113).

Institutional Review Board Statement: Not applicable.

Data Availability Statement: All mosquitoes (mounted on pins and/or microscope slides) collected in Querétaro State were deposited in the Culicidae Collection (CC-UL), Parasitology Department, Universidad Autónoma Agraria Antonio Narro unidad laguna (Autonomous Agrarian University Antonio Narro laguna unit), Torreón, Coahuila, Mexico.

Acknowledgments: We are grateful to the mosquito collection staff Salvador Morales-Avitia, Adelfo Sánchez-Trinidad, Félix Ordóñez-Sánchez, Juan J. Castro-Xochitla, Oscar L. Galindo-Soto, Federico Ortega-Lozano, Isis J. Morales-Avitia, Guillermo Morillón-Borjón, and Antonio A. Sánchez-García for their valuable collaboration during our collection trips; Jorge A. Alvarado-Zapata and Imelda Correa-Tinajero from the Biological Station "Quin" for their hospitality in Arcila, Querétaro; Eva I. Conde-Sánchez from Secretaría de Salud del estado de Querétaro (Health Secretary of Querétaro State); Herón Huerta-Jiménez from Instituto de Diagnóstico y Referencia Epidemiológica (Institute of Diagnosis and Epidemiological Reference) for providing some mosquito records collected in Querétaro state, and Thomas J. Zavortink for the revision of the English language and suggestions, which have improved the quality of this paper. Aldo I. Ortega-Morales would like to thank the National Council of Science and Technology (CONACyT) for founding some collection trips, Project: "Nomenclatural Catalog of the mosquitoes (Diptera: Culicidae) present in the Neo-volcanic Axis of some States of Mexico" (Grant QH1113). Luis M. Hernández-Triana would like to thank the Department of Environment and Rural Affairs (DEFRA) and the Devolved Administrations of Scotland and Wales who co-founded Contract C, project SV3045.

Conflicts of Interest: The authors declare no conflict of interest.

References

1. Muñoz-Cabrera, L.O.; Ibáñez-Bernal, S.; Corona-Vargas, M.C. Los mosquitos (Diptera: Culicidae) de Tlaxcala, México. I: Lista comentada de especies. *Folia Entomol. Mex.* **2006**, *45*, 223–271.
2. Ortega-Morales, A.I.; Mis-Ávila, P.C.; Elizondo-Quiroga, A.; Harbach, R.E.; Siller-Rodríguez, Q.K.; Fernández-Salas, I. The mosquitoes of Quintana Roo state, Mexico (Diptera: Culicidae). *Acta Zool. Mex.* **2010**, *26*, 33–46.
3. Ordóñez-Sánchez, F.; Sánchez-Trinidad, A.; Mis-Ávila, P.; Canul-Amaro, G.; Fernández-Salas, I.; Ortega-Morales, A.I. Nuevos registros de mosquitos (Diptera: Culicidae) en algunas localidades de Campeche y Quintana Roo. *Entomol. Mex.* **2013**, *12*, 850–854.
4. Chan-Chablé, R.J.; Ortega-Morales, A.I.; Martínez-Arce, A. First record of *Psorophora albipes* in Quintana Roo, Mexico. *J. Am. Mosq. Control Assoc.* **2016**, *32*, 237–239. [CrossRef]
5. Ortega-Morales, A.I.; Casas-Martínez, M.; Bond, G.; Harbach, R.E. First records of *Psorophora cilipes* and *Culex theobaldi* in Quintana Roo, Mexico. *J. Am. Mosq. Control Assoc.* **2018**, *34*, 124–127. [CrossRef]
6. Chan-Chablé, R.J.; Martínez-Arce, A.; Mis-Ávila, P.C.; Ortega-Morales, A.I. DNA barcodes and evidence of cryptic diversity of anthropophagous mosquitoes in Quintana Roo, Mexico. *Ecol. Evol.* **2019**, *9*, 4692–4705. [CrossRef]
7. Chan-Chablé, R.J.; Martínez-Arce, A.; Ortega-Morales, A.I.; Mis-Ávila, P.C. New records and updated checklist of mosquito species in Quintana Roo, Mexico, using DNA-Barcoding. *J. Am. Mosq. Control Assoc.* **2020**, *36*, 264–268. [CrossRef]
8. Canto-Mis, K.L.; Chan-Chablé, R.J.; Gómez-Rivera, A.S.; López-Sosa, X.Y.; González-Acosta, C.; Correa-Morales, F.; Mis-Ávila, P.C. Nuevos registros de distribución para *Uranotaenia sapphirina* (Osten Sacken, 1868) (Diptera: Culicidae) en Quintana Roo, México. *Rev. Chil. Entomol.* **2021**, *47*, 613–617.
9. Canto-Mis, K.L.; Gómez-Rivera, A.S.; Chan-Chablé, R.J.; González-Acosta, C.; Correa-Morales, F.; Mis-Ávila, P.C. Primer registro de distribución para *Psorophora varipes* (Coquillett, 1904) (Diptera: Culicidae) en Quintana Roo, México. *Rev. Chil. Entomol.* **2022**, *48*, 293–297.
10. Ibáñez-Bernal, S.; Mendoza-Palmero, F.S.; Hernández-Xoliot, R.A. Mosquitos (Insecta: Diptera: Culicidae). In *La Biodiversidad de Veracruz Estudio de Estado, Vol. II*; Cruz-Angón, A., Lorca-Hernández, F.G., Hernández-Ortiz, V., Morales-Mavil., J.E., Eds.; Comisión Nacional para el Conocimiento y Uso de la Biodiversidad CONABIO: Mexico City, México, 2011; pp. 399–403, Appendix 5–8.
11. Mendez-Andrade, A.; Rivera-García, K.D.; Ibáñez-Bernal, S. Notes on the *Toxorhynchites* of Mexico: Redescription of *Tx. moctezuma* (Dyar & Knab) and new record for *Tx. grandiosus* (Williston) in Veracruz (Diptera: Culicidae). *Zootaxa* **2019**, *4576*, 140–150.
12. Rivera-García, K.D.; Méndez-Andrade, A.; Ibáñez-Bernal, S. *Trichoprosopon mixtli* sp. nov., a new sabethini species (Diptera: Culicidae) from a Mexican cloud forest, with an assessment of the genus and keys for the identification pk known species. *Zootaxa* **2023**, *5254*, 94–116. [CrossRef] [PubMed]
13. Ortega-Morales, A.I.; Zavortink, T.J.; Huerta-Jiménez, H.; Sánchez-Rámos, F.J.; Valdés-Perezgasga, M.T.; Reyes-Villanueva, F.; Siller-Rodríguez, Q.; Fernández-Salas, I. Mosquito records from Mexico: The mosquitoes (Diptera: Culicidae) of Tamaulipas state. *J. Med. Entomol.* **2015**, *52*, 171–184. [CrossRef] [PubMed]
14. Ortega-Morales, A.I.; Zavortink, T.J.; Huerta-Jiménez, H.; Ibáñez-Bernal, S.; Siller-Rodríguez, Q.K. The mosquitoes (Diptera: Culicidae) of Hidalgo state, Mexico. *Acta Trop.* **2019**, *189*, 94–103. [CrossRef] [PubMed]
15. Ortega-Morales, A.I.; Pérez-Paredes, M.G.; Siller-Rodríguez, Q.K.; Moreno-García, M.; González-Acosta, C.; Correa-Morales, F. First record of *Aedes gabriel* in Hidalgo state, Mexico. *J. Am. Mosq. Control Assoc.* **2019**, *35*, 51–54. [CrossRef]
16. Ortega-Morales, A.I.; Zavortink, T.J.; Garza-Hernández, J.A.; Siller-Rodríguez, Q.K.; Fernández-Salas, I. The mosquitoes (Diptera: Culicidae) of Nuevo León, Mexico, with descriptions of two new species. *PLoS ONE* **2019**, *14*, e0217694. [CrossRef]
17. Villegas-Ramírez, H.M.; Ortega-Morales, A.I.; Flores-Suárez, A.; Fernández-Salas, I.; Ponce-García, G. First record of *Aedes podographicus* in Nuevo León state, Mexico. *J. Am. Mosq. Control Assoc.* **2021**, *37*, 87–89. [CrossRef]
18. Ortega-Morales, A.I.; Morillón-Borjón, G.; Morales-Avitia, I.J.; Sánchez-García, A.A.; Ayala-Sulca, Y.O.; Sánchez-Rámos, F.J. First record of *Wyeomyia mitchellii* in Nuevo León, Mexico. *J. Am. Mosq. Control Assoc.* **2022**, *38*, 216–218. [CrossRef]
19. Torres-Chable, O.M.; Baak-Baak, C.M.; Cigarroa-Toledo, N.; Zaragoza-Vera, C.V.; Arjona-Jiménez, G.; Moreno-Perez, L.G.; Machain-Williams, C.; García-Rejon, J. Mosquito fauna un home environments of Tabasco, Mexico. *Southwest. Entomol.* **2017**, *42*, 969–982. [CrossRef]

20. Ortega-Morales, A.I.; Méndez-López, R.; Garza-Hernández, J.A.; González-Álvarez, V.H.; Ruiz-Arrondo, I.; Huerta-Jiménez, H.; Rodríguez-Martínez, L.M.; Rodríguez-Pérez, M.A. The mosquitoes (Diptera: Culicidae) of Tabasco, Mexico. *J. Vector Ecol.* **2019**, *44*, 57–67. [CrossRef]
21. Rodríguez-Martínez, L.M.; Yzquierdo-Gómez, P.; González-Acosta, C.; Correa-Morales, F. First record of *Aedes* (*Ochlerotatus*) *fulvus* in Tabasco and distribution notes of other *Aedes* in Mexico. *Southwest. Entomol.* **2020**, *45*, 263–268. [CrossRef]
22. Ortega-Morales, A.I.; Rodríguez-Martínez, L.M.; Méndez-Alvarado, W.; Garza-Hernández, J.A.; López-Hernández, I.; Medrano-Santillana, M.; González-Acosta, C.; Correa-Morales, F. The distribution of *Uranotaenia sapphirina* and *Ur. socialis* in Tabasco, southern Mexico. *J. Am. Mosq. Control Assoc.* **2022**, *38*, 141–147. [CrossRef] [PubMed]
23. Dampf, A. Distribución y ciclo annual de *Uranotaenia syntheta* Dyar & Shannon en México y descripción del hipopigio masculine (Insecta, Diptera). *Rev. Soc. Mex. Hist. Nat.* **1943**, *4*, 147–170.
24. Dávalos-Becerril, E.; Correa-Morales, F.; González-Acosta, C.; Santos-Luna, R.; Peralta-Rodríguez, J.; Pérez-Rentería, C.; Ordóñez-Álvarez, R.; Huerta, H.; Carmona-Perez, M.; Díaz-Quiñonez, J.A.; et al. Urban and semi-urban mosquitoes of Mexico City: A risk for endemic mosquito borne-disease transmission. *PLoS ONE* **2019**, *14*, e0212987. [CrossRef] [PubMed]
25. Adeniran, A.A.; Hernández-Triana, L.M.; Ortega-Morales, A.I.; Garza-Hernández, J.A.; de la Cruz-Ramos, J.; Chan-Chablé, R.J.; Vázquez-Marroquín, R.; Huerta-Jiménez, H.; Nikolova, N.I.; Fooks, A.R.; et al. Identification of mosquitoes (Diptera: Culicidae) from Mexico State, Mexico using morphology and *COI* DNA barcoding. *Acta Trop.* **2021**, *213*, 105730. [CrossRef] [PubMed]
26. Vargas, L.; Martínez-Palacios, A. Anofelinos mexicanos, taxonomía y distribución. In *Anofelinos Mexicanos Taxonomia y Distribución*; Secretaría de Salud: Ciudad de México, Mexico, 1956; p. 181.
27. Vargas, L. Especies y distribución de mosquitos mexicanos no Anofelinos (Insecta, Diptera). *Rev. Inst. Salubr. Enf. Trop.* **1956**, *1*, 19–36.
28. Díaz-Nájera, A.; Vargas, L. Mosquitos mexicanos: Distribución geográfica actualizada. *Rev. Inv. Salud Pública.* **1973**, *33*, 111–125.
29. Ibáñez-Bernal, S.; Martínez-Campos, C. Clave para la aidentificación de larvas de mosquitos comunes en las áreas urbanas y suburbanas de la República Mexicana (Diptera: Culicidae). *Folia Entomol. Mex.* **1994**, *92*, 43–73.
30. Sánchez-Trinidad, A.; Ordóñez-Sánchez, F.; Valdez-Perezgasga, M.T.; Sánchez-Ramos, F.J.; Zavortink, T.J.; Cortés-Guzman, A.J.; Ortega-Morales, A.I. Geographical distribution of the *Aedes* Triseriatus group (Diptera: Culicidae) in Mexico. *J. Vector Ecol.* **2014**, *39*, 134–137. [CrossRef]
31. Quezada-Yaguachi, W.E.; Alquisira-Domínguez, M.; Vázquez-Anzúres, M.J.; Rebollo-Salinas, D.; Rescalvo-Luna, L.D.; Medina-Castañeda, A.; González-Acosta, C.; Correa-Morales, F.; Viveros-Santos, V.; Moreno-García, M. Nuevo registro de *Haemagogus* (*Haemagogus*) *equinus* Theobald, 1903 y otros mosquitos (Diptera: Culicidae) recolectados con ovitrampas en Jalpan de Serra, Querétaro, México. *Rev. Chil. Entomol.* **2023**, *49*, 73–81.
32. Ortega-Morales, A.I.; Moreno-García, M.; González-Acosta, C.; Correa-Morales, F. Mosquito surveillance in Mexico: The use of ovitraps for *Aedes aegypti*, *Ae. albopictus*, and non-target species. *Fla. Entomol.* **2018**, *101*, 623–626. [CrossRef]
33. Ortega-Morales, A.I.; Pérez-rentería, C.; Ordóñez-Álvarez, J.; Salazar, J.A.; Dzul-Manzanilla, F.; Correa-Morales, F.; Huerta-Jiménez, H. Update on the dispersal of *Aedes albopictus* in Mexico: 1988–2021. *Front. Trop. Dis.* **2022**, *2*, 72. [CrossRef]
34. INEGI. Instituto Nacional de Estadística y Geografía. 2018. Available online: https://www.inegi.org.mx/app/cuadroentidad/Qro/2018/01/1_4 (accessed on 15 February 2023).
35. Harbach, R.E.; Knight, K.L. *Taxonomist's Glossary of Mosquito Anatomy*; Plexus Publishing, Inc.: Marlton, NJ, USA, 1980; ISBN 9780937548004.
36. Knight, K.L.; Stone, A. *A Catalog of the Mosquitoes of the World (Diptera: Culicidae)*; Thomas Say Foundation: College Park, MD, USA, 1977.
37. Wilkerson, R.C.; Linton, Y.M.; Fonseca, D.M.; Schultz, T.R.; Price, D.C.; Strickman, D.A. Making mosquito taxonomy useful: A stable classification of Tribe Aedini that balances utility with current knowledge of evolutionary relationships. *PLoS ONE* **2015**, *10*, e0133602. [CrossRef]
38. WRBU. Walter Reed Biosystematics Unit, Smithsonian Institution. 2005. Available online: http://www.wrbu.org (accessed on 15 November 2022).
39. Folmer, O.; Black, M.; Hoeh, W.; Lutz, T.; Vrijenhoek, R. DNA Primers for amplification of mitochondrial Cytochrome c Oxidase Subunit I from diverse Metazoan Invertebrtes. *Mol. Mar. Biol. Biotechnol.* **1994**, *3*, 294–299.
40. Prosser, W.J.S.; deWaard, J.R.; Miller, S.C.; Hebert, P.D.N. DNA barcodes from century-old type specimens using next generation sequencing. *Mol. Ecol. Res.* **2016**, *16*, 487–489. [CrossRef] [PubMed]
41. Tamura, K.; Stoecher, G.; Peterson, N.; Filipski, A.; Kumar, S. MEGA6: Molecular evolutionary genetics analysis version 6.0. *Mol. Biol. Evol.* **2013**, *30*, 2725–2729. [CrossRef] [PubMed]
42. Hebert, P.N.D.; Cywinska, A.; Ball, S.L.; deWaard, J.R. Biological identification through DNA barcodes. *Proc. R. Soc. Biol. Sci.* **2003**, *270*, 213–321. [CrossRef]
43. Hebert, P.N.D.; Ratnasingham, S.; deWaard, J.R. Barcoding animal life: Cytochrome *c* oxidase subunit 1 divergences among closely related species. *Proc. R. Soc. Biol. Sci.* **2003**, *270*, 596–599. [CrossRef]
44. Hall, T. BioEdit: A user-friendly biological sequence alignment editor analysis program for Windows 95/98/NT. *Nucelic Acids Symp. Ser.* **1999**, *41*, 95–98.
45. Larkin, M.A.; Blackshields, G.; Brown, N.P.; Chenna, R.; McGettigan, P.A.; McWilliam, H.; Valentin, F.; Wallace, I.M.; Wilm, A.; Lopez, R.; et al. Clustal W and Clustal X version 2.0. *Bioinformatics* **2007**, *23*, 2947–2948. [CrossRef]

46. Hernández-Triana, L.M.; Brugman, V.A.; Nikolova, N.I.; Ruiz-Arrondo, I.; Barrero, E.; Thorne, T.; de Marco, M.F.; Kruger, A.; Lumley, S.; Johnson, N.; et al. DNA barcoding of British mosquitoes (Diptera, Culicidae) to support species identification, discovery of cryptic genetic diversity and monitoring invasive species. *Zookeys* **2019**, *832*, 57–76. [CrossRef]
47. Laurito, M.; Briscoe, A.G.; Almirón, W.R.; Harbach, R.E. Systematics of the *Culex coronator* complex (Diptera: Culicidae): Morphological and molecular assessment. *Zool. J. Linn. Soc.* **2018**, *182*, 735–757. [CrossRef]
48. Clark-Gil, S.; Darsie, R.F. The mosquitoes of Guatemala, their identification, distribution and bionomics, with keys to adult female and larvae. *Mosq. Syst.* **1983**, *15*, 151–294.
49. Lane, J.; Cerquiera, N.L. Os Sabetíneos de América (Diptera, Culicidae). *Arq. Zool. São Paulo* **1942**, *9*, 473–849.
50. Wilkerson, R.C.; Linton, T.-M.; Strickman, D. *Mosquitoes of the World*; Johns Hopkins University Press: Baltimore, MD, USA, 2021; Volume 2.
51. Vargas, L.; Martínez-Palacios, A. Descripción de *Wyeomyia* (*Wyeomyia*) *stonei*, n. sp., y notas sobre otros Sabethini de Mexico. *Rev. Inst. Salubr. Enf. Trop.* **1953**, *4*, 293–307.
52. Díaz-Nájera, A. Mosquitos tropicales de Mexico. *Rev. Invest. Salud Publ.* **1966**, *1*, 57–63.
53. Viveros-Santos, V.; Hernández-Triana, L.M.; Ibáñez-Bernal, S.; Ortega-Morales, A.I.; Nikolova, N.I.; Pairot, P.; Fooks, A.R.; Casas-Martínez, M. Integrated approaches for the identification of mosquitoes (Diptera: Culicidae) from the volcanoes of Central America Physiographic subprovence of the state of Chiapas, Mexico. *Vector Borne Zoonotic Dis.* **2022**, *22*, 120–137. [CrossRef]
54. Vargas, L. Notas sobre mosquitos nuevos para Mexico. *Rev. Inst. Salubr. Enf. Trop.* **1939**, *1*, 101–104.
55. Díaz-Nájera, A. Variaciones morfológicas en larvas de dos especies de *Trichoprosopon* y datos sobre su distribución. *Rev. Inst. Salubr. Enf. Trop.* **1961**, *3–4*, 201–219.
56. Hernández-Triana, L.M.; Garza-Hernández, J.A.; Ortega Morales, A.I.; Prosser, S.W.; Hebert, P.D.; Nikolova, N.I.; Barrero, E.; de Luna-Santillana, E.D.J.; González-Alvarez, V.H.; Mendez-López, R.; et al. An integrated molecular approach to untangling host-vector-pathogen interactions in mosquitoes (Diptera: Culicidae) from sylvan communities in Mexico. *Front. Vet. Sci.* **2021**, *7*, 564791. [CrossRef]
57. Talaga, S.; Leroy, C.; Guidez, A.; Dusfour, I.; Girod, R.; Dejean, A.; Murienne, J. DNA reference libraries of French Guianese mosquitoes for barcoding and metabarcoding. *PLoS ONE* **2017**, *12*, e0176993. [CrossRef]

diversity

MDPI

Article

Anopheles maculipennis Complex in The Netherlands: First Record of *Anopheles daciae* (Diptera: Culicidae)

A. Ibáñez-Justicia [1,*,†], Nathalie Smitz [2,*,†], Rody Blom [3], Ann Vanderheyden [4], Frans Jacobs [1], Kenny Meganck [2], Sophie Gombeer [4], Thierry Backeljau [4,5], Constantianus J. M. Koenraadt [3], J. S. Griep [3], Marc De Meyer [2] and Arjan Stroo [1]

1 Centre for Monitoring of Vectors (CMV), Netherlands Institute for Vectors, Invasive Plants and Plant Health (NIVIP), Netherlands Food and Consumer Product Safety Authority (NVWA), Geertjesweg 15, 6706EA Wageningen, The Netherlands
2 Royal Museum for Central Africa (RMCA), Barcoding Facility for Organisms and Tissues of Policy Concern (BopCo), Leuvensesteenweg 13-17, 3080 Tervuren, Belgium
3 Laboratory of Entomology, Plant Sciences Group, Wageningen University & Research, Droevendaalsesteeg 1, 6708PB Wageningen, The Netherlands
4 Barcoding Facility for Organisms and Tissues of Policy Concern (BopCo), Royal Belgian Institute of Natural Sciences (RBINS), Vautierstraat 29, 1000 Brussels, Belgium
5 Evolutionary Ecology Group, University of Antwerp, Universiteitsplein 1, 2610 Antwerp, Belgium
* Correspondence: a.ibanezjusticia@nvwa.nl (A.I.-J.); nathalie.smitz@africamuseum.be (N.S.); Tel.: +31-(0)-882231269 (A.I.-J.); +32-(0)-27695855 (N.S.)
† These authors contributed equally to this work.

Abstract: Despite their past importance as vectors of indigenous malaria, the species composition and spatial distribution of the members of the *Anopheles maculipennis* complex have been studied to a limited extent in the Netherlands. Therefore, this investigation focuses on the distribution of the members of this complex in the Netherlands, including *Anopheles daciae*, which has recently been found in countries bordering the Netherlands. In the framework of a national mosquito surveillance between 2010 and 2021, a total of 541 specimens of *An. maculipennis* s.l. were analyzed from 161 locations covering the entire territory. In addition, 89 specimens were analyzed from overwintering sites during the winter of 2020/2021. All individual mosquitoes were identified to species-level using Sanger sequencing of the ribosomal internal transcribed spacer 2. To characterize the habitat of *An. maculipennis* s.l. in the Netherlands, land cover use data was extracted in a 1 km buffer area around each finding location. For populations collected in summers between 2010 and 2021, the most frequent species was *An. messeae*, present in 88.19% of the locations, followed by *An. maculipennis* s.s. (11.80%), *An. atroparvus* (3.72%) and *An. daciae* (3.72%). *Anopheles daciae* was found in the southern inland areas of the country. Furthermore, *An. messeae* and *An. daciae* occurred in sympatry at overwintering sites. This study provides relevant information on the occurrence of species of the *Anopheles maculipennis* complex in the Netherlands, contributing to a better estimation of the risk of mosquito-borne disease in the country.

Keywords: mosquitoes; DNA-based species identification; ribosomal internal transcribed spacer 2 (ITS2); malaria vector

Citation: Ibáñez-Justicia, A.; Smitz, N.; Blom, R.; Vanderheyden, A.; Jacobs, F.; Meganck, K.; Gombeer, S.; Backeljau, T.; Koenraadt, C.J.M.; Griep, J.S.; et al. *Anopheles maculipennis* Complex in The Netherlands: First Record of *Anopheles daciae* (Diptera: Culicidae). *Diversity* **2022**, *14*, 636. https://doi.org/10.3390/d14080636

Academic Editors: Carlos Barceló and Ignacio Ruiz-Arrondo

Received: 29 June 2022
Accepted: 25 July 2022
Published: 10 August 2022

1. Introduction

In the Netherlands, the last published checklist of mosquitoes (Diptera: Culicidae) included 35 indigenous species [1]. This list is not static and was recently updated with a new indigenous species, *Culiseta longiareolata* [2]. Four members of the *Anopheles maculipennis* complex (= s.l.) are reported as present in the published checklist [1], some of which are capable of carrying pathogens of medical importance, including malaria: *Anopheles atroparvus* van Thiel, 1927, *An. messeae* Falleroni, 1926, *An. maculipennis* sensu stricto Meigen, 1818, and *An. melanoon* Hackett, 1934. However, this latter species is not considered to occur in the Netherlands by [3], and *An. maculipennis* s.s. is considered uncommon in the country [4]. *Anopheles atroparvus* is reported as the only malaria vector (*Plasmodium vivax* and *Plasmodium malariae*) in the coastal areas of the Netherlands. Experimentally, *An. atroparvus* can also be an effective host and capable of transmitting *Plasmodium ovale* to humans [5]. A comparison of the anopheline species composition between 1935 and 1999 showed a prevalence shift from *An. atroparvus* to *An. messeae* in the Delta of the Rivers Rhine and Meuse, coinciding with the disappearance of indigenous malaria [6]. In a study of overwintering mosquitoes in several farms in the Netherlands, *An. messeae* individuals were more frequently found and *An. atroparvus* was less common [4]. A decline of *An. atroparvus* over the 20th century was recorded in multiple European countries and was assumed to be linked to major ecological changes, such as drainage practices, surface water pollution, loss of suitable resting sites for hibernation, etc. [6–8].

Despite their past importance as vectors of indigenous malaria and their potential role in the transmission of imported tropical malaria, leading to the reappearance of autochthonous malaria cases in Europe [9], the species composition and spatial distribution of the members of *An. maculipennis* s.l. in the Netherlands has been poorly studied. Individuals of *An. maculipennis* s.l. were found at 144 sampling sites during a nationwide inventory of indigenous mosquitoes, involving natural, rural and urban habitats [10]. However, this nationwide inventory did not include DNA-based species identifications to distinguish the members of the complex. Yet, the nuclear ribosomal internal transcribed spacer 2 (ITS2) flanked by portions of the conserved 5.8S and 28S rDNA is useful in this respect [11–13]. Since the members of the *An. maculipennis* complex are difficult to discriminate by morphological characteristics, their identification needs to be verified by ITS2 sequencing.

Anopheles daciae Linton, Nicolescu & Harbach, 2004, is a recently described species of the *An. maculipennis* complex that is distributed throughout continental Europe [14] and has been found over the past years in the countries bordering the Netherlands, including Germany [15,16], the United Kingdom [11], and Belgium [17]. Nevertheless, *An. daciae* has not yet been reported in the Netherlands. Therefore, the aim of this study was 1) to investigate the distribution of *An. maculipennis* s.l. members present in the Netherlands by applying ITS2 sequencing and 2) to find evidence of the presence of *An. daciae* in the Netherlands, where it is expected to occur, given its presence in neighboring countries.

2. Materials and Methods

2.1. Sampling

Mosquito specimens for the study were collected from different surveys. Most of the specimens (n = 531) were collected between the months of May and September using Mosquito Magnet Liberty Plus traps (WoodstreamTM Co., Lititz, PA, USA) using octenol, in the framework of the National Mosquito Survey [10]. This included 145 specimens collected in 2011, 74 in 2012, 146 in 2013, 72 in 2014, 88 in 2015, 3 in 2016, 12 in 2017 and one in 2021. Additional specimens were collected during Exotic Mosquito Surveys [18] using a variety of sampling methods such as BG-Sentinel traps (n = 4) or BG-Mosquitaire traps (n = 5) (Biogents AG, Regensburg, Germany) both using BG-Sweetscent, and larval sampling using aquarium nets (n = 1). In total, 541 specimens (540 adults and 1 larva) of *An. maculipennis* s.l. were collected at 161 locations covering the entire territory of the Netherlands.

In addition to these 541 specimens, a total of 89 *An. maculipennis* s.l. specimens were collected in February (n = 65) and March (n = 24) 2021 from six bunkers of the New Dutch Waterline, located in the municipality of West-Betuwe, the Netherlands. These bunkers are well-known overwintering sites for several mosquito species [19].

All specimens were transported to the laboratory and were morphologically identified to the *Anopheles maculipennis* complex level using the key of Becker et al. [20]. After identification, specimens were placed in sterile vials and kept frozen at −20 °C until further processing.

2.2. DNA-Based Species Identification

Individual DNA was extracted from a leg or a part of abdomen using the NucleoSpin® Tissue DNA extraction kit (Macherey-Nagel, Düren, Germany), following the manufacturer's protocols, with an elution volume of 70 µL. For the mosquitoes collected from overwintering sites in February and March 2021, DNA was extracted following the ammoniumhydroxide-protocol as described by [21]. The ITS2 fragment was amplified using the primers of [16], with thermal cycling conditions, PCR reactions and purification for sequencing following [17], except for the mosquitoes collected in overwintering sites in February and March 2021. For these latter, ITS2 fragment were amplified using MyTaq® HS Red mix (Bioline, UK) using the primers as described in [16] and the following thermal cycling conditions: 1 min at 95 °C, followed by 35 cycles of 95 °C for 15 s, 53 °C for 15 s, and 72 °C for 10 s. Forward and reverse strands were assembled and corrected with Geneious® Prime v.2019.2.3 (Biomatters Ltd., Auckland, New Zealand), after which consensus sequences were generated and trimmed to remove the primers and low-quality ends.

ITS2 consensus sequences were used as queries to search for most similar sequences in GenBank (NCBI, National Centre for Biotechnology), using the Basic Local Alignment Search Tool (https://blast.ncbi.nlm.nih.gov/Blast.cgi, accessed on 15 June 2022). To discriminate between *An. daciae* and *An. messeae*, aligned consensus sequences were visually checked for the presence of the five species-specific diagnostic sites [12]. Species-assigned consensus sequences were then aligned together with sequences of all other *Anopheles* species occurring in the Netherlands, namely *An. algeriensis* Theobald, 1903, *An. claviger* (Meigen, 1804), *An. plumbeus* Stephens, 1828 [22], and with outgroup sequences (namely *An. funestus* sensu stricto Giles, 1900, and *An. minimus* Theobald, 1901), using ClustalW in Geneious® Prime v.2019.2.3. We also included ITS2 sequences of *An. melanoon* Hackett, 1934, which is part of *An. maculipennis* s.l. and was reported to occur in the Netherlands in the past [22] but was not identified in previous reports [3]. Conspecific identical sequences were removed from the database to retain unique ITS2 sequences in the final alignment. Using the web application FindModel (http://hiv.lanl.gov/content/sequence/findmodel/findmodel.html, accessed on 15 June 2022), the Kimura 2-parameter was identified as the best evolution model describing our data [23,24]. A rooted maximum likelihood tree (ML) was constructed using MEGA X v.10.0.5 (Kumar et al., 2018), with branch support assessed by 1000 bootstrap replicates. Average interspecific K2P distances were calculated with the R v.3.6.1 package Spider v3.6.2 [25,26].

2.3. Habitat Characterization

To characterize the habitat of *An. maculipennis* s.l. in the Netherlands, a 1 km buffer area was created around each finding location (excluding overwintering locations). Using this pre-defined buffer zone in ArcGIS v.10.7.1 [27] and the 2018 version of the raster file of the Corine Land Cover [28], the areas covered by the five main Land Cover Classes were extracted: artificial or urban areas, agricultural areas, forest and seminatural areas, wetlands, and water bodies. The expected number of specimens per Land Cover Class was calculated using the following formula: (total number of specimens per species × percentage Land Cover in the Netherlands)/100. Habitat association per species was verified by comparing observed number of specimens per species per Land Cover Class with expected number of specimens per species per Land Cover Class using a Fisher's exact test. All statistical analyses were conducted in RStudio v1.4.1717 [25].

3. Results

The ITS2 fragment was scored in the 541 specimens collected between May and September. Of these, 496 specimens were assigned as *An. messeae*, 25 as *An. maculipennis* s.s., 11 as *An. atroparvus*, and nine as *An. daciae*. Of the 89 mosquitoes collected from the six bunkers, 82 specimens were identified as *An. messeae* (92.13%), and seven were identified as *An. daciae* (8.99%).

No ITS2 sequences were shared between the four species of *An. maculipennis* s.l. collected in the Netherlands, with each of the four species involving one unique species-specific haplotype (Figure 1). Average interspecific K2P distances ranged from 0.687 to 8.258% (Table 1). Double peaks at two of the five supposedly diagnostic sites discriminating *An. messeae* from *An. daciae* were observed in three ITS2 sequences of *An. daciae*, namely position 218 (A/T) (n = 2) and 220 (C/T) (n = 1) (site numbering following [12]). Such ambiguities were not recorded in the 578 *An. messeae* ITS2 sequences (214 (T), 218 (T), 220 (C), 416 (G), and 436 (G)).

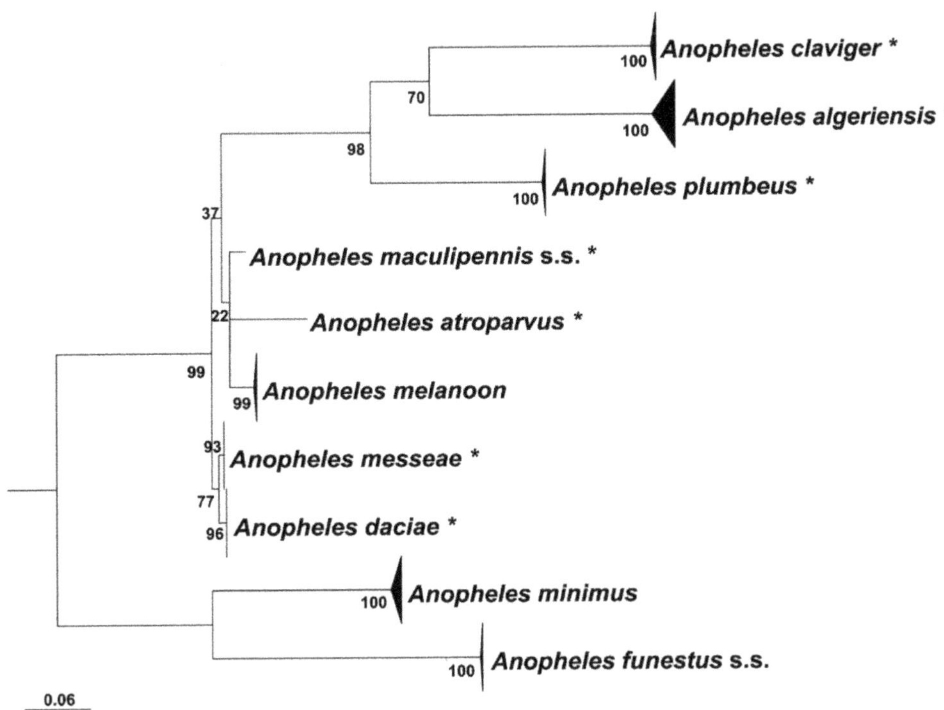

Figure 1. Condensed ITS2 ML-tree (K2P model) of five members of *Anopheles maculipennis* s.l. (*An. maculipennis* s.s.; *An. messeae; An. atroparvus; An. daciae; An. melanoon*), four of which were collected in the Netherlands in the present study, including *An. plumbeus, An. claviger* and *An. algeriensis* occurring in the Netherlands, and *An. funestus* sensu stricto and *An. minimus* as outgroups (GenBank accession numbers: KP298399, KP298400, OK570292, OK570315). Duplicate sequences per species were excluded, with the ITS2 databases for *An. messeae* and *An. daciae* including a few sequences displaying ambiguous sites. Numbers at nodes are bootstrap support values. * species reported in the Netherlands.

Excluding the sampled overwintering sites (i.e., the six bunkers), the most frequent species captured was *An. messeae*, being present in 80.19% of the locations, followed by *An. maculipennis* s.s. (11.80%), *An. atroparvus* (3.72%) and *An. daciae* (3.72%) (Figure 2, Table S1). *Anopheles messeae* was found in a total of 142 locations, being found in sympatry with *An. maculipennis* s.s. at four locations and with *An. atroparvus* at five locations. *Anopheles daciae* was found in sympatry with *An. maculipennis* s.s. at three out of the six locations where it was identified (Figure 2). Using ITS2, the present investigation provides the first solid evidence of the occurrence of *An. daciae* in the Netherlands. The 16 identified specimens were captured only in 2011, 2012, 2013 and 2021, in the southern part of the country (Figure 2). At overwintering collection sites, the seven *An. daciae* were collected in

February from five out of six bunkers; no *An. daciae* were identified in March. *Anopheles maculipennis* s.s. and *An. atroparvus* were not found at the overwintering sites.

Table 1. Descriptive statistics of the genetic diversity of ITS2 within *Anopheles maculipennis* s.l. in the Netherlands, including the average interspecific K2P distances among sequences (excluding conspecific identical sequences).

	n	N_H	N_P	Average Interspecific K2P (%) ± Range (%)
An. atroparvus	11	1	0	8.258 ± 0.000
An. daciae	9 + **7**	1	0 *	0.881 ± 0.327
An. maculipennis s.s.	25	1	0	3.114 ± 0.000
An. messeae	496 + **82**	1	0	0.687 ± 0.003

n: sample size. N_H: number of haplotypes. N_P: number of polymorphic nucleotide sites. * ambiguities recorded at two of the five species-diagnostic sites. Numbers in bold are specimens collected at overwintering sites (2020/2021).

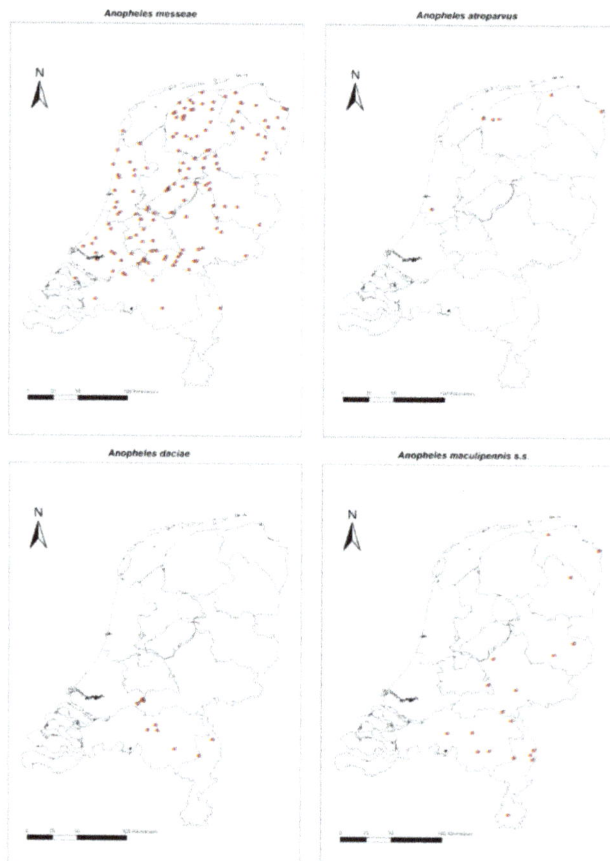

Figure 2. Distribution of *Anopheles maculipennis* s.l. in the Netherlands (details in Table S1) identified using DNA-based techniques: *Anopheles messeae*, *Anopheles atroparvus*, *Anopheles daciae* and *Anopheles maculipennis* sensu stricto.

Except for *An. maculipennis* s.s., preferred Land Cover Class for *An. messeae*, *An. daciae* and *An. atroparvus* are areas with predominant agricultural use (Figure 3). *Anopheles maculipennis* s.s. was found in Land Cover Classes artificial habitats (e.g., industrial or residential areas) and agricultural areas in similar proportions. *Anopheles atroparvus* was not found in Land Cover Classes forests or seminatural areas. *Anopheles atroparvus* was collected in the north of the country, at six locations near the coast, where it is more probable to encounter mixing seawater and fresh water. A significant difference between expected and observed distributions per Land Cover Class was observed for *An. messeae* ($p < 0.001$), indicating a significantly higher occurrence at artificial Land Cover Classes than expected. For *An. maculipennis* s.s. ($p = 0.164$), *An. daciae* ($p = 1$) and *An. atroparvus* ($p = 1$), no significant differences between expected and observed distributions were found (Figure 3).

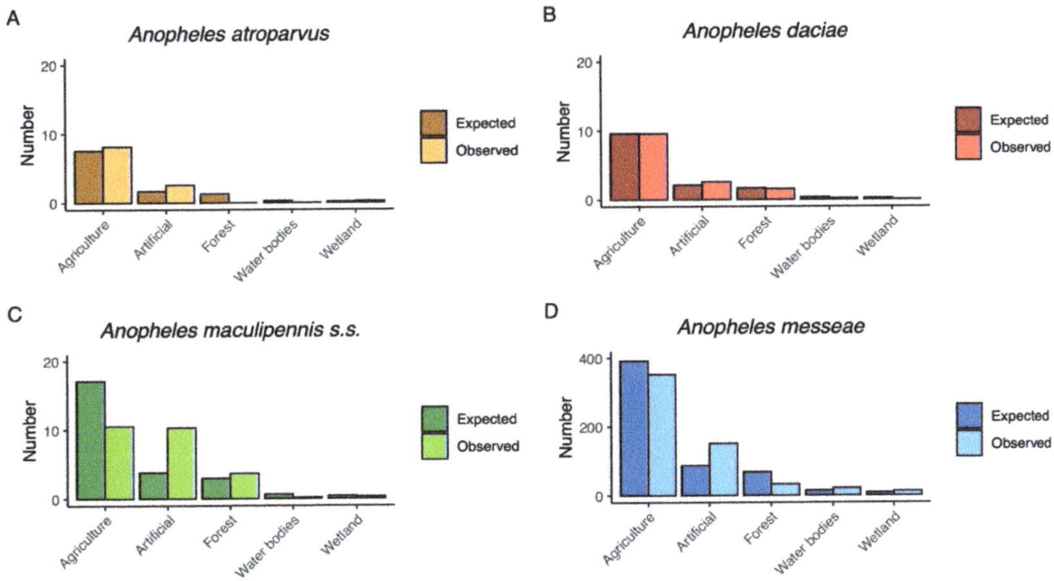

Figure 3. Expected distribution versus observed distribution per Land Cover Class of (**A**). *An. atroparvus* ($n = 11$), (**B**). *An. daciae* ($n = 16$), (**C**). *An. maculipennis* s.s. ($n = 25$) and (**D**). *An. messeae* ($n = 478$). Significant differences between observed and expected distributions were observed for *An. messeae* only (Fisher's exact test, $p < 0.001$).

4. Discussion and Conclusions

In the Netherlands, four species of *Anopheles maculipennis* s.l. were identified: *An. messeae*, *An. maculipennis* s.s., *An. atroparvus* and *An. daciae*. Our study presents the first report of *An. daciae* in the Netherlands using ITS2 sequencing.

Each identified species of the complex involved a single ITS2 haplotype, with smallest average interspecific K2P distances between *An. daciae* and *An. messeae* (0.7%). However, ambiguous sites at some of the species-diagnostic positions were observed [17,29–31]. Double peaks in chromatograms can result from slight differences among ITS2 copies (heterozygosity) and their regular occurrence in specimens from different countries, surveys and years, suggests that only two sites are diagnostic between *An. daciae* and *An. messeae*, namely positions 416 (A/G) and 436 (C/G). For the remainder, phylogenetic DNA sequence analyses of ND4, ND5, COI and Hunchback gene fragments provide no support for the distinction of *An. daciae* and *An. messeae* [17,31], while other taxonomically diagnostic features between these nominal species are still poorly investigated (e.g., hybrid incompatibility, morphology, ecology, cytotaxonomy, etc.). Therefore, it has been proposed to regard *An. daciae* as a species inquirenda (i.e., a species of doubtful identity [32]) [17,33].

Our study shows that *An. messeae* is the species with the widest distribution in the Netherlands compared to any other species of *An. maculipennis* s.l. Similarly to a study in Germany [14], *An. messeae* was the most frequent species in the analyzed samples. In the Netherlands, the species was most commonly found in agricultural areas. Important and ecologically relevant features of the Dutch agricultural landscape are the drainage ditches, which are an important refuge for biodiversity in agricultural landscapes [34] and represent a preferred breeding site for *Anopheles* mosquitoes. However, when comparing the distribution of *An. messeae* with the overall distribution of Land Cover Classes in the Netherlands, the species seems to be associated with artificial Land Cover Classes, more than we would expect based on the land cover distribution of the entire country. Interestingly, the prevalence of *An. messeae* observed in the present study is comparable to that reported by [4], which focused on overwintering mosquitoes. In the latter study, *An. maculipennis* s.s. was not detected, and the presence of *An. daciae* was not investigated by ITS2 sequence data. *Anopheles atroparvus*, a common species breeding in brackish waters, was found at three sites in the coastal areas of the Netherlands during the study of [4].

Anopheles maculipennis s.s. is also widely distributed in the Netherlands, covering large areas from North to South, but it was not found near the coastal areas. Our data show that this species occurs in similar proportions in both urban and agricultural Land Cover Classes. In our study, no preference of *An. maculipennis* s.s. towards a specific class was identified. However, similar to *An. atroparvus* and *An. daciae*, the collected number of *An. maculipennis* s.s. specimens was too low for adequate analyses. As such, further research is needed to identify habitat preferences. In Belgium, *An. maculipennis* s.s. appears to be the most frequent and widespread species of the complex [17]. However, this observation was based on a survey of artificial breeding sites. Therefore, the higher prevalence of *An. maculipennis* s.s. is not surprising, since this species seems better adapted to artificial habitats compared to other species of *An. maculipennis* s.l. [35–37]. In the present study, most of the specimens were collected using adult traps at randomly generated locations across the country, including urban, rural and natural Land Cover Classes, thus preventing a sampling bias [10]. The occurrence of *An. maculipennis* s.s. in urban areas in our study (>40% of the sampling sites) seems to corroborate its preference for man-made habitats.

Except for *An. atroparvus* [6], the potential role of *An. messeae, An. maculipennis* s.s. and *An. daciae* in the historical transmission of malaria in the Netherlands is unknown. For *An. daciae*, the present highlighted species distribution does not fit with the historical areas where the malaria parasite occurred [6]. The main vector of malaria was *An. atroparvus* [6], a species dependent on brackish water. Nowadays, in comparison with the other members of the species complex, the species distribution of *An. atroparvus* indicates the presence of the species in scarce locations nearby coastal areas.

This study also shows that *An. messeae* and *An. daciae* live in sympatry during winter. It remains unclear, however, how the species within the *An. maculipennis* complex differ in overwintering strategies. Earlier studies have shown that *An. atroparvus* enters diapause in early winter, but occasionally continues its blood-feeding behaviour to maintain fat reserves, in contrast to *An. messeae*, which remains inactive throughout winter [38,39]. To our knowledge, this is the first study to find *An. daciae* overwintering in artificial shelters together with other mosquito species that are in diapause [19]. Furthermore, *An. daciae* was only found in the southern inland areas and occurring in areas with Land Cover Classes associated with agricultural activities.

This study provides accurate and unbiased information on the occurrence of the species of the *Anopheles maculipennis* complex in the Netherlands and shows that *An. messeae* is the most frequent species in the country during summers and can occur in sympatry with *An. daciae* in winters. Unbiased occurrence data are needed to develop mosquito species distribution models that can contribute to a better estimation of the risk of mosquito-borne diseases in the country. In addition, while the ITS2 gene fragment is an adequate tool for the identification of the species of the *Anopheles maculipennis* complex, exploring the species' whole genomes will further help elucidate the phylogenetic relationships between the complex members and support their taxonomic status.

Supplementary Materials: The following supporting information can be downloaded at: https://www.mdpi.com/article/10.3390/d14080636/s1, Table S1: Sampling locations (latitude, longitude) and methods (BGS: BG-Sentinel trap; BGM: BG-Mosquitaire trap; MMLP: Mosquito Magnet Liberty Plus traps), collection year, and DNA-based identification results of specimens collected in the Netherlands.

Author Contributions: Conceptualization, A.I.-J., A.S. and N.S.; methodology, A.I.-J., N.S., A.V., F.J., R.B., K.M. and S.G.; analyses, A.I.-J., N.S., A.V., R.B. and C.J.M.K.; data curation, A.I.-J., N.S., A.V., F.J., R.B., C.J.M.K., K.M. and S.G.; writing—original draft preparation, A.I.-J. and N.S.; writing—review and editing, A.V., F.J., R.B., K.M., S.G., T.B., C.J.M.K., J.S.G., M.D.M. and A.S.; visualization, A.I.-J., N.S., C.J.M.K. and R.B.; supervision, T.B., M.D.M. and A.S.; project administration and funding acquisition, T.B., M.D.M. and A.S. All authors have read and agreed to the published version of the manuscript.

Funding: Part of this work (overwintering mosquito sampling) is part of the research program One Health PACT with project number 109,986, which is (partly) financed by the Dutch Research Council (NWO). The Barcoding Facility for Organisms and Tissues of Policy Concern (BopCo—http://bopco.myspecies.info/, accessed on 15 June 2022) is financed by the Belgian Science Policy Office (Belspo).

Institutional Review Board Statement: Not applicable.

Data Availability Statement: The data that support the findings of this study are openly available in GenBank at https://www.ncbi.nlm.nih.gov/genbank/, accessed on 15 June 2022. Accession numbers: *An. atroparvus*: ON033651-ON033660 and ON033662; *An. daciae*: ON053456-ON053464 and ON407975-ON407981; *An. maculipennis* s.s.: ON033742-ON033766; *An. messeae*: ON053466-ON053930, ON053932-ON053962 and ON407982-ON408063.

Acknowledgments: We would like to thank our NVWA colleagues for their assistance in the laboratory and in the field.

Conflicts of Interest: The authors declare no conflict of interest.

References

1. Verdonschot, P.F.M.; Beuk, P.L.T. Family Culicidae. In *Checklist of the Diptera of the Netherlands*; Beuk, P.L.T., Ed.; KNNV Uitgeverij: Utrecht, The Netherlands, 2019. Available online: https://diptera-info.nl/news.php?fam=Culicidae (accessed on 13 June 2022).
2. Deblauwe, I.; Ibanez-Justicia, A.; De Wolf, K.; Smitz, N.; Schneider, A.; Stroo, A.; Jacobs, F.; Vanslembrouck, A.; Gombeer, S.; Dekoninck, W.; et al. First detections of *Culiseta longiareolata* (Diptera: Culicidae) in Belgium and the Netherlands. *J. Med. Entomol.* **2021**, *58*, 2524–2532. [CrossRef] [PubMed]
3. Takken, W.; Kager, P.A.; Verhave, J.P. Will malaria return to North-West Europe? In *Emerging Pests and Vector-Borne Diseases in Europe*; Takken, W., Knols, B.G.J., Eds.; Wageningen Academic Publishers: Wageningen, The Netherlands, 2007; pp. 23–24.
4. Scholte, E.J.; Mars, M.H.; Braks, M.; Den Hartog, W.; Ibañez-Justicia, A.; Koopmans, M.; Koenraadt, C.J.; De Vries, A.; Reusken, C. No evidence for the persistence of Schmallenberg virus in overwintering mosquitoes. *Med. Vet. Entomol.* **2013**, *28*, 110–115. [CrossRef] [PubMed]
5. Collins, W.E.; Jeffery, G.M. Plasmodium ovale: Parasite and disease. *Clin. Microbiol. Rev.* **2005**, *18*, 570–581. [CrossRef] [PubMed]
6. Takken, W.; Geene, R.; Adam, W.; Jetten, T.H.; van der Velden, J.A. Distribution and dynamics of larval populations of *Anopheles messeae* and *A. atroparvus* in the delta of the rivers Rhine and Meuse, The Netherlands. *Ambio* **2002**, *31*, 212–218. [CrossRef]
7. Rodhain, J.; Van Hoof, M.T. Recherches sur l'anophélisme en Belgique. *Ann. De La Société Belg. De Médecine Trop.* **1942**, *21*, 19–43.
8. Van Seventer, H.A. The disappearance of malaria in the Netherlands. *Ned. Tijdschr. Voor Geneeskd.* **1969**, *113*, 2055–2056.
9. ECDC. Multiple Reports of Locally-Acquired Malaria Infections in the EU. Available online: https://www.ecdc.europa.eu/sites/portal/files/documents/RRA-Malaria-EU-revised-September-2017_0.pdf (accessed on 20 May 2022).
10. Ibañez-Justicia, A.; Stroo, A.; Dik, M.; Beeuwkes, J.; Scholte, E.J. National mosquito (Diptera: Culicidae) survey in The Netherlands 2010-2013. *J. Med. Entomol.* **2015**, *52*, 185–198. [CrossRef]
11. Danabalan, R.; Monaghan, M.T.; Ponsonby, D.J.; Linton, Y.M. Occurrence and host preferences of *Anopheles maculipennis* group mosquitoes in England and Wales. *Med. Vet. Entomol.* **2014**, *28*, 169–178. [CrossRef]
12. Nicolescu, G.; Linton, Y.M.; Vladimirescu, A.; Howard, T.M.; Harbach, R.E. Mosquitoes of the *Anopheles maculipennis* group (Diptera: Culicidae) in Romania, with the discovery and formal recognition of a new species based on molecular and morphological evidence. *Bull. Entomol. Res.* **2004**, *94*, 525–535. [CrossRef]
13. Proft, J.; Maier, W.A.; Kampen, H. Identification of six sibling species of the *Anopheles maculipennis* complex (Diptera: Culicidae) by a polymerase chain reaction assay. *Parasitol. Res.* **1999**, *85*, 837–843. [CrossRef]

14. Bertola, M.; Mazzucato, M.; Pombi, M.; Montarsi, F. Updated occurrence and bionomics of potential malaria vectors in Europe: A systematic review (2000-2021). *Parasit Vectors* **2022**, *15*, 88. [CrossRef] [PubMed]
15. Luhken, R.; Czajka, C.; Steinke, S.; Jost, H.; Schmidt-Chanasit, J.; Pfitzner, W.; Becker, N.; Kiel, E.; Kruger, A.; Tannich, E. Distribution of individual members of the mosquito *Anopheles maculipennis* complex in Germany identified by newly developed real-time PCR assays. *Med. Vet. Entomol.* **2016**, *30*, 144–154. [CrossRef]
16. Weitzel, T.; Gauch, C.; Becker, N. Identification of *Anopheles daciae* in Germany through ITS2 sequencing. *Parasitol. Res.* **2012**, *111*, 2431–2438. [CrossRef] [PubMed]
17. Smitz, N.; De Wolf, K.; Gheysen, A.; Deblauwe, I.; Vanslembrouck, A.; Meganck, K.; De Witte, J.; Schneider, A.; Verle, I.; Dekoninck, W.; et al. DNA identification of species of the *Anopheles maculipennis* complex and first record of *An. daciae* in Belgium. *Med. Vet. Entomol.* **2021**, *35*, 442–450. [CrossRef]
18. Ibañez-Justicia, A.; Koenraadt, C.J.M.; Stroo, A.; van Lammeren, R.; Takken, W. Risk-based and adaptive invasive mosquito surveillance at Lucky bamboo and used tire importers in the Netherlands. *J. Am. Mosq. Control. Assoc.* **2020**, *36*, 89–98. [CrossRef] [PubMed]
19. Koenraadt, C.J.M.; Mohlmann, T.W.R.; Verhulst, N.O.; Spitzen, J.; Vogels, C.B.F. Effect of overwintering on survival and vector competence of the West Nile virus vector *Culex pipiens*. *Parasit Vectors* **2019**, *12*, 147. [CrossRef] [PubMed]
20. Becker, N.; Petric, D.; Zgomba, M.; Boase, C.; Madon, M.; Dahl, C.; Kaiser, A. *Mosquitoes and Their Control*; Kluwer Academic/Plenum Publishers: Heidelberg, Germany; Dordrecht, The Nertherlands; London, UK; New York, NY, USA, 2010.
21. Vogels, C.B.; van de Peppel, L.J.; van Vliet, A.J.; Westenberg, M.; Ibanez-Justicia, A.; Stroo, A.; Buijs, J.A.; Visser, T.M.; Koenraadt, C.J. Winter Activity and Aboveground Hybridization Between the Two Biotypes of the West Nile Virus Vector Culex pipiens. *Vector Borne Zoonotic Dis.* **2015**, *15*, 619–626. [CrossRef]
22. Robert, V.; Günay, F.; Le Goff, G.; Boussès, P.; Sulesco, T.; Khalin, A.; Medlock, J.M.; Kampen, H.; Petrić, D.; Schaffner, F. Distribution chart for Euro-Mediterranean mosquitoes (western Palaearctic region). *J. Eur. Mosq. Control Assoc.* **2019**, *37*, 1–28.
23. Posada, D.; Crandall, K.A. MODELTEST: Testing the model of DNA substitution. *Bioinformatics* **1998**, *14*, 817–818. [CrossRef]
24. Tao, N.; Richardson, R.; Bruno, W.; Kuiken, C. FindModel. 2016. Available online: http://hiv.lanl.gov/content/sequence/findmodel/findmodel.html (accessed on 18 February 2021).
25. RStudioTeam. RStudio: Integrated Development for R. RStudio. 2020 Boston, MA: PBC. Available online: https://rstudio.com/ (accessed on 26 February 2021).
26. Brown, S.D.; Collins, R.A.; Boyer, S.; Lefort, M.C.; Malumbres-Olarte, J.; Vink, C.J.; Cruickshank, R.H. Spider: An R package for the analysis of species identity and evolution, with particular reference to DNA barcoding. *Mol. Ecol. Resour.* **2012**, *12*, 562–565. [CrossRef]
27. ESRI. *ArcGIS Desktop: Release 10*; Environmental Systems Research Institute: Redlands, CA, USA, 2011.
28. Copernicus. CORINE Land Cover. Available online: https://land.copernicus.eu/pan-european/corine-land-cover (accessed on 17 April 2022).
29. Bezzhonova, O.V.; Goryacheva, I.I. Intragenomic heterogeneity of rDNA internal transcribed spacer 2 in *Anopheles messeae* (Diptera: Culicidae). *J. Med. Entomol.* **2008**, *45*, 337–341. [CrossRef] [PubMed]
30. Culverwell, C.L.; Vapalahti, O.P.; Harbach, R.E. *Anopheles daciae*, a new country record for Finland. *Med. Vet. Entomol.* **2020**, *34*, 145–150. [CrossRef] [PubMed]
31. Lilja, T.; Eklof, D.; Jaenson, T.G.T.; Lindstrom, A.; Terenius, O. Single nucleotide polymorphism analysis of the ITS2 region of two sympatric malaria mosquito species in Sweden: *Anopheles daciae* and *Anopheles messeae*. *Med. Vet. Entomol.* **2020**, *34*, 364–368. [CrossRef] [PubMed]
32. ICZN. *International Code of Zoological Nomenclature*, 4th ed.; International Trust for Zoological Nomenclature: London, UK, 1999; 306p.
33. Calzolari, M.; Desiato, R.; Albieri, A.; Bellavia, V.; Bertola, M.; Bonilauri, P.; Callegari, E.; Canziani, S.; Lelli, D.; Mosca, A.; et al. Mosquitoes of the Maculipennis complex in Northern Italy. *Sci. Rep.* **2021**, *11*, 6421. [CrossRef] [PubMed]
34. Mukete, B.; Vermaat, J.E.; van Straalen, N.M. Variability in microhabitats prevalence in low lying peat polder ditches of the Netherlands. *Appl. Ecol. Environ. Res.* **2013**, *11*, 629–643. [CrossRef]
35. Becker, N.; Petrić, D.; Zgomba, M.; Boase, C.; Madon, M.B.; Dahl, C.; Kaiser, A. *Mosquitoes: Identification, Ecology and Control*; Springer Nature Switzerland AG: Cham, Switzerland, 2020; 570p.
36. Dakić, Z.; Zoran, K.; Novica, S.; Mijomir, P.; Miloje, C.; Stanimirovic, Z.; Nikola, I.; Poluga, J.; Pavlovic, M. Ecology of *Anopheles* mosquitoes in Belgrade area. Estimating vector potential for malaria retransmission. *Acta Veterinaria* **2008**, *58*, 603–614.
37. Weyer, F. Die geographische Verbreitung der Rassen von *Anopheles maculipennis* in Deutschland. *Z. Für Parasitenkd.* **1938**, *10*, 437–463. [CrossRef]
38. Ramsdale, C.D.; Wilkes, T.J. Some aspects of overwintering in southern England of the mosquitoes *Anopheles atroparvus* and *Culiseta annulata* (Diptera: Culicidae). *Ecol. Entomol.* **1985**, *10*, 449–454. [CrossRef]
39. Weyer, F. Versuche zur Übertragung der affen-malaria durch stechmucken. *Arch. Schiff. f. Tropenhyg.* **1937**, *41*, 167–172.

diversity

MDPI

Article

DNA Identification and Diversity of the Vector Mosquitoes *Culex pipiens* s.s. and *Culex torrentium* in Belgium (Diptera: Culicidae)

Ann Vanderheyden [1,*], Nathalie Smitz [2,*], Katrien De Wolf [3,4], Isra Deblauwe [3], Wouter Dekoninck [5], Kenny Meganck [2], Sophie Gombeer [1], Adwine Vanslembrouck [3,5], Jacobus De Witte [3], Anna Schneider [3], Ingrid Verlé [3], Marc De Meyer [2], Thierry Backeljau [1,6], Ruth Müller [3] and Wim Van Bortel [3,7]

1 Royal Belgian Institute of Natural Sciences, Barcoding Facility for Organisms and Tissues of Policy Concern, Vautierstraat 29, 1000 Brussels, Belgium; sgombeer@naturalsciences.be (S.G.); tbackeljau@naturalsciences.be (T.B.)
2 Royal Museum for Central Africa, Barcoding Facility for Organisms and Tissues of Policy Concern, Leuvensesteenweg 17, 3080 Tervuren, Belgium; kenny.meganck@africamuseum.be (K.M.); marc.de.meyer@africamuseum.be (M.D.M.)
3 The Unit of Entomology, Department Biomedical Sciences, Institute of Tropical Medicine, Nationalestraat 155, 2000 Antwerp, Belgium; katrien.dewolf@ugent.be (K.D.W.); ideblauwe@itg.be (I.D.); avanslembrouck@itg.be (A.V.); jdewitte@itg.be (J.D.W.); aschneider@itg.be (A.S.); ingridverle@hotmail.com (I.V.); rmuller@itg.be (R.M.); wvanbortel@itg.be (W.V.B.)
4 Terrestrial Ecology Unit, Department of Biology, Ghent University, K.L. Ledeganckstraat 35, 9000 Ghent, Belgium
5 Royal Belgian Institute of Natural Sciences, Scientific Heritage Service, Vautierstraat 29, 1000 Brussels, Belgium; wdekoninck@naturalsciences.be
6 Evolutionary Ecology Group, University of Antwerp, Universiteitsplein 1, 2610 Antwerp, Belgium
7 Outbreak Research Team, Institute of Tropical Medicine, Nationalestraat 155, 2000 Antwerp, Belgium
* Correspondence: avanderheyden@naturalsciences.be (A.V.); nathalie.smitz@africamuseum.be (N.S.); Tel.: +32-(0)-2-627-4123 (A.V.); +32-(0)-2-769-5855 (N.S.)

Citation: Vanderheyden, A.; Smitz, N.; De Wolf, K.; Deblauwe, I.; Dekoninck, W.; Meganck, K.; Gombeer, S.; Vanslembrouck, A.; De Witte, J.; Schneider, A.; et al. DNA Identification and Diversity of the Vector Mosquitoes *Culex pipiens* s.s. and *Culex torrentium* in Belgium (Diptera: Culicidae). *Diversity* **2022**, *14*, 486. https://doi.org/10.3390/d14060486

Academic Editors: Carlos Barceló and Ignacio Ruiz-Arrondo

Received: 16 May 2022
Accepted: 13 June 2022
Published: 15 June 2022

Abstract: This survey reports on the DNA identification and occurrence of *Culex torrentium* and *Cx. pipiens* s.s. in Belgium. These native disease-vector mosquito species are morphologically difficult to separate, and the biotypes of *Cx. pipiens* s.s. are morphologically indistinguishable. *Culex torrentium* and *Cx. pipiens* s.s. were identified using the COI and ACE2 loci. We recorded 1248 *Cx. pipiens* s.s. and 401 *Cx. torrentium* specimens from 24 locations in Belgium (collected between 2017 and 2019). *Culex pipiens* biotypes *pipiens* and *molestus*, and their hybrids, were differentiated using fragment-size analysis of the CQ11 locus (956 *pipiens* and 227 *molestus* biotype specimens, 29 hybrids). Hybrids were observed at 13 out of 16 sympatric sites. These results confirm that both species are widespread in Belgium, but while *Cx. torrentium* revealed many COI haplotypes, *Cx. pipiens* s.s. showed only one abundant haplotype. This latter observation may either reflect a recent population-wide demographic or range expansion, or a recent bottleneck, possibly linked to a *Wolbachia* infection. Finally, new evidence is provided for the asymmetric but limited introgression of the *molestus* biotype into the *pipiens* biotype.

Keywords: *Culex pipiens* biotypes *pipiens*/*molestus*; hybrids; disease vectors; DNA-based identification; cytochrome c oxidase I (COI); fragment-size analyses (ACE2; CQ11)

1. Introduction

Culex pipiens s.l. is a complex of three species, viz. *Cx. australicus* Dobrotworsky & Drummond, 1953; *Cx. pipiens* s.s. Linnaeus, 1758; and *Cx. quinquefasciatus* Say, 1823. The latter species is common in (sub)tropical regions with no known established populations in Europe [1], but has been introduced in the Netherlands with airplane traffic [2]. In contrast, *Culex australicus* is endemic to Australia. In Europe, hybrids between *Cx. quinquefasciatus* and *Cx. pipiens* s.s. were characterized in southern coastal regions and the Mediterranean region by applying DNA methods [3]. Despite morphological similarities with *Cx. pipiens* s.s. [4], *Culex torrentium* Martini, 1925 is no longer considered as belonging to the *Cx. pipiens* species complex [5,6]. Nevertheless, both species occur throughout Europe; *Cx. torrentium* is more common in northern Europe and at high elevations further south, whereas *Cx. pipiens* s.s. is more common in the south, but the exact species distribution limits are still unclear [7]. The two species occur in sympatry and are native in Belgium, where *Cx. pipiens* s.s. appears to be more abundant and widespread [8–13]. Within *Cx. pipiens* s.s., two biotypes are recognized, viz. *Cx. pipiens* biotype *pipiens* Linnaeus, 1758, and *Cx. pipiens* biotype *molestus* Forskål, 1775 [14].

The identification of *Cx. torrentium* and *Cx. pipiens* s.s. is difficult as the two species differ by a few subtle morphological characteristics only [4,15]. The biotypes of *Cx. pipiens* s.s. are morphologically indistinguishable [14], but they show four key behavioural differences [16–19]. Females of *Cx. pipiens* biotype *pipiens* need a bloodmeal to produce their first batch of viable eggs, prefer feeding on birds, breed in open spaces, and overwinter in a state of diapause. In contrast, females of *Cx. pipiens* biotype *molestus* can produce a first batch of viable eggs without a bloodmeal, prefer feeding on mammals, can breed in confined mating spaces, and do not overwinter in a state of diapause. In temperate regions of Europe, including Belgium, both biotypes co-occur in open aboveground spaces, but *Cx. pipiens* biotype *molestus* has a preference for confined spaces such as cellars, cesspits, human-made basements, or subways, where these mosquitoes mate and remain active throughout the year [20,21]. Hybrids between biotypes have been reported [13,19,22], displaying a combination of the behavioural traits of both biotypes [20,23]. However, since hybrids are less frequent than expected under random mating, the biotypes may show some degree of reproductive isolation [24,25]. Across the Mediterranean basin, populations in open spaces are genetically more homogenous, with individuals displaying mixed biotype ancestry and a mix of the four key behavioural traits [20]. Thus, the genetic differentiation between biotypes decreases gradually from north to south across the western Palearctic. This may be linked to less severe winters, allowing the non-diapausing *molestus* biotype to survive in open-space environments and admix [20].

In Europe, *Culex pipiens* s.s. is the principal vector for West Nile Virus (WNV), and several other arboviruses [26]. *Culex pipiens* s.s. biotype hybrids with an opportunistic feeding behaviour seem to transmit WNV between birds and humans more easily than non-hybrid *Cx. pipiens* biotypes [23,27–29]. *Culex torrentium*, in turn, is an important vector for Sindbis virus (SINV) in Sweden [30]. However, *Cx. torrentium* also has a high potential to transmit WNV [31,32]. In view of the recent outbreaks of WNV infections in Germany and the Netherlands [33], it is important to closely monitor competent *Culex* vectors.

The distinction between *Cx. torrentium*, *Cx. pipiens* s.s., and *Cx. pipiens* biotypes, has not been investigated systematically. Hence, the distribution and identity of these taxa is still poorly known in most European countries [34]. However, this information is essential to establish reliable risk projection and control programmes, particularly for the early detection of WNV vectors and their potential spread in Europe [35,36]. Therefore, the present paper reports on the DNA-based identification, occurrence, and diversity of *Cx. torrentium* and *Cx. pipiens* s.s., as well as *Cx. pipiens* biotypes *molestus* and *pipiens* and their hybrids, in Belgium.

2. Materials and Methods

2.1. Sampling

Adult and larval mosquitoes were collected in 2017 (August–November), 2018, and 2019 (both April–November) in the framework of the MEMO project (Monitoring of Exotic MOsquito species in Belgium [11]). Thirty-one potential points of entry (PoEs) for exotic mosquitoes in Belgium were surveyed using different sampling and trapping methods [11]. The PoEs included ports and airports, used tire and lucky bamboo import companies, parking lots along highways, wholesale markets, a flower auction, an allotment garden, an industrial area, and cemeteries along the border with Germany. Specimens were morphologically identified as *Cx. pipiens* s.l./*Cx. torrentium* using the keys of [4,37]. A random subset of 1689 *Cx. pipiens* s.l./*Cx. torrentium* specimens were selected for DNA-based identification (Table S1), using the sample_frac function of the dplyr package in R v4.03 [38].

2.2. DNA Extraction and COI Amplification

DNA was extracted from legs (adults) or abdomen (larvae) using the NucleoSpin® Tissue DNA extraction kit (Macherey-Nagel), following the manufacturer's protocol, but with an elution volume of 70 µL. Remaining parts of the specimens and dried DNA extracts are stored at the Royal Belgian Institute of Natural Sciences (Collection Identifier: IG34179). The universal primers LCO1490 and HCO2198 [39] were used to amplify the mitochondrial cytochrome c oxidase subunit I (COI) barcode region (658 bp) [40]. If this was unsuccessful, the C1N-2191 and C1J-1718 primer combination [41] was used to amplify a 472 bp fragment of the COI barcode region. All PCR mixtures, cycling conditions, purification, and sequencing details are as described by [42] (Table S2). Raw sequences were trimmed, corrected, translated into amino acids, and assembled using Geneious® v.10.0.4 (Biomatters Ltd., Auckland, New Zealand). A consensus sequence was generated for each specimen.

2.3. Fragment-Size Analyses

To distinguish between *Cx. pipiens* s.s. and *Cx. torrentium*, a fragment of the Acetylcholinesterase-2 locus (ACE2) was amplified in a 10 µL PCR reaction volume as described by [43] (Table S2). This method also allows for detection of the eventual presence of introduced exotic *Cx. quinquefasciatus*. Using the forward primer B1246s and the reverse primers ACEpip, ACEquin, and ACEtorr, species-specific fragment sizes were produced, viz. 610 bp for *Cx. pipiens* s.s., 416 bp for *Cx. torrentium*, and 274 bp for *Cx. quinquefasciatus* [43]. PCR products were checked on a 2.5% agarose gel (45 min; 90 V).

To identify the two *Cx. pipiens* s.s. biotypes and their hybrids, the CQ11 microsatellite locus was amplified using the forward primer CQ11F2 and the reverse primers pipCQ11R and molCQ11R, following [44] (Table S2). PCR products were checked on a 2.5% agarose gel (45 min; 90V), with a band at 200 bp for *Cx. pipiens* biotype *pipiens*, and at 250 bp for *Cx. pipiens* biotype *molestus*. Hybrids showed both bands. Such hybrids were subsequently re-extracted and re-amplified for the CQ11 locus to exclude possible DNA contamination and confirm their status by visualisation of the two bands.

2.4. COI Data Analyses

The species identification engine of BOLD was used (www.boldsystems.org, accessed on 24 February 2020) with the species-level barcode records option to find the closest matching reference sequence. A Neighbour-Joining (NJ) tree was constructed to examine the clustering support of each *Culex* species occurring in Belgium [12,45] (Geneious® v10.0.4. (Biomatters Ltd., Auckland, New Zealand), Tamura-Nei distance model, 1000 bootstrap replicates). To do so, all publicly available COI sequences (http://www.boldsystems.org/index.php/databases, 16 March 2020) for these *Culex* species were aligned, using ClustalW in Geneious® v10.0.4 (Biomatters Ltd., Auckland, New Zealand), with the newly generated COI sequences in this study. COI sequences of four species of the genus *Coquillettidia* Dyar, 1905 were included as outgroup (GenBank accession numbers: GQ165785, GQ165801, GQ165802, and GQ165803). The alignment was checked for stop codons and trimmed to retain 658 bp. Sequences of less than 300 bp and conspecific identical sequences were discarded.

Pairwise differences in COI nucleotide frequencies between species, biotypes, and biotype hybrids were evaluated using Wright's *F*-statistics in Arlequin v3.5 (1000 random permutations for significance, with subsequent standard Bonferroni correction) [46]. Haplotype frequencies, mean number of pairwise nucleotide differences (*k*) and nucleotide diversity (*Pi*) were also estimated with Arlequin v3.5, excluding sequences with ambiguous sites.

2.5. Habitat Characterization: Land-Cover Classes

The percentage of Corine Land-Cover (CLC) classes (© European Union, Copernicus Land Monitoring Service 2021, European Environment Agency (EEA)) was calculated in a 2.5 km buffer zone around each sampling location. The latest raster file (CLC 2018) was used, and calculation was performed in Q-GIS and R v4.03. The levels were grouped into five main CLC classes, i.e., artificial or urban areas, agricultural areas, forest and seminatural areas, wetlands, and water bodies.

3. Results

In total, 34,401 specimens from 27 out of 31 PoEs were morphologically identified as *Cx. pipiens* s.l./*Cx. torrentium* of which 1689 specimens from 24 sites were selected for DNA-based verification. Of these, 573 were adults and 1113 were larvae. Adults were collected using a Mosquito Magnet trap (N = 242; 42.2%), Frommer Updraft Gravid trap (N = 59; 10.3%) and BG-Sentinel trap (N = 272; 47.5%). The four PoEs where these species were not collected were only surveyed using oviposition traps. Based on the BOLD similarity percentages, the COI NJ-tree (Figure S1), and the ACE2 fragment sizes (Figure S2), 401 specimens were identified as *Cx. torrentium* (N_{adult} = 40; N_{larva} = 361—Table S1), and 1248 as *Cx. pipiens* s.s. Thirty-seven specimens did not provide ACE2 results and were therefore considered as *Cx. pipiens* s.l./*Cx. torrentium*. Three sequences were of too low quality for identification. The ACE2 fragment-size analysis provided no evidence of *Cx. quinquefasciatus*. In the NJ-tree, *Cx. torrentium* forms a cluster with 74.9% bootstrap support inside the *Cx. pipiens* s.s./*Cx. torrentium* group (Figure S1). The *Cx. torrentium* cluster includes all generated and downloaded (BOLD) COI sequences.

Based on the CQ11 fragment-size analysis 956 specimens were assigned to *Cx. pipiens* biotype *pipiens* (N_{adult} = 315 (33%); N_{larva} = 641 (67%)) and 227 specimens to *Cx. pipiens* biotype *molestus* (N_{adult} = 187 (82%); N_{larva} = 40 (18%)) (Figure S3, Table S1). More adults of the *molestus* biotype were collected than larvae, and the *pipiens* biotype. Additionally, 29 specimens were identified as hybrids between both biotypes (N_{adult} = 8 (28%); N_{larva} = 21 (72%)), while the biotypes of 36 *Cx. pipiens* s.s. sequences were not determined due to missing CQ11 results. These sequences, together with those identified as *Cx. pipiens* s.l./*Cx. torrentium* (N = 37), were excluded from further analyses. The abundance of each taxon at each sampling site is shown in Figure 1. In most sites where both biotypes co-occur (N = 16), crossbreeding was identified, with hybrids detected at 13 sampling locations (Figure 1, Table S3). Sites where biotypes co-occurred included environments dominated by urban (Kallo, Charleroi, Zeebrugge and Zaventem), agricultural (Villers-Le-Bouillet, Vrasene, Frameries, Aische-en-Refail, Büllingen and Natoye), and forest and seminatural (Eupen, Dilsen-Stokkem, Houyet and Maasmechelen) areas (Figure 2). At these sites, larval stages of both biotypes were collected on the same days in the same types of larval habitats, viz. used tires, drainage holes, plastic containers, and cemetery flower vases, on multiple occasions. Once they were also found together in a large artificial pond. *Culex pipiens* s.s. and *Cx. torrentium* were collected on the same days in the following same types of larval habitats: used tires, drainage holes, cemetery flower vases, plastic sheets, and metal and plastic containers. The new COI sequences were deposited in GenBank (accession numbers: *Cx. torrentium*—OM749168-OM749568; *Cx. pipiens* biotype *pipiens*—OM748139-OM749094; *Cx. pipiens* biotype *molestus*—OM747912-OM748138; *Cx. pipiens* biotype *pipiens* X *Cx. pipiens* biotype *molestus*—OM747883-OM747911; *Cx. pipiens* s.s.: OM749132-OM749167; *Cx. pipiens* s.l./*Cx. torrentium*—OM749095-OM749131).

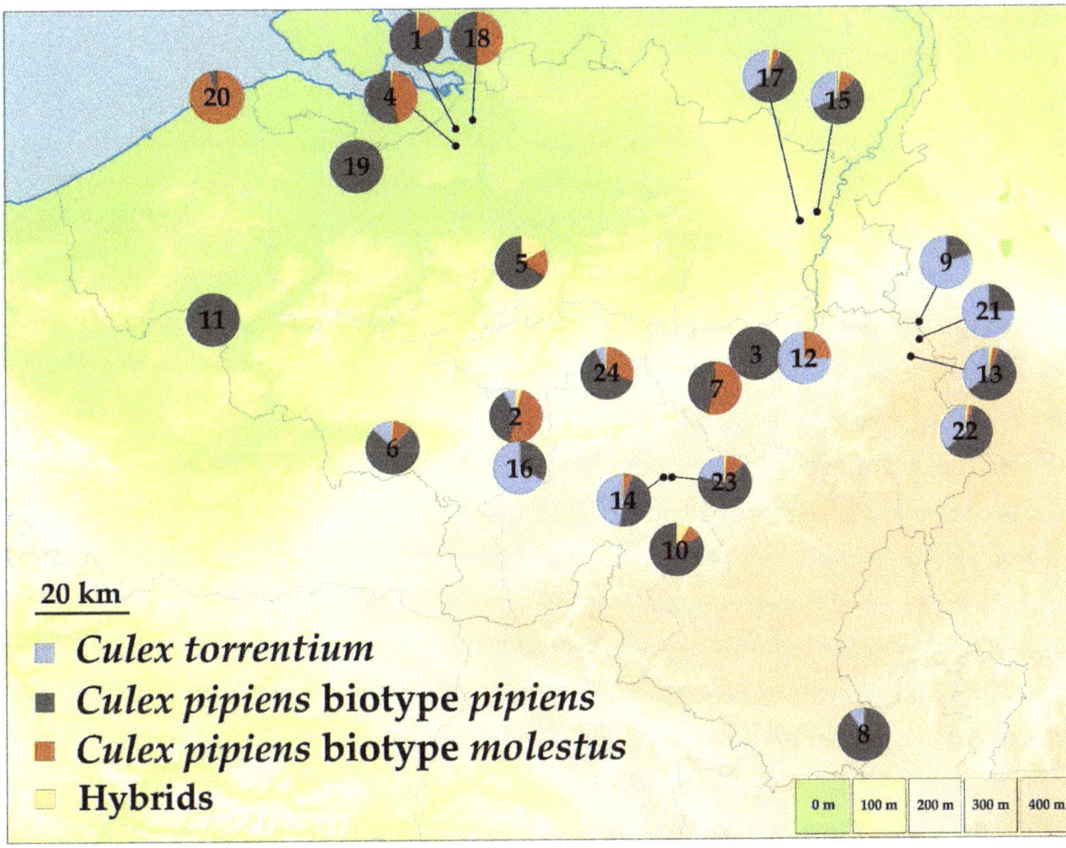

Figure 1. Distribution of *Culex* specimens identified using DNA-based techniques and collected during the MEMO survey 2017–2019 [11]. Hybrids = *Cx. pipiens* biotype *pipiens* X *Cx. pipiens* biotype *molestus*. The numbers indicated on the pie charts are the collection site numbers used in Table S3.

The pairwise F_{ST} values, i.e., a measure of the average genetic differentiation, between *Cx. pipiens* biotype *pipiens* and *Cx. pipiens* biotype *molestus*, were significantly different from zero, which was also the case between the latter and the hybrids (Table 1). The hybrids and *Cx. pipiens* biotype *pipiens* showed the smallest average pairwise nucleotide differences and the lowest average nucleotide diversities (Table 2), despite *Cx. pipiens* biotype *pipiens* being the most widespread taxon in this survey (Figure 1; Table S4). *Culex torrentium* showed higher average pairwise nucleotide differences and nucleotide diversities than *Cx. pipiens* biotype *pipiens* (Table 2).

The most common COI haplotypes in *Cx. pipiens* s.s. were H1 (698 out of 1248 sequences, including 19 hybrids, 91 *Cx. pipiens* biotype *molestus*, and 588 *Cx. pipiens* biotype *pipiens*) and H2 (84 out of 1248 sequences, including 82 *Cx. pipiens* biotype *molestus* and 2 *Cx. pipiens* biotype *pipiens*) (Figure 3). The most common COI haplotype in *Cx. torrentium* was H3 (125 out of 401 sequences) (Figure 3). The haplotype of 509 sequences could not be identified because of ambiguous sites or short fragment lengths ($N_{Cx.\ torrentium}$ = 122; $N_{Cx.\ pipiens\ biotype\ pipiens}$ = 330; $N_{Cx.\ pipiens\ biotype\ molestus}$ = 49; $N_{hybrids}$ = 8).

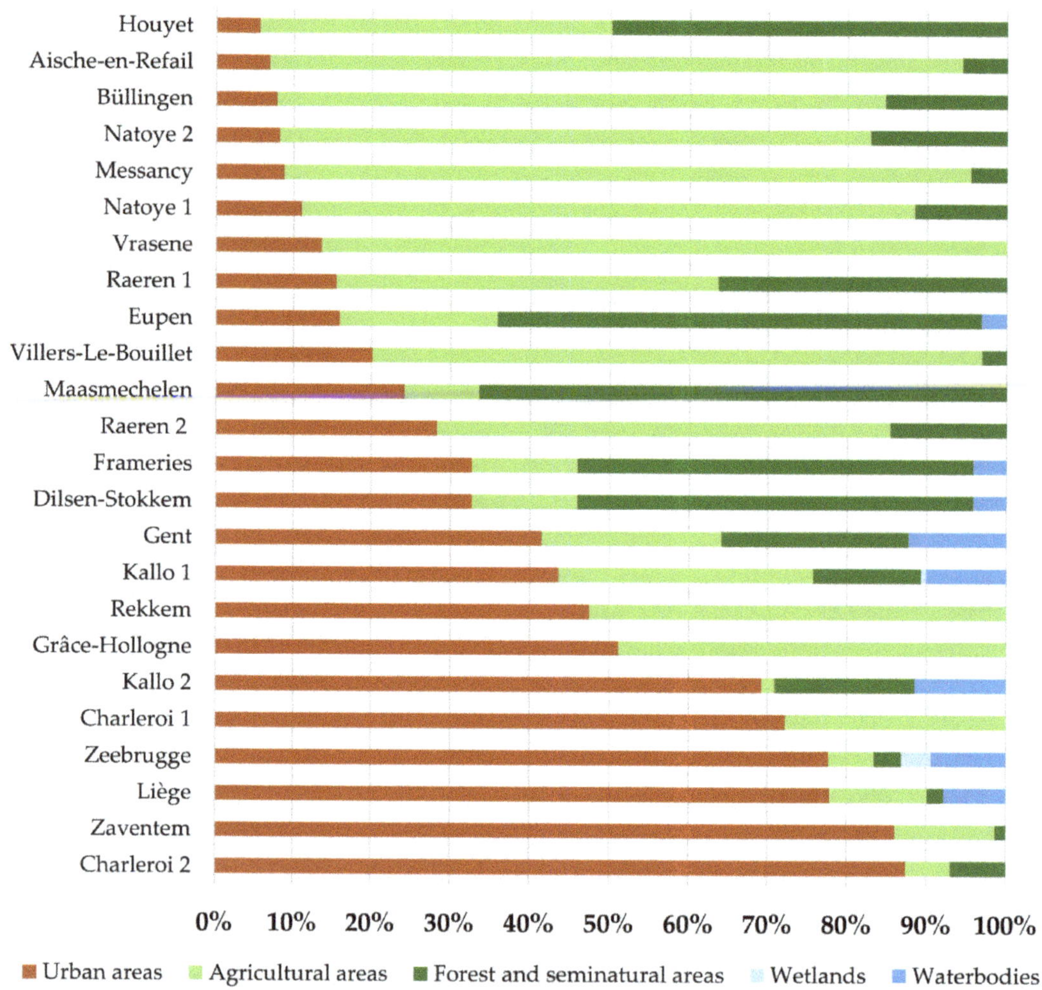

Figure 2. The percentage of main Corine Land-Cover classes in a 2.5 km buffer zone around the sampling locations (group levels based on five classes).

Table 1. Pairwise F_{ST} estimates between biotypes and biotype hybrids of *Culex pipiens* s.s. based on COI sequences, calculated using Arlequin v3.5. Hybrids = *Cx. pipiens* biotype *molestus* X *Cx. pipiens* biotype *pipiens*. Significant values after standard Bonferroni correction marked by an asterisk ($p < 0.0005$).

F_{ST}	Hybrids	Cx. *pipiens* Biotype *molestus*	Cx. *pipiens* Biotype *pipiens*
Hybrids	0	-	-
Cx. pipiens biotype *molestus*	0.2172 *	0	-
Cx. pipiens biotype *pipiens*	0.0164	0.4737 *	0

* = significant values (after Bonferroni correction).

Table 2. COI sequence diversity of *Culex pipiens* s.s. biotypes and *Cx. torrentium*, calculated using Arlequin v3.5. Hybrids = *Cx. pipiens* biotype *molestus* x *Cx. pipiens* biotype *pipiens*, N_{COI} = number of generated COI sequences, N_{Hap} = number of haplotypes, $N_{UnsharedHap}$ = number of haplotypes specific to the taxon, S = number of polymorphic sites, *k* = average pairwise nucleotide differences, *Pi* = nucleotide diversity.

	Hybrids	*Cx. pipiens* Biotype *molestus*	*Cx. pipiens* Biotype *pipiens*	*Cx. torrentium*
N_{COI}	29	227	956	401
N_{Hap}	3	6	24	50
$N_{UnsharedHap}$	1	3	20	50
S	3	8	19	31
k	0.2069 ± 0.2606	0.5458 ± 0.4522	0.0703 ± 0.1418	0.8990 ± 0.6275
Pi	0.0006 ± 0.0008	0.0014 ± 0.0012	0.0002 ± 0.0004	0.0024 ± 0.0018

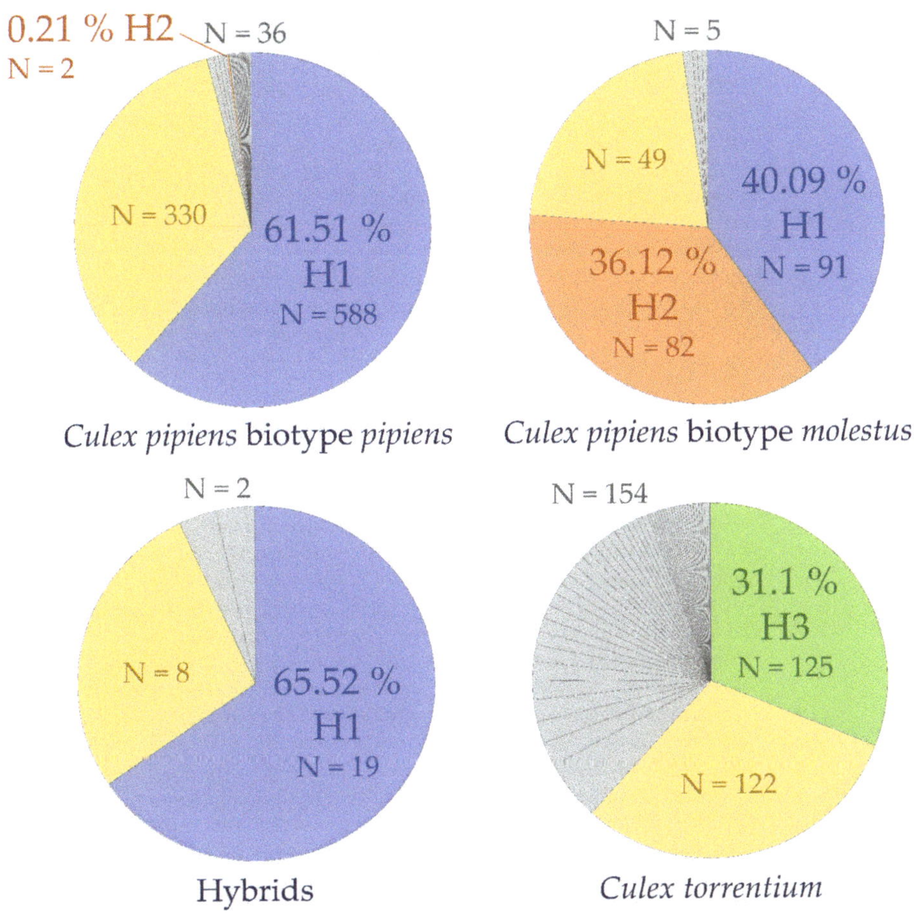

Figure 3. COI haplotype frequencies calculated with Arlequin v.3.5. [46]. Hybrids = *Culex pipiens* biotype *pipiens* X *Cx. pipiens* biotype *molestus*, blue = *Cx. pipiens* s.s. haplotype H1, orange = *Cx. pipiens* s.s. haplotype H2, green = *Cx. torrentium* haplotype H3, yellow = sequences to which no haplotypes could be assigned, grey = haplotypes with a frequency < 6%.

4. Discussion

4.1. Species Occurrence in Belgium

In line with previous studies [8,9,13], *Cx. pipiens* s.s. appears to be more common in Belgium than *Cx. torrentium*. Likewise, *Culex pipiens* biotype *pipiens* is more common and widespread in Belgium than *Cx. pipiens* biotype *molestus*, as the latter comprises only 13% of the specimens (Table S4). However, industrial areas were overrepresented in this survey for the early detection of exotic *Aedes* species [11]; thus, the sampling may have been biased. As such, the *molestus* biotype, with its preference for hypogean habitats [21] and highly eutrophic waters in confined mating spaces [20,21], may be underrepresented, as these environments were not surveyed during the MEMO project [11]. Thus, more targeted surveys are needed to determine the actual prevalence and distribution of *Cx. pipiens* biotype *molestus* in Belgium. Nevertheless, this study confirms the co-occurrence of both biotypes in urban, agricultural, and forest and seminatural habitats [13,47–50] (Figure 2). Despite the sympatric occurrence of both biotypes, only few hybrid specimens were found (1.7%; Tables S3 and S4), i.e., less than in Germany (4.2%) [19], Portugal (8–10%) [47], and Italy (14.4%) [49]. These low hybridisation rates suggest at least partial reproductive/ecological isolation between biotypes [20,47], with some rare haplotypes identified as specific to one biotype (Table 2, see next section). The sympatric co-occurrence of both biotypes and their hybrids with their opportunistic feeding behaviour [47,51] hints at the potential danger of viral transmissions from birds to humans (i.e., to act as bridge for disease vectors). However, the low frequency of hybrids likely limits their potential epidemiological role in WNV outbreaks.

While *Cx. pipiens* s.s. and *Cx. torrentium* are sympatric in some areas, the latter species was not collected in the north of Belgium (Figure 1). Again, this may be a sampling artefact, since the species was collected all over Belgium during the nationwide MODIRISK mosquito survey (2007–2010) [9,48]. Both species were equally observed in different habitats and have adapted to a life in human neighbourhoods [52], with *Cx. torrentium* and *Cx. pipiens* s.s. larvae often found in small artificial and nutrient-rich bodies of water [7,52]. Thus, both species are widespread in Belgium, but their exact distribution limits in Europe remain to be determined.

4.2. COI Haplotype Composition and Genetic Variability

Belgian *Cx. torrentium* showed a higher COI variability than *Cx. pipiens* s.s., which is in line with [52,53]. However, we found no evidence of further sub-structuring or taxonomic differentiation within *Cx. torrentium*, while previous studies reported some morphological variability within the species [53,54]. The limited intraspecific variation within *Cx. pipiens* s.s. is consistent with [54,55]. Hence, COI haplotype H1 had a prevalence of 61.51% in *Cx. pipiens* biotype *pipiens* and of 40.09% in *Cx. pipiens* biotye *molestus*. Similar prevalences were reported by [52] (H1 = haplotype 1), [3] (H1 = haplotype A/C), and [56] (H1 = haplotype H). As such, haplotype H1 is widespread and most frequent in northern temperate climates (Germany, Japan, North America, and Russia) [3,52,56]. Situations in which populations show limited genetic variation and consist of a highly frequent haplotype, jointly with a few rare haplotypes, can be explained by either a recent population-wide demographic or range expansion, or a recent bottleneck, possibly in combination with a *Wolbachia* infection [52]. Such a *Wolbachia* infection can severely reduce mitochondrial diversity [52,57,58]. This might, in part, explain the limited COI diversity in *Cx. pipiens* s.s., which shows *Wolbachia* infection rates of >90% [52,59–61], whereas COI diversity might have been retained in *Cx. torrentium*, within which *Wolbachia* infections appear to be very rare [59,62].

In Belgium, haplotype H2 was almost exclusively found in *Cx. pipiens* biotype *molestus*. This is somewhat unexpected, as this haplotype is rarely found in temperate climates, but associated with (sub)tropical climates ([3] H2 = haplotype E/E1; [56] H2 = haplotype C). Elsewhere, haplotype H2 is prevalent in *Cx. quinquefasciatus* (42%) and its hybrids with *Cx. pipiens* s.s. (32%) [3]. Hence, COI haplotypes in *Cx. pipiens* s.l. are not species-specific [56].

Currently, the biotypes *pipiens* and *molestus* of *Cx. pipiens* s.s. are regarded as different monophyletic evolutionary units undergoing incipient ecological speciation, so that they may be distinct phylogenetic entities [22,27,47,63–65]. This was supported by the significant F_{ST} estimates found in the present study. The different mating behaviours of both biotypes was considered as an initial factor of a sympatric speciation process [47]. The limited level of hybridisation is not bidirectional, with a mainly male-mediated introgression from *molestus* to the *pipiens* biotype [47], which explains the prevalence of H1 (typical of *pipiens*) and absence of H2 (typical of *molestus*) in hybrids. This asymmetric introgression may reflect a mating strategy wherein stenogamous *molestus* males mate with both *molestus* and *pipiens* females in above-ground habitats, while *pipiens* males mate (via specialised swarming behaviour) in open spaces and, therefore, have a higher disposition to mate with *pipiens* females [47]. An experimental study revealed at least one reproductive isolating mechanism, with females actively avoiding copulation with males of the other biotype, and *pipiens* females being unsuccessful in receiving *molestus* males' sperm [66].

Supplementary Materials: The following supporting information can be downloaded at: https://www.mdpi.com/article/10.3390/d14060486/s1, Figure S1: Neighbour-Joining tree based on the COI sequences available on BOLD/GenBank for *Culex* species present in Belgium [12,45] including the sequences generated in this study, and identified based on the ACE2 fragment-size analysis. Bootstrap values are indicated at the branches. Sequences were collapsed in species clusters. N = total number of unique sequences included, following Geneious® v10.0.4; Figure S2: Example of ACE2 fragment-size analysis on a 2.5% agarose gel. PCR multiplex including the primers B1246s (Forward), ACEpip (Reverse), ACEtorr (Reverse), and ACEquin (Reverse). Cycling conditions are provided in Table S2. P = *Cx. pipiens* s.s. (610 bp); T = *Cx. torrentium* (416 bp); Figure S3: Example of CQ11 fragment-size analysis on a 2.5% agarose gel. PCR multiplex including the primers CQ11F2 (Forward), molCQ11R (Reverse) and pipCQ11R (Reverse). Cycling conditions are provided in Table S2. PP = *Culex pipiens* biotype *pipiens* (200 bp); PM = *Culex pipiens* biotype *molestus* (250 bp); H = *Cx. pipiens* biotype *pipiens* X *Cx. pipiens* biotype *molestus* (200 bp and 250 bp); Table S1: Detailed list of DNA-based identified specimens, including their life stage at collection and the trapping method; Table S2: Summary of PCR cycling conditions for the amplification of the COI, ACE2, and CQ11 loci; Table S3: Map codes, municipalities, and coordinates of sampling localities with taxon occurrence; Table S4: Overview of the COI sequencing success per year. N = number of specimens. Reference [67] are cited in the Supplementary Materials.

Author Contributions: Conceptualization: N.S., I.D., K.D.W., W.D., R.M. and W.V.B.; Methodology: N.S., K.D.W., I.D., A.V. (Adwine Vanslembrouck), J.D.W., A.S., I.V. and W.V.B.; Formal Analysis and Software: N.S., A.V. (Ann Vanderheyden), K.M. and S.G.; Data Curation: W.D., N.S. and I.D.; Writing—Original Draft Preparation: A.V. (Ann Vanderheyden) and N.S.; Writing—Review and Editing: I.D., K.D.W., W.D., K.M., S.G., A.V. (Adwine Vanslembrouck), J.D.W., A.S., I.V., M.D.M., T.B., R.M. and W.V.B.; Visualization: A.V. (Ann Vanderheyden) and N.S.; Supervision: I.D., W.V.B., R.M., T.B. and M.D.M.; Project Administration and Funding Acquisition: I.D., W.V.B., R.M., T.B. and M.D.M. All authors have read and agreed to the published version of the manuscript.

Funding: This work is part of the MEMO project (tender: CES-2016-02), funded by the Flemish, Walloon and Brussels regional governments and the Federal Public Service (FPS) Public Health, Food Chain Safety and Environment in the context of the National Environment and Health Action Plan (NEHAP) (Belgium). The Barcoding Facility for Organisms and Tissues of Policy Concern (BopCo-http://bopco.myspecies.info/, accessed 15 May 2022) is financed by the Belgian Science Policy Office (Belspo). The Outbreak Research Team of the Institute of Tropical Medicine is funded by the Department of Economy, Science and Innovation of the Flemish government.

Institutional Review Board Statement: Not applicable.

Data Availability Statement: The data presented in this study are openly available in GenBank at https://www.ncbi.nlm.nih.gov/genbank/, accessed 15 May 2022, reference number: OM747883-OM749568.

Acknowledgments: We would like to thank all cooperating companies for giving access to their private property during mosquito sampling surveys. We would like to thank the laboratory and technical staff at the Institute of Tropical Medicine Antwerp (ITM), at the Royal Belgian Institute of Natural Sciences (RBINS), and at the Royal Museum for Central Africa (RMCA).

Conflicts of Interest: The authors declare no conflict of interest.

References

1. ECDC. Culex pipiens Factsheet. 2020. Available online: https://www.ecdc.europa.eu/en/all-topics-z/disease-vectors/facts/mosquito-factsheets/culex-pipiens-factsheet-experts (accessed on 16 February 2022).
2. Scholte, E.; Ibáñez-Justicia, A.; Stroo, A.; Zeeuw, J.D.; Hartog, W.D.; Reusken, C. Mosquito collections on incoming intercontinental flights at Schiphol International airport, the Netherlands, 2010–2011. *J. Eur. Mosq. Control Assoc.* **2014**, *32*, 17–21.
3. Shaikevich, E.V.; Vinogradova, E.B.; Bouattour, A.; Gouveia de Almeida, A.P. Genetic diversity of *Culex pipiens* mosquitoes in distinct populations from Europe: Contribution of *Cx. quinquefasciatus* in Mediterranean populations. *Parasites Vectors* **2016**, *9*, 47. [CrossRef] [PubMed]
4. Becker, N.; Petric, D.; Zgomba, M.; Boase, C.; Madon, M.; Dahl, C.; Kaiser, A. *Mosquitoes and Their Control*; Springer: Heidelberg, Germany; New York, NY, USA, 2010.
5. Miller, B.R.; Crabtree, M.B.; Savage, H.M. Phylogeny of fourteen *Culex* mosquito species, including the *Culex pipiens* complex, inferred from the internal transcribed spacers of ribosomal DNA. *Insect Mol. Biol.* **1996**, *5*, 93–107. [CrossRef]
6. Harbach, R.E. *Culex pipiens*: Species versus species complex taxonomic history and perspective. *J. Am. Mosq. Control Assoc.* **2012**, *28*, 10–23. [CrossRef]
7. Hesson, J.C.; Rettich, F.; Merdić, E.; Vignjević, G.; Östman, Ö.; Schäfer, M.; Schaffner, F.; Foussadier, R.; Besnard, G.; Medlock, J.; et al. The arbovirus vector *Culex torrentium* is more prevalent than *Culex pipiens* in northern and central Europe. *Med. Vet. Entomol.* **2014**, *28*, 179–186. [CrossRef]
8. Dekoninck, W.; Hendrickx, F.; Versteirt, V.; Coosemans, M.; De Clercq, E.M.; Hendrickx, G.; Hance, T.; Grootaert, P. Changes in species richness and spatial distribution of mosquitoes (Diptera: Culicidae) inferred from museum specimen records and a recent inventory: A case study from Belgium suggests recent expanded distribution of arbovirus and malaria vectors. *J. Med. Entomol.* **2013**, *50*, 237–243. [CrossRef]
9. Versteirt, V.; Boyer, S.; Damiens, D.; De Clercq, E.; Dekoninck, W.; Ducheyne, E.; Grootaert, P.; Garros, C.; Hance, T.; Hendrickx, G. Nationwide inventory of mosquito biodiversity (Diptera: Culicidae) in Belgium, Europe. *Bull. Èntomol. Res.* **2013**, *103*, 193–203. [CrossRef] [PubMed]
10. Deblauwe, I.; Sohier, C.; Schaffner, F.; Rakotoarivony, L.M.; Coosemans, M. Implementation of surveillance of invasive mosquitoes in Belgium according to the ECDC guidelines. *Parasites Vectors* **2014**, *7*, 201. [CrossRef]
11. Deblauwe, I.; De Wolf, K.; Smitz, N.; Vanslembrouck, A.; Schneider, A.; De Witte, J.; Verlé, I.; Dekoninck, W.; De Meyer, M.; Backeljau, T.; et al. *Monitoring of Exotic Mosquitoes in Belgium (MEMO): Final Report Phase 7 Part 1: MEMO Results*; Institute of Tropical Medicine: Antwerp, Belgium, 2020.
12. Boukraa, S.; Dekoninck, W.; Versteirt, V.; Schaffner, F.; Coosemans, M.; Haubruge, E.; Francis, F. Updated checklist of the mosquitoes (Diptera: Culicidae) of Belgium. *J. Vector Ecol.* **2015**, *40*, 398–407. [CrossRef] [PubMed]
13. Boukraa, S.; de La Grandiere, M.A.; Bawin, T.; Raharimalala, F.N.; Zimmer, J.Y.; Haubruge, E. Diversity and ecology survey of mosquitoes potential vectors in Belgian equestrian farms: A threat prevention of mosquito-borne equine arboviruses. *Prev. Vet. Med.* **2016**, *124*, 58–68. [CrossRef]
14. Harbach, R.E.; Harrison, B.A.; Gad, A.M. *Culex (Culex) molestus* Forskål (Diptera: Culicidae): Neotype designation, description, variation, and taxonomic status. *Proc. Entomol. Soc. Wash.* **1984**, *86*, 521–542.
15. Razygraev, A.V.; Sulesco, T.M. The Use of the Bayes Factor for identification of *Culex pipiens* and *C. torrentium* (Diptera: Culicidae) based on morphometric wing characters. *Èntomol. Rev.* **2020**, *100*, 220–227. [CrossRef]
16. Byrne, K.; Nichols, R.A. *Culex pipiens* in London Underground tunnels: Differentiation between surface and subterranean populations. *Heredity* **1999**, *82*, 7–15. [CrossRef]
17. Lundström, J. Mosquito-borne viruses in western Europe: A review. *J. Vector Ecol.* **1999**, *24*, 1–39. [PubMed]
18. Vinogradova, E.B. *Culex Pipiens Pipiens Mosquitoes: Taxonomy, Distribution, Ecology, Physiology, Genetics, Applied Importance and control*; Pensoft: Sofia, Bulgaria, 2000.
19. Rudolf, M.; Czajka, C.; Börstler, J.; Melaun, C.; Jöst, H.; von Thien, H.; Badusche, M.; Becker, N.; Schmidt-Chanasit, J.; Krüger, A.; et al. First nationwide surveillance of *Culex pipiens* Complex and *Culex torrentium* mosquitoes demonstrated the presence of *Culex pipiens* biotype *pipiens/molestus* hybrids in Germany. *PLoS ONE* **2013**, *8*, e71832. [CrossRef]
20. Haba, Y.; McBride, L. Origin and status of *Culex pipiens* mosquito ecotypes. *Curr. Biol.* **2022**, *32*, R237–R246. [CrossRef]
21. Becker, N.; Jöst, A.; Weitzel, T. The *Culex pipiens* complex in Europe. *J. Am. Mosq. Control Assoc.* **2012**, *28*, 53–67. [CrossRef] [PubMed]
22. Yurchenko, A.A.; Masri, R.A.; Khrabrova, N.V.; Sibataev, A.K.; Fritz, M.L.; Sharakhova, M.V. Genomic differentiation and intercontinental population structure of mosquito vectors *Culex pipiens pipiens* and *Culex pipiens molestus*. *Sci. Rep.* **2020**, *10*, 7504. [CrossRef] [PubMed]

23. Fritz, M.L.; Walker, E.D.; Miller, J.R.; Severson, D.W.; Dworkin, I. Divergent host preferences of above- and below-ground *Culex pipiens* mosquitoes and their hybrid offspring. *Med. Vet. Entomol.* **2015**, *29*, 115–123. [CrossRef]
24. Osório, H.C.; Zé-Zé, L.; Amaro, F.; Nunes, A.; Alves, M.J. Sympatric occurrence of *Culex pipiens* (Diptera, Culicidae) biotypes *pipiens, molestus* and their hybrids in Portugal, Western Europe: Feeding patterns and habitat determinants. *Med. Vet. Entomol.* **2014**, *28*, 103–109. [CrossRef] [PubMed]
25. Gomes, B.; Sousa, C.A.; Vicente, J.L.; Pinho, L.; Calderón, I.; Arez, E.; Almeida, A.P.; Donnelly, M.J.; Pinto, J. Feeding patterns of *molestus* and *pipiens* forms of *Culex pipiens* (Diptera: Culicidae) in a region of high hybridization. *Parasites Vectors* **2013**, *6*, 93. [CrossRef] [PubMed]
26. Brugman, V.A.; Hernández-Triana, L.M.; Medlock, J.M.; Fooks, A.R.; Carpenter, S.; Johnson, N. The role of *Culex pipiens* L. (Diptera: Culicidae) in virus transmission in Europe. *Int. J. Environ. Res. Public Health* **2018**, *15*, 389. [CrossRef] [PubMed]
27. Fonseca, D.M.; Keyghobadi, N.; Malcolm, C.A.; Mehmet, C.; Schaffner, F.; Mogi, M.; Fleischer, R.C.; Wilkerson, R.C. Emerging Vectors in the *Culex pipiens* Complex. *Science* **2004**, *303*, 1535–1538. [CrossRef]
28. Gomes, B.; Kioulos, E.; Papa, A.; Almeida, A.P.G.; Vontas, J.; Pinto, J. Distribution and hybridization of *Culex pipiens* forms in Greece during the West Nile virus outbreak of 2010. *Infect. Genet. Evol.* **2013**, *16*, 218–225. [CrossRef] [PubMed]
29. Ciota, A.T.; Chin, P.A.; Kramer, L.D. The effect of hybridization of *Culex pipiens* complex mosquitoes on transmission of West Nile virus. *Parasites Vectors* **2013**, *6*, 305. [CrossRef]
30. Hesson, J.C.; Verner-Carlsson, J.; Larsson, A.; Ahmed, R.; Lundkvist, Å.; Lundström, J.O. *Culex torrentium* mosquito role as major enzootic vector defined by rate of Sindbis virus infection, Sweden, 2009. *Emerg. Infect. Dis.* **2015**, *21*, 875. [CrossRef] [PubMed]
31. Leggewie, M.; Badusche, M.; Rudolf, M.; Jansen, S.; Börstler, J.; Krumkamp, R.; Huber, K.; Krüger, A.; Schmidt-Chanasit, J.; Tannich, E.; et al. *Culex pipiens* and *Culex torrentium* populations from Central Europe are susceptible to West Nile virus infection. *One Health* **2016**, *2*, 88–94. [CrossRef] [PubMed]
32. Jansen, S.; Heitmann, A.; Lühken, R.; Leggewie, M.; Helms, M.; Badusche, M.; Rossini, G.; Schmidt-Chanasit, J.; Tannich, E. *Culex torrentium*: A potent vector for the transmission of west nile virus in central Europe. *Viruses* **2019**, *11*, 492. [CrossRef]
33. ECDC. Surveillance and Disease Data for West Nile Virus Infections. 2020. Available online: https://www.ecdc.europa.eu/en/west-nile-fever/surveillance-and-disease-data (accessed on 21 February 2022).
34. Weitzel, T.; Braun, K.; Collado, A.; Jöst, A.; Becker, N. Distribution and frequency of *Culex pipiens* and *Culex torrentium* (Culicidae) in Europe and diagnostic allozyme markers. *Eur. Mosq. Bull.* **2011**, *29*, 22–37.
35. Calistri, P.; Giovannini, A.; Hubalek, Z.; Ionescu, A.; Monaco, F.; Savini, G.; Lelli, R. Epidemiology of West Nile in Europe and in the Mediterranean Basin. *Open Virol. J.* **2010**, *4*, 29. [CrossRef]
36. García-Carrasco, J.M.; Muñoz, A.R.; Olivero, J.; Segura, M.; Real, R. Predicting the spatio-temporal spread of West Nile virus in Europe. *PLoS Negl. Trop. Dis.* **2021**, *15*, e0009022. [CrossRef] [PubMed]
37. Gunay, F.; Picard, M.; Robert, V. MosKeyTool, An Interactive Identification Key for Mosquitoes of Euro-Mediterranean. 2018. Available online: http://medilabsecure.com/moskeytool (accessed on 5 August 2020).
38. Wickham, H.; François, R.; Henry, L.; Müller, K. A Grammar of Data Manipulation dplyr 1.0.8. 2020. Available online: https://dplyr.tidyverse.org/index.html (accessed on 1 April 2022).
39. Folmer, O.; Black, M.; Hoeh, W.; Lutz, R.; Vrijenhoek, R. DNA primers for amplification of mitochondrial cytochrome c oxidase subunit I from diverse metazoan invertebrates. *Mol. Mar. Biol. Biotechnol.* **1994**, *3*, 294–299. [PubMed]
40. Hebert, P.; Cywinska, A.; Ball, S.; DeWaard, J. Biological identifications through DNA barcodes. *Proc. R. Soc. London Ser. B Biol. Sci.* **2003**, *270*, 313–321. [CrossRef]
41. Simon, C.; Frati, F.; Beckenback, A.; Crespi, B.; Liu, H.; Flook, P. Evolution, weighting, and phylogenetic utility of mitochondrial gene sequences and a compilation of conserved polymerase chain reaction primers. *Ann. Èntomol. Soc. Am.* **1994**, *87*, 653–701. [CrossRef]
42. Ibáñez-Justicia, A.; Smitz, N.; den Hartog, W.; van de Vossenberg, B.; De Wolf, K.; Deblauwe, I.; Van Bortel, W.; Jacobs, F.; Vaux, A.G.C.; Medlock, J.M.; et al. Detection of exotic mosquito species (Diptera: Culicidae) at international airports in Europe. *Int. J. Environ. Res. Public Health* **2020**, *17*, 3450. [CrossRef] [PubMed]
43. Smith, J.L.; Fonseca, D.M. Rapid assays for identification of members of the *Culex (Culex) pipiens* complex, their hybrids, and other sibling species (Diptera: Culicidae). *Am. J. Trop. Med. Hyg.* **2004**, *70*, 339–345. [CrossRef] [PubMed]
44. Bahnck, C.M.; Fonseca, D.M. Rapid assay to identify the two genetic forms of *Culex (Culex) pipiens* L. (Diptera: Culicidae) and hybrid populations. *Am. J. Trop. Med. Hyg.* **2006**, *75*, 251–255. [CrossRef]
45. De Wolf, K.; Vanderheyden, A.; Deblauwe, I.; Smitz, N.; Gombeer, S.; Vanslembrouck, A.; Meganck, K.; Dekoninck, W.; DE Meyer, M.; Backeljau, T.; et al. First record of the West Nile virus bridge vector *Culex modestus* Ficalbi (Diptera: Culicidae) in Belgium, validated by DNA barcoding. *Zootaxa* **2021**, *4920*, 131–139. [CrossRef]
46. Excoffier, L.; Lischer, H.E.L. Arlequin suite ver 3.5: A new series of programs to perform population genetics analyses under Linux and Windows. *Mol. Ecol. Resour.* **2010**, *10*, 564–567. [CrossRef]
47. Gomes, B.; Sousa, C.A.; Novo, M.T.; Freitas, F.B.; Alves, R.; Côrte-Real, A.R.; Salgueiro, P.; Donnelly, M.J.; Almeida, A.P.; Pinto, J. Asymmetric introgression between sympatric molestus and pipiens forms of *Culex pipiens* (Diptera: Culicidae) in the Comporta region, Portugal. *BMC Evol. Biol.* **2009**, *9*, 262. [CrossRef] [PubMed]

48. Versteirt, V.; De Clercq, E.; Dekoninck, W.; Damiens, D.; Ayrinhac, A.; Jacobs, F.; Bortel, W.V. MODIRISK: Mosquito Vectors of Disease: Spatial Biodiversity, Drivers of Change, and Risk. Antwerp, Belgium. 2011. Available online: https://www.belspo.be/belspo/SSD/science/Reports/MODIRISK_Phase1Summary.pdf (accessed on 31 January 2017).
49. Di Luca, M.; Toma, L.; Boccolini, D.; Severini, F.; La Rosa, G.; Minelli, G.; Bongiorno, G.; Montarsi, F.; Arnoldi, D.; Capelli, G.; et al. Ecological Distribution and CQ11 Genetic Structure of *Culex pipiens* Complex (Diptera: Culicidae) in Italy. *PLoS ONE* **2016**, *11*, e0146476. [CrossRef]
50. Zittra, C.; Flechl, E.; Kothmayer, M.; Vitecek, S.; Rossiter, H.; Zechmeister, T.; Fuehrer, H.-P. Ecological characterization and molecular differentiation of *Culex pipiens* complex taxa and *Culex torrentium* in eastern Austria. *Parasites Vectors* **2016**, *9*, 197. [CrossRef] [PubMed]
51. Huang, S.; Hamer, G.L.; Molaei, G.; Walker, E.D.; Goldberg, T.L.; Kitron, U.D.; Andreadis, T.G. Genetic variation associated with mammalian feeding in *Culex pipiens* from a West Nile virus epidemic region in Chicago, Illinois. *Vector-Borne Zoonotic Dis.* **2009**, *9*, 637–642. [CrossRef] [PubMed]
52. Werblow, A.; Klimpel, S.; Bolius, S.; Dorresteijn, A.W.C.; Sauer, J.; Melaun, C. Population structure and distribution patterns of the sibling mosquito species *Culex pipiens* and *Culex torrentium* (Diptera: Culicidae) reveal different evolutionary paths. *PLoS ONE* **2014**, *9*, e102158. [CrossRef] [PubMed]
53. Werblow, A.; Bolius, S.; Dorresteijn, A.W.C.; Melaun, C.; Klimpel, S. Diversity of *Culex torrentium* Martini, 1925—A potential vector of arboviruses and filaria in Europe. *Parasitol. Res.* **2013**, *112*, 2495–2501. [CrossRef]
54. Fedorova, M.V.; Shaikevich, E.V. Morphological and molecular-genetic distinctions between adult mosquitoes *Culex torrentium* Martini and *C. pipiens* Linnaeus (Diptera, Culicidae) from Moscow Province. *Èntomol. Rev.* **2007**, *87*, 127–135. [CrossRef]
55. Guillemaud, T.; Pasteur, N.; Rousset, F. Contrasting levels of variability between cytoplasmic genomes and incompatibility types in the mosquito *Culex pipiens*. *Proc. R. Soc. London Ser. B Biol. Sci.* **1997**, *264*, 245–251. [CrossRef]
56. Koosha, M.; Oshaghi, M.A.; Sedaghat, M.M.; Vatandoost, H.; Azari-Hamidian, S.; Abai, M.R.; Hanafi-Bojd, A.A.; Mohtarami, F. Sequence analysis of mtDNA COI barcode region revealed three haplotypes within *Culex pipiens* assemblage. *Exp. Parasitol.* **2017**, *181*, 102–110. [CrossRef]
57. Cariou, M.; Duret, L.; Charlat, S. The global impact of *Wolbachia* on mitochondrial diversity and evolution. *J. Evol. Biol.* **2017**, *30*, 2204–2210. [CrossRef]
58. Hurst, G.D.D.; Jiggins, F.M. Problems with mitochondrial DNA as a marker in population, phylogeographic and phylogenetic studies: The effects of inherited symbionts. *Proc. R. Soc. B Boil. Sci.* **2005**, *272*, 1525–1534. [CrossRef]
59. Bergman, A.; Hesson, J.C. *Wolbachia* prevalence in the vector species *Culex pipiens* and *Culex torrentium* in a Sindbis virus-endemic region of Sweden. *Parasites Vectors* **2021**, *14*, 428. [CrossRef] [PubMed]
60. Duron, O.; Lagnel, J.; Raymond, M.; Bourtzis, K.; Fort, P.; Weill, M. Transposable element polymorphism of *Wolbachia* in the mosquito *Culex pipiens*: Evidence of genetic diversity, superinfection and recombination. *Mol. Ecol.* **2005**, *14*, 1561–1573. [CrossRef] [PubMed]
61. Vinogradova, E.B.; Fedorova, M.V.; Shaikevich, E.V. Endosymbiotic bacterium *Wolbachia pipientis* in synanthropic populations of the mosquito *Culex pipiens pipiens* L. (Diptera, Culicidae). *Dokl. Biol. Sci.* **2003**, *389*, 172–175. [CrossRef] [PubMed]
62. Leggewie, M.; Krumkamp, R.; Badusche, M.; Heitmann, A.; Jansen, S.; Schmidt-Chanasit, J.; Tannich, E.; Becker, S.C. *Culex torrentium* mosquitoes from Germany are negative for *Wolbachia*. *Med. Vet. Entomol.* **2018**, *32*, 115–120. [CrossRef] [PubMed]
63. Shaikevich, E. PCR-RFLP of the COI gene reliably differentiates *Cx. pipiens*, *Cx. pipiens f. molestus* and *Cx. torrentium* of the *Pipiens* Complex. *Eur. Mosq. Bull.* **2007**, *23*, 25–30.
64. Gomes, B.; Wilding, C.S.; Weetman, D.; Sousa, C.A.; Novo, M.T.; Savage, H.M.; Almeida, A.P.G.; Pinto, J.; Donnelly, M.J. Limited genomic divergence between intraspecific forms of *Culex pipiens* under different ecological pressures. *BMC Evol. Biol.* **2015**, *15*, 197. [CrossRef]
65. Kothera, L.; Godsey, M.; Mutebi, J.P.; Savage, H.M. A comparison of aboveground and belowground populations of *Culex pipiens* (Diptera: Culicidae) mosquitoes in Chicago, Illinois, and New York City, New York, using microsatellites. *J. Med. Entomol.* **2010**, *47*, 805–813. [CrossRef] [PubMed]
66. Kim, S.; Trocke, S.; Sim, C. Comparative studies of stenogamous behaviour in the mosquito *Culex pipiens* complex. *Med. Vet. Entomol.* **2018**, *32*, 427–435. [CrossRef]
67. Bourguet, D.; Fonseca, D.; Vourch, G.; Dubois, M.; Chandre, F.; Severini, C.; Raymond, M. The acetylcholinesterase gene Ace: A diagnostic marker for the *pipiens* and *quinquefasciatus* forms of the *Culex pipiens* complex. *J. Am. Mosq. Control Assoc.* **1998**, *14*, 390–396. [PubMed]

diversity

Article

Morphological and Molecular Characterization Using Genitalia and *CoxI* Barcode Sequence Analysis of Afrotropical Mosquitoes with Arbovirus Vector Potential

Eddyson Montalvo-Sabino [1,10], Ana Paula Abílio [2], Milehna Mara Guarido [3,6], Vera Valadas [1], Maria Teresa Novo [1], Ayubo Kampango [2,4], Carla Alexandra Sousa [1], José Fafetine [5], Marietjie Venter [3], Peter N. Thompson [6], Leo Braack [7,8], Anthony John Cornel [9], Ricardo Parreira [1] and António Paulo Gouveia de Almeida [1,*]

1 Global Health and Tropical Medicine (GHTM), Institute of Hygiene and Tropical Medicine (IHMT), NOVA University Lisbon (NOVA), 1349-008 Lisboa, Portugal
2 Instituto Nacional de Saúde (INS), Marracuene 3943, Mozambique
3 Zoonotic Arbo- and Respiratory Virus Program, Centre for Viral Zoonoses, Department Medical Virology, Faculty of Health Sciences, University of Pretoria, Pretoria 0031, South Africa
4 Department of Zoology and Entomology, University of Pretoria, Pretoria 0002, South Africa
5 Centro de Biotecnologia, Universidade Eduardo Mondlane (UEM), Maputo 3453, Mozambique
6 Department of Production Animal Studies, Faculty of Veterinary Science, University of Pretoria, Onderstepoort 0110, South Africa
7 Malaria Consortium, Mahidol University, Bangkok 73170, Thailand
8 UP Institute for Sustainable Malaria Control, University of Pretoria, Pretoria 0031, South Africa
9 Mosquito Control Research Laboratory, Department of Entomology and Nematology and Vector Genetics Laboratory, Department of Pathology, Microbiology, and Immunology, University of California at Davis, 9240 South Riverbend Avenue, Parlier, CA 93648, USA
10 Escuela Académico Profesional de Medicina Veterinaria, Facultad de Medicina Veterinaria y Zootecnia, Universidad Nacional Hermilio Valdizán, Huanuco 10003, Peru
* Correspondence: palmeida@ihmt.unl.pt

Citation: Montalvo-Sabino, E.; Abílio, A.P.; Guarido, M.M.; Valadas, V.; Novo, M.T.; Kampango, A.; Sousa, C.A.; Fafetine, J.; Venter, M.; Thompson, P.N.; et al. Morphological and Molecular Characterization Using Genitalia and *CoxI* Barcode Sequence Analysis of Afrotropical Mosquitoes with Arbovirus Vector Potential. *Diversity* **2022**, *14*, 940. https://doi.org/10.3390/d14110940

Academic Editors: Michael Wink, Carlos Barceló and Ignacio Ruiz-Arrondo

Received: 26 September 2022
Accepted: 31 October 2022
Published: 2 November 2022

Abstract: Potential arboviral Afrotropical mosquito vectors are underrepresented in public databases of *CoxI* barcode sequences. Furthermore, available *CoxI* sequences for many species are often not associated with voucher specimens to match the corresponding fine morphological characterization of specimens. Hence, this study focused on the characterization of Culicine mosquitoes from South Africa, Mozambique, and Angola and their classification using a complementary approach including a morphological analysis of specimens' genitalia and phylogenetic study based on the analysis of *CoxI* barcode sequences using maximum likelihood and Bayesian phylogenetic inference methods, alongside Median-Joining Network and PCOORD analyses. Overall, 800 mosquitoes (652 males and 148 females) from 67 species, were analyzed. Genitalia from 663 specimens allowed the identification of 55 species of 10 genera. A total of 247 *CoxI* partial gene sequences corresponding to 65 species were obtained, 11 of which (*Aedes capensis*, *Ae. mucidus*, *Culex andersoni*, *Cx. telesilla*, *Cx. inconspicuosus*, *Eretmapodites subsimplicipes*, *Er. quinquevittatus*, *Ficalbia uniformis*, *Mimomyia hispida*, *Uranotaenia alboabdominalis*, and *Ur. mashonaensis*) are, to the best of our knowledge, provided here for the first time. The presence of *Cx. pipiens* ecotypes *molestus* and *pipiens* and their hybrids, as well as *Cx. infula*, is newly reported in the Afrotropical region. The rates of correct sequence identification using BOLD and BLASTn (≥95% identity) were 64% and 53%, respectively. Phylogenetic analysis revealed that, except for subgenus *Eumelanomyia* of *Culex*, there was support for tribes Aedini, Culicini, Ficalbiini, and Mansoniini. A divergence >2% was observed in conspecific sequences, e.g., *Aedeomyia africana*, *Ae. cumminsii*, *Ae. unilineatus*, *Ae. metallicus*, *Ae. furcifer*, *Ae. caballus*, and *Mansonia uniformis*. Conversely, sequences from groups and species complexes, namely, *Ae. simpsoni*, *Ae. mcintoshi*, *Cx. bitaeniorhynchus*, *Cx. simpsoni*, and *Cx. pipiens* were insufficiently separated. A contribution has been made to the barcode library of Afrotropical mosquitoes with associated genitalia morphological identifications.

Keywords: *CoxI*; barcode; Culicidae; mosquitoes; mosquito genitalia; systematics; phylogenetics; arboviruses

1. Introduction

Many mosquito species are important vectors of pathogens, including arboviruses, which can cause various febrile, neurological, and hemorrhagic diseases and, therefore, pose a considerable burden on human health and health systems [1]. While currently, the most important arboviruses transmitted by mosquitoes are dengue (DENV), Zika (ZIKV), Chikungunya (CHIKV), and yellow fever (YFV), outbreaks caused by West Nile (WNV), Rift Valley fever (RVFV), and Japanese Encephalitis (JEV) viruses have also been reported in recent years, becoming emerging health problems [2,3].

Mosquitoes (Diptera: Culicidae) are widely distributed throughout the world (except for Antarctica), with 3570 valid species and 130 subspecies thus far documented [1]. The correct identification of mosquito species that may be involved in pathogen transmission is the first step in the surveillance and control of mosquito-borne diseases and has been based on morphological analysis of mainly adult specimens, but also fourth instar larvae [4–8]. Furthermore, several mosquito species can only be identified based on morphological differences in the male genitalia (and occasionally on other male-specific structures), rendering the identification of their female counterparts sometimes unsolved. Nevertheless, the characteristics of male genitalia are structural, allowing accurate and reliable species identification, in addition to being less susceptible to general body damage that is so common in field samples. However, genitalia dissection is a fine and tedious process that requires specific and specialized training [9]. Furthermore, some mosquitoes form closely related, morphologically indistinguishable, cryptic species complexes, with each species having ecological and host preferences and reproductive isolation, constituting biological individual taxa. To overcome the difficulty in their identification, nucleic acids-based molecular identification methods are used for, for example, members of multiple *Anopheles* species complexes [10] and *Culex (Culex) pipiens* complex members [11].

So far, despite the medical importance of diseases such as dengue, yellow fever, West Nile fever, Zika, and Rift Valley fever, studies aimed at the molecular identification of vectors of arboviruses of African origin [12–16] are limited compared to those regarding the analysis of malaria vectors, or even arbovirus vectors of non-African origin [17–28]. The molecular identification of many species occurring in countries such as South Africa, Mozambique, and Angola that have high mosquito and viral richness are not available [12,29–32]. Cytochrome c oxidase subunit I (*CoxI*) barcode sequences of many Afrotropical mosquito vectors of arboviruses are lacking due to the underrepresentation of specimens in the largest public genomic sequence databases most frequently searched (BOLD and GenBank). Examination of the global representation of *CoxI* barcode Culicidae species sequences in BOLD clearly reveals the underrepresentation of African-derived taxa (https://www.boldsystems.org/index.php/Public_SearchTerms, accessed on 12 May 2022). Furthermore, it is essential to have reliable and comprehensively annotated reference databases of verified sequences that can be used for comparison for species identification [20]. Phylogenetic analyses based on some GenBank/BOLD records have suggested that some partial genomic sequences obtained from mosquitoes have been incorrectly assigned, a type of error that has already been identified in studies based on the *CoxI* marker [25] and internal transcribed spacer of nuclear ribosomal DNA (*ITS*) [33].

The objectives of this work were: (i) to morphologically characterize Afrotropical mosquitoes of the Culicinae subfamily, focusing on the analysis of genitalia of adult specimens, in order to have morphological vouchers associated with a matching mitochondrial *CoxI* sequence to be obtained sequently; (ii) to perform a phylogenetic reconstruction that would allow the identification of the sequences obtained in this work, but that would also (iii) shed light on the agreement between phylogenetic tree topology and the current morphology-based taxonomic arrangement.

2. Materials and Methods

2.1. Mosquito Sampling and Preparation of Male Mosquito Genitalia

The mosquito collection analyzed in this work represented a convenience sample comprising specimens previously collected in three countries in southern Africa (Mozambique, South Africa, and Angola; Supplementary Material-SIV (File S-IV), Figure S1) between 2014 and 2018, within the scope of various scientific projects related to arbovirus detection and epidemiology assessments of arboviruses. After collection and subsequent transportation to the Institute of Hygiene and Tropical Medicine | NOVA University Lisbon (IHMT | NOVA), these mosquitoes, listed in Supplementary Material-I (File S-I), Table S1, were kept dehydrated in silica gel tubes at room temperature.

All mosquitoes were classified according to species, species complex, or species group (where possible) based on the analysis of their morphological features, following the keys of Edwards [4], Jupp [5], and Harbach [34]. The classification of the Aedini tribe followed that of Wilkerson et al. [35] and taxa nomenclature as in https://mosquito-taxonomic-inventory.myspecies.info/valid-species-list# (accessed on 12 May 2022).

The genitalia of all male, and some female, specimens were dissected and slide-mounted for careful examination. The terminal part of the mosquito abdomen was sectioned at the level of segment VII/VIII and immersed in Marc André's solution for a minimum of 7 days at room temperature. Afterward, mosquito genitalia were placed on a slide with a drop of a polyvinyl-chloral-formo-phenol medium, dissected under a stereomicroscope, and covered with a coverslip [14]. Analysis of the different structures of the mosquito genitalia and (sometimes) of maxillary palps, were carried out using an Olympus microscope (BX5,1) and their identification and naming of parts followed the nomenclature of Harbach and Knight [9]. Photographs were taken with an Olympus SC30 digital camera and processed with the Zerene Stacker program (https://www.zerenesystems.com/, accessed on 12 May 2022). In *Culex* subgenus *Oculeomyia*, we relied on the description by Sirivanakarn [36] and Harbach [34] to confirm the identification based on the genitalia.

2.2. DNA Extraction, Partial Amplification of CoxI, and Culex Pipiens Complex Molecular Identification

Total genomic DNA was extracted from mosquito legs and abdomens, as previously described [14]. The barcode N-terminal region of the *CoxI* gene was amplified using the specific primers (LCO1490 and HCO2198), using reaction conditions described by Folmer et al. [37]. The amplified products of 658 bp were visualized under UV illumination after electrophoresis in 2% agarose gels. Whenever a specific amplification product was not observed, to obtain a *CoxI*-specific amplicon, an alternative strategy was used. This entailed the use of primers LCO1490 and TL2-N-3014 and the thermal profile previously described by Tchouassi et al. [38]. In case unsuccessful amplifications prevailed, a final attempt called for the design of new primers using multiple alignments of *CoxI* nucleotide sequences downloaded from the GenBank genomic database. These sequences were aligned using MAFFT v7 (https://mafft.cbrc.jp/alignment/server/, accessed 10 November 2021), and these alignments served as a starting point for the design of degenerate primers using the primer design-M tool (https://bio.tools/primerdesign-m, accessed 10 November 2021). The chosen primers (C_degF 5′-ACWTTATAYTTYATTTTYGG-3′ and C_degR 5′-GTTARWARTAT-WGTAATWGC-3′) were used at a final concentration of 500 nM in 20 µL PCR reactions containing 10 µL NZYTaq II 2x Green Master Mix (NYTech, Portugal), 2 µL of a 1:10 dilution of the original DNA extract, and 6 µL of nuclease-free water. The amplification conditions included one denaturation step at 95 °C for 5 min, followed by 40 cycles of amplification (denaturation: 30 s at 95 °C; annealing: 40 s at 43 °C; extension: 1 min at 72 °C) and a final extension step for 5 min at 72 °C.

A multiplex PCR assay that targets species-specific polymorphisms at the intron-2 of the acetylcholinesterase gene intron-2 (*Ace-2*) sequence of *Cx. pipiens* Linnaeus, 1758 and *Cx. quinquefasciatus* Say, 1823 was carried out with primers B1246s, ACEpip, and ACEquin, as described by Smith and Fonseca [11], yielding a PCR product of 610 bp for *Cx. pipiens* and 274 bp for *Cx. quinquefasciatus*. Differentiation of the *Cx. pipiens* ecotype *molestus* and *Cx. pipiens* ecotype *pipiens* followed the analysis of the CQ11 microsatellite flanking region, described by Bahnck and Fonseca [39], yielding a PCR-product approximately 200 bp in size for *Cx. pipiens* form *pipiens* and 250 bp for form *molestus*.

2.3. Amplicon Sequencing and Nucleotide Sequence Analyses

The amplified PCR products corresponding to partial sequences of the *CoxI* gene from each of the analyzed samples were purified and sequenced by the Sanger method (STABVida, Lda. 2825-182 Caparica, Portugal) using primers LCO1490 or C_deg_F, and the respective reverse primers when the obtained chromatogram lacked in quality. The sequences obtained were edited using the Chromas tool version 2.6.6 (https://technelysium.com.au/wp/, accessed on 10 November 2021). Low-quality sequences were excluded during the editing process. In these cases, a new amplification was performed from the same DNA extract. The purification and sequencing of the obtained amplification products were also repeated, as described above. All amplification products were sequenced, which ranged from 399–661 nucleotides. However, for phylogenetic and divergence analysis, only sequences greater than 500 nucleotides were considered.

The search for homologous sequences available in publicly accessible genomic databases (GenBank/ENA/DDBJ) was performed both with the BLASTn tool (https://blast.ncbi.nlm.nih.gov/Blast.cgi, accessed on 10 November 2021) and the taxonomy search engine in the BOLDSystems v4 database (https://www.boldsystems.org/index.php/IDS_OpenIdEngine, accessed on November 2021). These same tools were used for the identification/confirmation of the identity of our sequences. Multiple sequence alignments were constructed using the G-INS-i iterative refinement method as implemented in MAFFT v7. The obtained alignments were treated with Gblocks (http://phylogeny.lirmm.fr/phylo_cgi/one_task.cgi?task_type=gblocks, accessed 10 November 2021) after selecting the most permissive editing options. The evaluation of the phylogenetic signal of all used sequence datasets was carried out using the likelihood-mapping method, as implemented in the TREE-PUZZLE software [40].

For the phylogenetic sequence analysis, two different approaches were explored: the Maximum Likelihood optimization criterion (ML) and a Bayesian phylogenetic inference-based approach. For both, the first step of the analysis involved the choice of the best nucleotide substitution model to be used (GTR + Γ, GTR + I or GTR + Γ + I models: GTR-General Time Reversal; Γ-Gamma distribution; proportion I of invariant sites), using the IQtree software [41], which was also used for ML phylogenetic reconstruction. The topological support of the branches in the obtained trees was assessed with bootstrap analysis and an approximate likelihood ratio test [aLRT], also implemented in Iqtree. In either case, 1000 replicates of the original sequence data were used, and bootstrap or aLRT values ≥ 75 (% of the total number of replicates) were considered as indicating strong topological support.

Bayesian phylogenetic analyses were carried out using BEAST v1.10.4 software [42], using the same sequence data sets and evolutionary models adopted for the ML analyses. The Bayesian analyses consisted of two independent Markov chain Monte Carlo (MCMC) runs until 1×10^8 states had been sampled at every 10,000th MCMC step (10% of which were later discarded as burn-in). In all cases, chain convergence was assessed using Tracer software v1.7.1 (http://beast.bio.ed.ac.uk/tracer, accessed on 10 November 2021), which was also used to check for an adequate effective sample size (ESS) higher than 200 (after the removal of the burn-in). The tree distribution was summarized using TreeAnnotator software v1.8.3 as a maximum clade credibility (MCC) tree, using median heights as the node heights in the tree. All the phylogenetic trees were visualized using FigTree v1.4.2

software (http://tree.bio.ed.ac.uk/software/figtree/, accessed 10 November 2021). At specific branches, a posterior probability value of ≥0.80 was considered as indicating strong topological support. In all trees, the sequence of the species *An. neomaculipalpus* Curry, 1931 (KM592986.1) was used as the outgroup. The trees obtained with maximum credibility (product of the Bayesian analysis) were selected to depict a topological organization of the branches more compatible with *a priori* taxonomic expectations. Specific branches were labeled with one to three "*" signs, according to the number of phylogenetic construction methods/tests that confirmed such topology (aLRT and bootstrap/ML + posterior probability/Bayesian analyses). The original trees can be found in Supplementary Material-III (File S-III).

The average intraspecific and interspecific genetic variation were calculated using genetic distances corrected with the Kimura 2-parameter model (K2P), as implemented in the MEGA X software.

Median Joining networks analysis was performed using SplitsTree5 5.0.0_alpha application with default options [43] for computing unrooted phylogenetic networks from alignments of sequences. The Neighbor Net method [44] was used (default options) to obtain compatible splits, and the Splits Network Algorithm method [45] was used (default options) to obtain split networks. Principal coordinates analysis was also carried out using the software available on the platform (https://www.hiv.lanl.gov/content/sequence/PCOORD/PCOORD.html, accessed on 10 April 2022).

3. Results

3.1. Mosquito Identification: Morphological and Molecular

A total of 800 mosquitoes, comprising 652 (81.5%) males and 148 (18.5%) females, were analyzed in this study. These included 73 specimens from Angola, 515 from South Africa, and 212 from Mozambique, representing 67 species belonging to 10 genera: *Aedeomyia* (2), *Aedes* (28), *Coquillettidia* (3), *Culex* (24), *Eretmapodites* (2), *Ficalbia* (1), *Lutzia* (1), *Mansonia* (2), *Mimomyia* (2), and *Uranotaenia* (2) (File S-I, Table S1).

Of these, genitalia from 652 male and 11 female specimens were dissected and their analysis confirmed the identification of 55 species (File S-IV, Table S1), a photographic record of which can be found in Supplementary Material-II (File S-II). The respective slides are deposited in the IHMT | NOVA Insect Collection.

From a subsample of genitalia-confirmed male *Cx. pipiens* and *Cx. quinquefasciatus* and females from the *Cx. pipiens* complex, *Ace2* multiplex PCR allowed us to confirm 8 specimens (2 females and 6 males) as *Cx. pipiens* and 18 as *Cx. quinquefasciatus* (7 females and 11 males; File S-IV, Figure S2a). No hybrids were identified and four samples failed to react. Multiplex PCR for the CQ11 microsatellite flanking region identified one of the *Cx. pipiens* as the *pipiens* ecotype (EM305), two as the *molestus* ecotype (EM326 and EM332), and four as hybrids of the two ecotypes (EM300, EM302, EM303, EM304), while no amplification product was obtained for one male *Cx. pipiens* (EM306) (File S-IV, Figure S2b).

The amplification of the *CoxI* gene was successful in 247/333 specimens (74.2%). The majority (n = 184) of the *CoxI* amplicons were obtained with the Folmer et al. [37] protocol, while the remaining 63 sequences were obtained either with the Tchouassi et al. protocol (n = 14) [38] or using the degenerate primers/protocol here described (n = 49). A total of 65 species were identified through molecular analysis (File S-IV, Table S2). Not all species could be identified by both methods as in some, no males were available, and in others, no amplification was obtained, leaving the total number of species identified by either method as 67. Only 64% of the sequences obtained were correctly identified by the BOLD tool, i.e., corresponding to the genitalia-confirmed species, and 53% shared ≥95% identity with a given species-specific sequence using the BLASTn tool. For eleven of these species, and as far as we could ascertain, partial *CoxI* sequences are provided here for the first time. These species include *Ae. (Albuginosus) capensis* Edwards, 1924; *Ae. (Mucidus) mucidus* (Karsch, 1887); *Cx. (Culex) andersoni* Edwards, 1914; *Cx. (Cux.) telesilla* de Meillon and Lavoipierre, 1945; *Cx. (Eumelanomyia) inconspicuosus* (Theobald, 1908); *Er. subsimplicipes*

Edwards, 1914; *Er. quinquevittatus* Theobald, 1901; *Fi. uniformis* (Theobald, 1904); *Mi. (Mim) hispida* (Theobald, 1910); *Ur. (Uranotaenia) alboabdominalis* Theobald, 1910; and *Ur. (Pseudoficalbia) mashonaensis* Theobald, 1901 (the male genitalia of which are presented in Figure 1).

Figure 1. Genitalia of male mosquito species whose partial sequence of the *CoxI* gene was obtained for the first time in this study: (**a**,**b**) *Ae. (Alb.)* capensis, (**c**,**d**) *Ae. (Muc.) mucidus*, (**e**) *Er. subsimplicipes*, (**f**) *Er. quinquevittatus*, (**g**,**h**) *Cx. (Cux.) andersoni*, (**i**,**j**) *Cx. (Cux.) telesilla*, (**k**,**l**) *Cx. (Eum.) inconspicuosus*, (**m**) *Fi. uniformis*, (**n**) *Mi. hispida*, (**o**) *Ur. (Ura.) alboabdominalis*, (**p**) *Ur. (Pfc.) mashonaensis*. Most photographs represent the whole genitalia, with the exception of (**b**) detail of gonostylus, (**d**) detail of basal dorsomesal lobe, claspettes, and proteger, (**g**,**i**,**k**) phallosome, and (**h**,**j**,**l**) gonocoxite with gonostylus.

3.2. *Mosquito Identification Using Phylogenetic Reconstruction*

3.2.1. Genus *Aedeomyia*

Aedeomyia sequences were grouped phylogenetically according to their subgenera (Figure 2). *Aedeomyia (Aedeomyia) africana* Neveu-Lemaire, 1906 from Mozambique (File S-II Figure S1) was grouped according to a conspecific sequence from Malawi and another from Madagascar, *Ad. (Ady) madagascarica* Brunhes, Boussès & da Cunha Ramos, 2011. Those from Kenya formed their own clade with a divergence of 6.9% $\pm$ 1.3 between the two *Ad. africana* clades (Figure 2 and Supplementary Material File File S-IV, Table S3). *Aedeomyia (Lepiothauma) furfurea* (Enderlein, 1923), both from Mozambique and South Africa, formed a strong clade. The divergence between these two species was >10%. Networks and PCOORD analyses agreed with that topology (File S-IV, Figure S3).

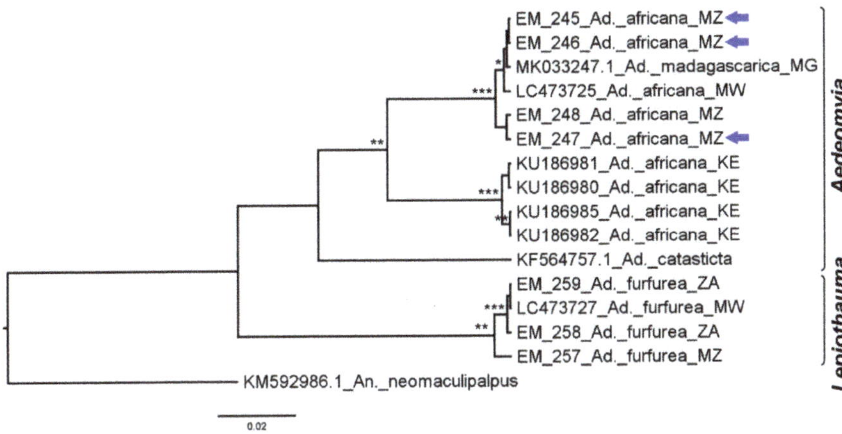

Figure 2. Phylogenetic analysis of 15 partial *coxI* nucleotide sequences from *Aedeomyia* mosquitoes. At specific branches, the number of * indicates the tree topological support revealed by the different phylogenetic reconstruction methods, assuming relevant bootstrap and aLRT values above 75% and posterior probability values above 0.80. The sequences obtained in this work are designated with the "EM" code, while those with associated genitalia are indicated with horizontal blue arrows. Reference sequences downloaded from the public databases are shown by their respective access codes (Boldsystems) or accession numbers (GenBank), as well as the country of origin [South Africa (ZA), Madagascar (MG), Malawi (MW), Mozambique (MZ), Kenya (KE)]. Vertical lines mark the *Aedeomyia* and *Lepiothauma* subgenera.

3.2.2. Genus *Aedes*

Aedes sequences formed two main clusters, with species within subgenera *Mucidus* and *Ochlerotatus* forming a cluster separated from species representing all the other *Aedes* subgenera (Figure 3).

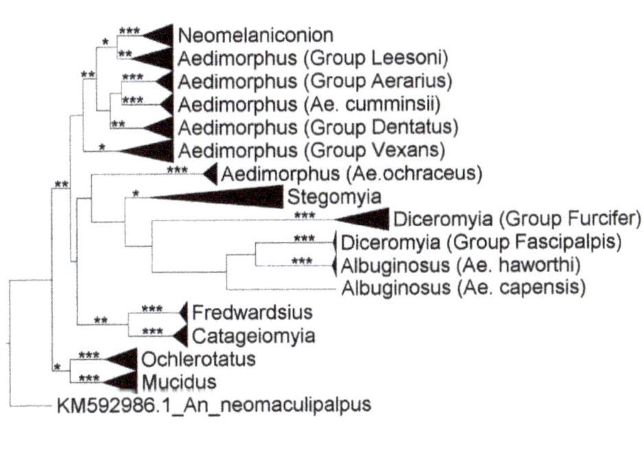

Figure 3. Phylogenetic analysis of 172 partial *coxI* nucleotide sequences from *Aedes* mosquitoes. At specific branches, the number of * indicates the tree topological support revealed by the different phylogenetic reconstruction methods, assuming relevant bootstrap and aLRT values above 75% and posterior probability values above 0.80. In the collapsed branches are the species of the subgenera and/or informal groups of the subgenera.

Species in the subgenus *Neomelaniconion* formed a single, monophyletic, strongly supported clade (Figure 4) in which *Ae. (Neo.) mcintoshi* Huang, 1985 (File S-II Figure S16), *Ae. (Neo.) unidentatus* McIntosh, 1971, and *Ae. (Neo.) circumluteolus* (Theobald, 1908) (File S-II Figure S15) were grouped in a clade with a variation of 1.2% $\pm$ 0.3 that overlapped the interspecific divergence (1.1–1.4%) (File S-IV, Table S4). *Aedes* (*Neo.*) *lineatopennis* (Ludlow, 1905) formed a sister clade, showing a divergence with the other species $\geq$ 5.4%. Networks and PCOORD analyses supported these results (File S-IV, Figure S4).

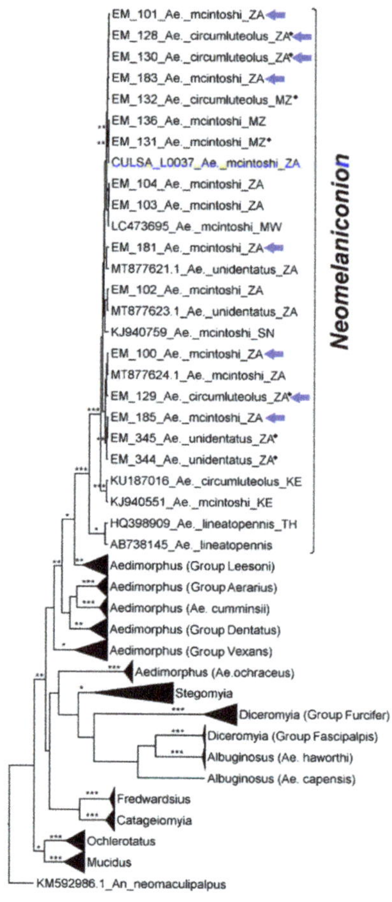

Figure 4. Phylogenetic analysis of 172 partial *coxI* nucleotide sequences from mosquitoes of the genus *Aedes*, presenting the subgenus *Neomelaniconion*. At specific branches, the number of * indicates the tree topological support revealed by the different phylogenetic reconstruction methods, assuming relevant bootstrap and aLRT values above 75% and posterior probability values above 0.80. The sequences obtained in this work are indicated with the "EM" code, and those with associated genitalia are indicated by horizontal blue arrows. The sequences downloaded from GenBank and Boldsystems are indicated by their respective accession numbers and access codes (respectively); the symbol "." after the code of our sequence indicates that said sequence was not identified by Boldsystems. Their country of origin [South Africa (ZA), Malawi (MW), Mozambique (MZ), Kenya (KE), Senegal (SN), Thailand (TH)] are also indicated. The vertical line marks the subgenus *Neomelaniconion* and the collapsed branches indicate the species of the subgenera and/or informal groups of the subgenera.

Most sequences of taxa in the subgenus *Aedimorphus* formed a polyphyletic clade, separating into subclades according to their morphologically based groupings designated by McIntosh [46] (Figure 5). However, *Ae. (Adm.) cumminsii* (Theobald, 1903) sequences from Kenya, Guinea, and Senegal shared an inter-group variation that ranged from 0.7–2.3%, according to the origin (File S-IV, Table S5), forming a clade distant from the conspecific sequences from South Africa, which joined the Dentatus group, *Ae. (Adm.) dentatus* (Theobald, 1904) (File S-II Figure S3) and *Ae. (Adm.) pachyurus* Edwards, 1936, to which they belong. The divergence between these two groups of *Ae. cumminsii* was ≥7.4%. Networks and PCOORD analyses corroborated this finding (File S-IV, Figure S5).

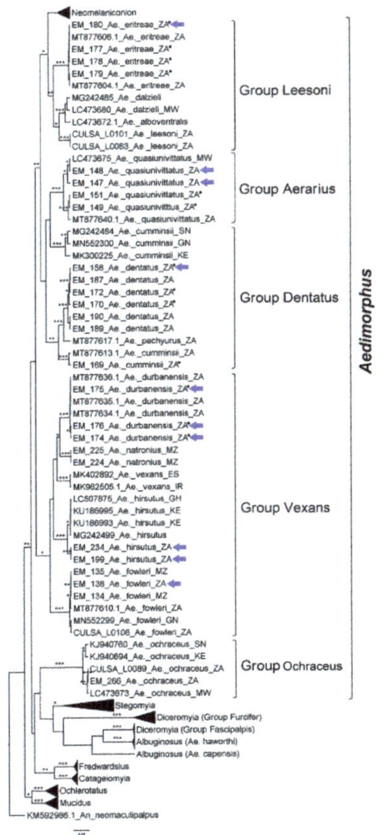

Figure 5. Phylogenetic analysis of 172 partial *coxI* nucleotide sequences from mosquitoes of the genus *Aedes*, presenting the subgenus *Aedimorphus*. At specific branches, the number of * indicates the tree topology support revealed by the different phylogenetic reconstruction methods, assuming relevant bootstrap and aLRT values above 75% and posterior probability values above 0.80. The sequences obtained in this work have the "EM" code and those with associated genitalia are indicated by horizontal blue arrows. The sequences downloaded from GenBank and Boldsystems are indicated by their respective accession numbers and access codes (respectively); the symbol "." after the code of our sequence indicates that said sequence was not identified by Boldsystems. Their country of origin [South Africa (ZA), Spain (ES), Ghana (GH), Guinea (GN), Iran (IR), Malawi (MW), Mozambique (MZ), Kenya (KE), Senegal (SN)] are also indicated. The vertical lines mark the informal groups and the subgenus *Aedimorphus*; the collapsed branches are the species of the subgenera and/or informal groups of the subgenera.

The *Stegomyia* subgenus formed a monophyletic clade in which most species formed well-supported species group clades (Figure 6). One exception was within the Simpsoni group where *Ae.* (*Stg.*) *simpsoni* (Theobald, 1905) and *Ae.* (*Stg.*) *bromeliae* (Theobald, 1911) formed a single clade with a variation of 1.1% ± 0.3, while the interspecific divergence of the species in the clade was 1.3% ± 0.4. *Ae.* (*Stg.*) *unilineatus* (Theobald, 1906) formed two monophyletic sister clades comprising sequences from either South Africa (File S-II Figure S20) or Pakistan, with a global intraspecific variation of 3.4% ± 0.6 and an inter-clade variation of 5.3% ± 0.9. Similarly, sequences from *Ae.* (*Stg.*) *metallicus* (Edwards, 1912) (Figure S-II S19) formed two monophyletic sister clades, with an inter-clade distance of 7.4% ± 1.0. These results were corroborated by the networks and PCOORD analyses, evidencing the near lack of separation of *simpsoni/bromeliae*, wider separation for the two groups of *unilineatus*, and even greater separation for *metallicus* (File S-IV, Figure S6).

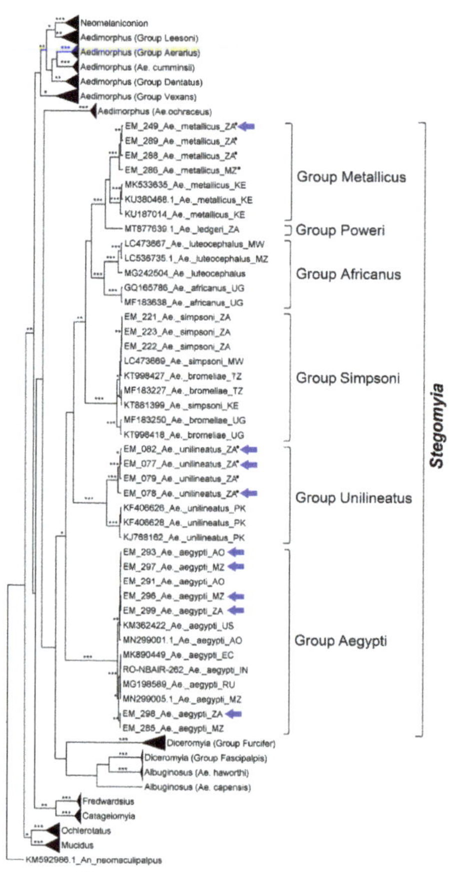

Figure 6. Phylogenetic analysis of 172 partial *coxI* nucleotide sequences from mosquitoes of the genus *Aedes*, presenting the subgenus *Stegomyia*. At specific branches, the number of * indicates the tree topological support revealed by the different phylogenetic reconstruction methods, assuming relevant bootstrap and aLRT values above 75% and posterior probability values above 0.80. The sequences obtained in this work are indicated with the "EM" code and those with associated genitalia are indicated by horizontal blue arrows. The sequences downloaded from GenBank and Boldsystems are indicated by their respective accession numbers and access codes (respectively); the symbol "." after the code of our sequence indicates that said sequence was not identified by Boldsystems. Their country of origin [South Africa (ZA), Angola (AO), United States (US), Ecuador (EC), Russian

Federation (RU), India (IN), Malawi (MW), Mozambique (MZ), Pakistan (PK), Kenya (KE), Tanzania (TZ), Uganda (UG)] are also indicated. The vertical lines mark the informal groups and the subgenus Stegomyia; the collapsed branches are the species of the subgenera and/or informal groups of the subgenera.

Subgenus *Diceromyia* was paraphyletic, but the two species included—*Ae. (Dic.) furcifer* (Edwards, 1913) and *Ae. (Dic.) fascipalpis* (Edwards, 1912), both represented by specimens from South Africa (File S-II Figures S11 and S12)—formed species-specific clades with strong support (Figure 7), confirmed in networks and PCOORD analyses (File S-IV, Figure S7).

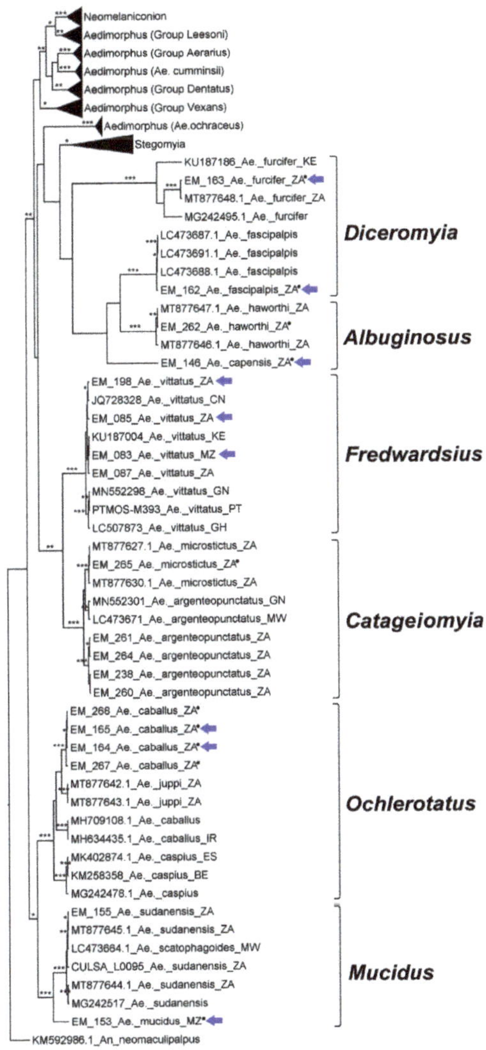

Figure 7. Phylogenetic analysis of 172 partial *coxI* nucleotide sequences from mosquitoes of the genus *Aedes*, presenting the subgenera *Diceromyia*, *Albuginosus*, *Fredwardsius*, *Catageiomyia*, *Ochlerotatus*, and

Mucidus. At specific branches, the number of * indicates the tree topological support revealed by the different phylogenetic reconstruction methods, assuming relevant bootstrap and aLRT values above 75% and posterior probability values above 0.80. The sequences obtained in this work are indicated with the "EM" code and those with associated genitalia are indicated by horizontal blue arrows. The sequences downloaded from GenBank and Boldsystems are indicated by their respective accession numbers and access codes (respectively); the symbol "." after the code of our sequence indicates that said sequence was not identified by Boldsystems. Their country of origin [South Africa (ZA), Belgium (BE), China (CN), Spain (ES), Ghana (GH), Guinea (GN), India (IN), Iran (IR), Malawi (MW), Mozambique (MZ), Portugal (PT), Kenya (KE)] are also indicated. The vertical lines mark the subgenera shown; the collapsed branches are the species of the subgenera and/or informal groups of the subgenera.

Ochlerotatus sequences formed a strong clade, with equally strong paraphyletic subclades; in these, *Ae. (Och.) caballus* (Theobald, 1912) (File S-II Figure S17) and *Ae. (Och.) juppi* McIntosh, 1973, both from South Africa, segregated into closer subclades, while *Ae. caballus* from Iran formed a separate cluster (Figure 7). Intraspecific variance within each of the three groups was low ($\leq$0.5% $\pm$ 0.2); interspecific divergence between *Ae. caballus* and *Ae. juppi* from SA was 2.8% $\pm$ 0.7, and *Ae. caballus* from Iran had a divergence $\geq$ 3.6% $\pm$ 0.8 to either *Ae. caballus* or *Ae. juppi* from SA (File S-IV, Table S6). Networks and PCOORD analyses also placed *Ae. caballus* and *Ae. juppi* from SA closer to one another and farther apart from *Ae. caballus* from Iran (File S-IV, Figure S8).

The clade defining the subgenus *Mucidus* was strongly supported. *Aedes* (*Muc.*) *sudanensis* (Theobald, 1908) and *Ae.* (*Muc.*) *scatophagoides* (Theobald, 1901) were grouped in a single monophyletic cluster with an intra-clade variation of 0.9% $\pm$ 0.3 and an interspecific divergence of 0.6% $\pm$ 0.2. The *Ae. mucidus* sequence from a Mozambique specimen segregated away from all *Ae. scatophagoides* with a divergence of 7.5% $\pm$ 1.2. Similarly, networks and PCOORD analyses placed *Ae. mucidus* sequences far from the *sudanensis* and *scatophagoides*, which were either pooled in an unsolved group or distributed along a single "dimension" without segregation (File S-IV, Figure S9).

3.2.3. Genus *Eretmapodites*

Eretmapodites sequences formed monophyletic clades separating the various species analyzed. Based on morphological features of male genitalia, *Er. intermedius*, *Er. subsimilicipes* (File S-II Figure S22), and *Er. chrysogaster* were very similar and considered members of the "Chysogaster group," and separated quite distinctly from a clade consisting of a sequence of *Er. quinquevittatus* from Mozambique (File S-II Figure S21), which had quite different male genitalia and adult scutal patterns and a GenBank sequence denoted as *Er. silvestris* Ingram and de Meillon, 1927 (Figure 8). *Eretmapodites subsimplicipes* showed no intraspecific variation and diverged from *Er. quinquevittatus* by 9.3% $\pm$ 1.5 (File S-IV, Table S7). Similar results were obtained with networks and PCOORD analyses (File S-IV, Figure S10).

3.2.4. Genera *Culex* and *Lutzia*

Sequences from the genus *Culex* segregated into a highly polyphyletic topology, where most species of subgenus *Culex* segregated into clusters intermingled with members of other subgenera. Two major clades with support of one of the three methods were formed; the first contained species of the subgenus *Culex*, namely, some members of the groups Pipiens, Sitiens, the subgroup Vishnui, and the subgenus *Oculeomyia*; the second clade contained species of the subgenus *Culex*, namely, members of the Pipiens and Duttoni groups and the subgenus *Culiciomyia*. Other separate minor clades, without support among one another, were formed by species of subgroups Sitiens and Decens and the subgenus *Eumelanomyia*, with the genus *Lutzia* as a monophyletic clade (Figure 9).

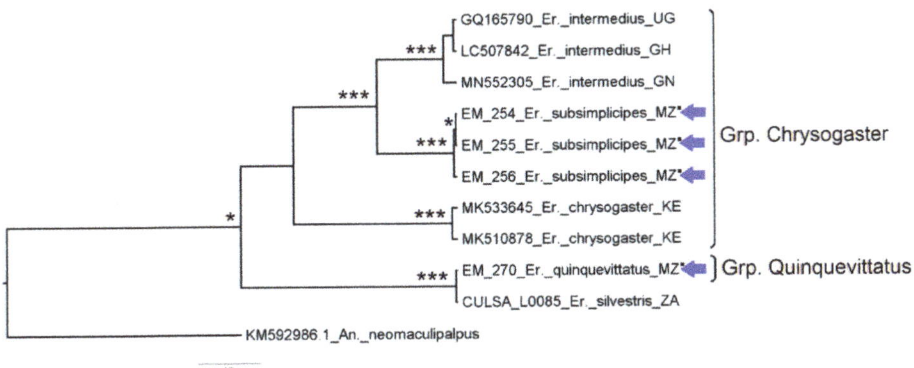

Figure 8. Phylogenetic analysis of 10 partial *coxI* nucleotide sequences from mosquitoes of the genus *Eretmapodites*. At specific branches, the number of * indicates the tree topological support revealed by the different phylogenetic reconstruction methods, assuming relevant bootstrap and aLRT values above 75% and posterior probability values above 0.80. The sequences obtained in this work are indicated with the "EM" code and those with associated genitalia are indicated by horizontal blue arrows; the "Grp" indicated group is marked by vertical lines. The sequences downloaded from GenBank and Boldsystems are indicated by their respective accession numbers and access codes (respectively); the symbol "." after the code of our sequence indicates that said sequence was not identified by Boldsystems. Their country of origin [South Africa (ZA), Ghana (GH), Guinea (GN), Mozambique (MZ), Kenya (KE), Uganda (UG)] are also indicated.

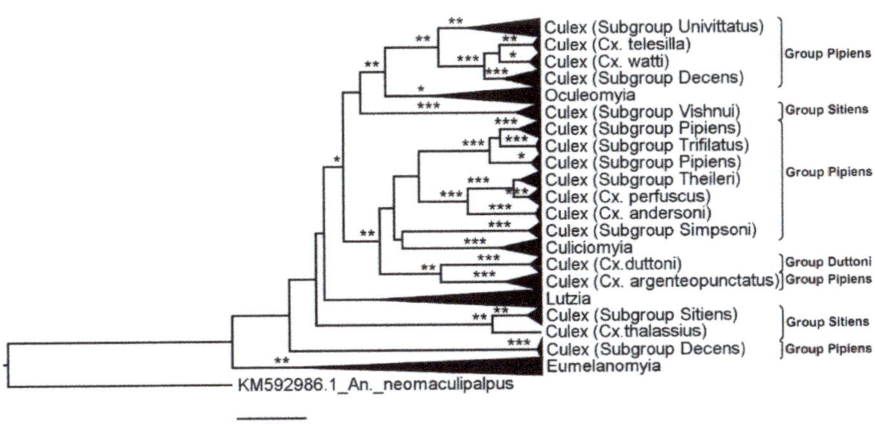

Figure 9. Phylogenetic analysis of 170 partial *coxI* nucleotide sequences from *Culex* and *Lutzia* mosquitoes. At specific branches, the number of * indicates the tree topological support revealed by the different phylogenetic reconstruction methods, assuming relevant bootstrap and aLRT values above 75% and posterior probability values above 0.80. In the collapsed branches there are species of the genus *Lutzia* and the subgenus and informal subgroups of the genus *Culex*; the vertical lines mark the informal groups.

The Univittatus subgroup formed a strongly supported clade, within which *Cx. (Cux.) univittatus* Theobald, 1901, *Cx. (Cux.) neavei* Theobald, 1906, and *Cx. (Cux.) perexiguus* Theobald, 1903, (File S-II Figures S27–S29) segregated into well-supported monophyletic clades (Figure 10). Sequences of *Cx. perexiguus* from South Africa and Mozambique clustered with sequences from other African countries, Europe, and the Middle East, with a divergence of 0.5% ± 0.2 between *Cx. perexiguus* from Europe and the Middle East and those from Africa (File S-IV, Table S8). *Culex univittatus* from Africa were segregated from those of European origin. Networks and PCOORD analyses confirmed these results (File S-IV, Figure S11).

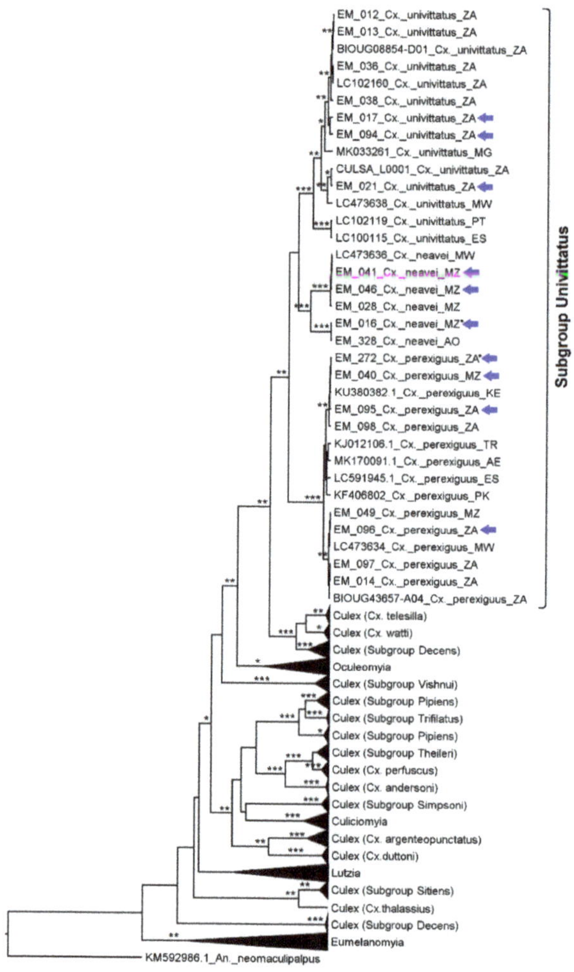

Figure 10. Phylogenetic analysis of 170 partial *coxI* nucleotide sequences from *Culex* and *Lutzia* mosquitoes, presenting the subgroup Univittatus. At specific branches, the number of * indicates the tree topological support revealed by the different phylogenetic reconstruction methods, assuming relevant bootstrap and aLRT values above 75% and posterior probability values above 0.80. The sequences obtained in this work are indicated by the "EM" code and those with associated genitalia are indicated by horizontal blue arrows. The sequences downloaded from GenBank and Boldsystems are indicated by their respective accession numbers and access codes (respectively); the symbol "." After the code of our sequence indicates that said sequence was not identified by Boldsystems. Their country of origin [South Africa (ZA), Angola (AO), United Arab Emirates (AE), Spain (ES), Madagascar (MG), Malawi (MW), Mozambique (MZ), Pakistan (PK), Portugal (PT), Kenya (KE), Turkey (TR)] are also indicated. The vertical line marks the subgroup Univittatus; the collapsed branches are the species of the genus *Lutzia* and subgenera and/or informal groups of the genus *Culex*.

Species of the pipiens complex, *Cx. quinquefasciatus*, and *Cx. pipiens* (File S-II Figures S31 and S32), and all those molecularly typed as *Cx. quinquefasciatus, Cx. pipiens (pipiens* ecotype plus hybrids of the *pipiens* and *molestus* ecotypes), formed a strongly supported clade (Figure 11). This monophyletic clade included *Cx. (Cux.) trifilatus* Edwards, 1914 (File S-II Figure S36), specimens of *Cx. pipiens* ecotype *molestus*, and one that could not be confirmed molecularly. The intra-clade variation supporting *Cx. quinquefasciatus, Cx. pipiens* (*pipiens* ecotype plus *pipiens-molestus* hybrids), and *Cx. trifilatus* was 1.6% ± 0.3, while the *molestus* ecotype diverged >2% in relation to the *pipiens* ecotype and *Cx. quinquefasciatus*, and *Cx. trifilatus* diverged ≥ 2.9% from any of the Pipiens subgroup members (File S-IV, Table S9). Similar results were obtained with networks and PCOORD analyses (File S-IV, Figure S12).

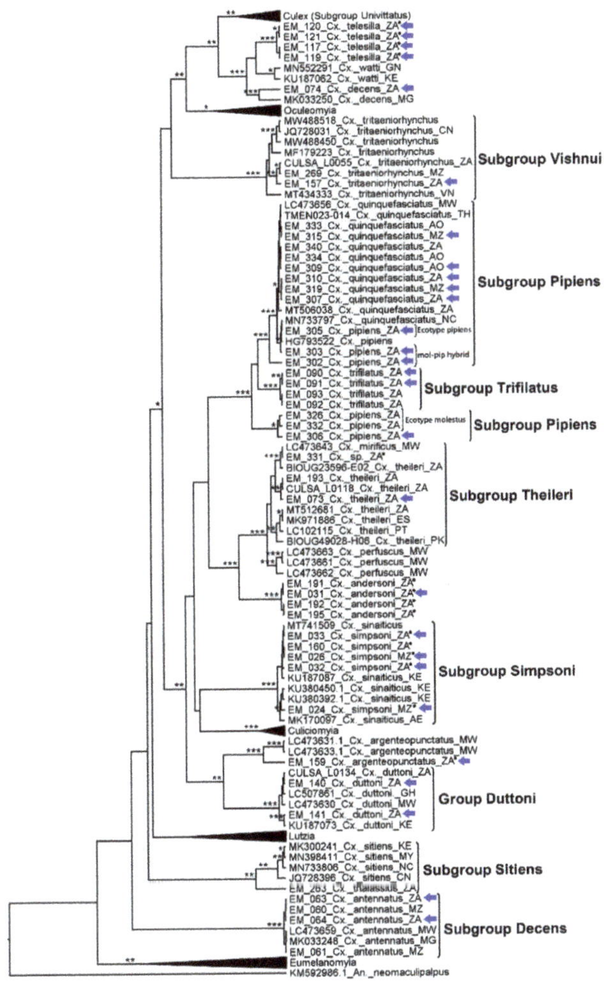

Figure 11. Phylogenetic analysis of 170 partial *coxI* nucleotide sequences from *Culex* and *Lutzia* mosquitoes, presenting the informal groups of the *Culex* subgenus. At specific branches, the number of * indicates the tree topological support revealed by the different phylogenetic reconstruction methods, assuming relevant bootstrap and aLRT values above 75% and posterior probability values

above 0.80. The sequences obtained in this work are indicated by the "EM" code and those with associated genitalia are indicated by horizontal blue arrows. The sequences downloaded from GenBank and Boldsystems are indicated by their respective accession numbers and access codes (respectively); the symbol "." after the code of our sequence indicates that said sequence was not identified by Boldsystems. Their country of origin [South Africa (ZA), Angola (AO), China (CN), Ghana (GH), Guinea (GN), United Arab Emirates (AE), Spain (ES), New Caledonia (NC), Madagascar (MG), Malaysia (MY), Malawi (MW), Mozambique (MZ), Pakistan (PK), Portugal (PT), Kenya (KE), Thailand (TH), Vietnam (VN)] are also indicated. The vertical lines mark the informal subgroups; the collapsed branches are the species of the genus Lutzia and subgenera and/or informal groups of the genus *Culex*, in addition to the ecotypes of *Cx. pipiens*.

Sequences from South African *Cx. (Cux.) theileri* Theobald, 1903 (File S-II Figure S35), another from a female originally identified (by us) as *Cx.* sp. (EM331) and *Cx. (Cux.) mirificus* Edwards, 1913 from Malawi (sharing 100% *CoxI* identity with EM331), formed sister clades with other sequences of *Cx. theileri* from Spain, Portugal, and Pakistan, with an intraspecific variation of 0.8% $\pm$ 0.2 (Figure 11). These were joined by *Cx. (Cux.) perfuscus* Edwards, 1914 and *Cx. andersoni* (File S-II Figure S24), forming a larger, well-supported monophyletic clade, a pattern that was supported by networks and PCOORD analyses (File S-IV, Figure S13).

Culex spp. of the subgroup Simpsoni formed a strongly supported clade in which the sequences of *Cx. (Cux.) simpsoni* Theobald, 1905 from this study, which had been morphologically confirmed through the male genitalia (File S-II Figure S33), did not segregate from the sequences of *Cx. (Cux.) sinaiticus* Kirkpatrick, 1925 from GenBank (Figure 11). Intraclade, intraspecific and interspecies divergence values overlapped, ranging from 0.2% to 0.4% ($\pm$0.1–0.2). These species were neither segregated by networks nor PCOORD analyses (File S-IV, Figure S14).

Subgenus *Oculeomyia* formed a monophyletic clade with branch support in only one of three methods (Figure 12); *Culex (Ocu.) bitaeniorhynchus* Giles, 1901 (File S-II File S44), *Cx. (Ocu.) infula* Theobald, 1901 (File S-II Figure S43), *Cx. (Ocu.) annulioris* Theobald, 1901 (File S-II Figure S41), and *Cx. (Ocu.) poicilipes* (Theobald, 1903) (File S-II Figure S42) sequences formed sister clades. However, the clades containing *Cx. bitaeniorhynchus* and *Cx. infula* were not species-specific; rather, sequences were grouped according to geographic origin, separating African specimens from ones originating in Asia. Hence, to unravel the relation of these taxa, we performed a further phylogenetic reconstruction with a larger data set (File S-IV, Figure S15). Similarly, African sequences obtained in this work deviated from the large clade formed by sequences from Asia and the Middle East, without separation of *Cx. bitaeniorhynchus* and *Cx. infula*. The distance between the various groups of sequences from the various countries of origin, or of different species, did not surpass 3%, and the divergence of these clades ranged between 2.0–2.7% (File S-IV, Table S10a,b). Networks and PCOORD analyses (File S-IV, Figure S16) still failed to separate *Cx bitaeniorhynchus* from *Cx. infula*.

The subgenus *Culiciomyia* was grouped into a defined clade with strong branch support, where *Cx. (Cui.) cinereus* Theobald, 1901 and *Cx. (Cui.) nebulosus* Theobald, 1901 (File S-II Figures S38 and S39) formed equally strong monophyletic clades (Figure 12) with low intraspecific variation for each branch ($\leq$0.4%), diverging by 3.6% $\pm$ 0.8.

Eumelanomyia sequences were grouped in an external clade of the remaining *Culex* subgenera (Figure 12), with *Cx. inconspicuosus* from South Africa (File S-II Figure S40) forming a strong clade with an intraspecific variation of 0.8% $\pm$ 0.3.

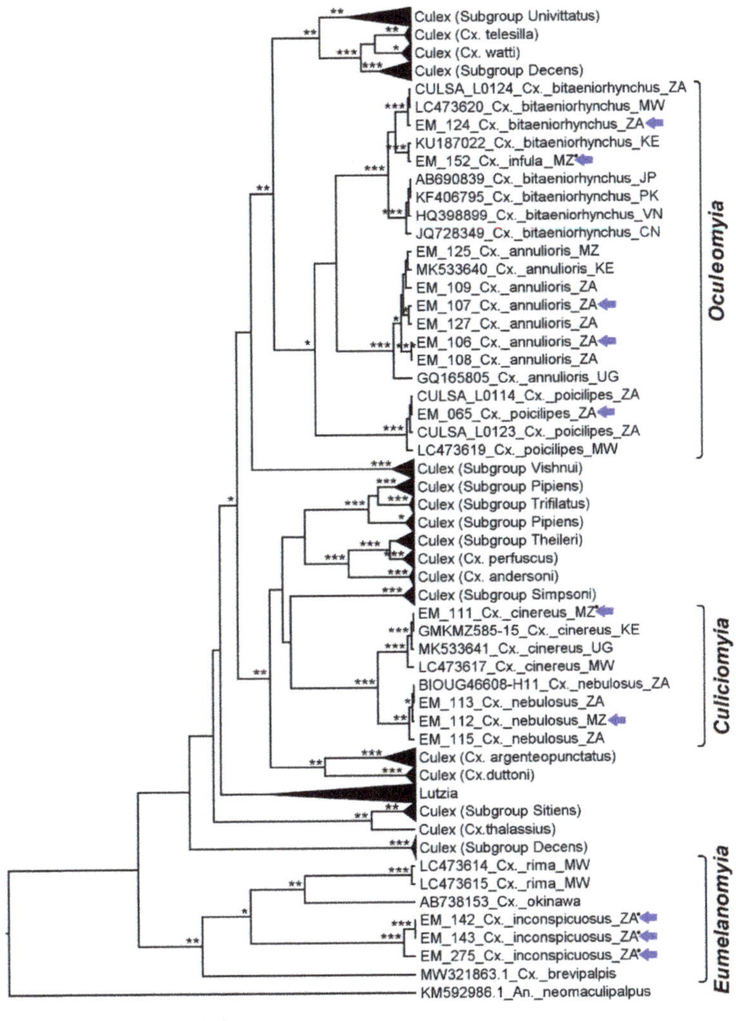

Figure 12. Phylogenetic analysis of 170 partial *coxI* nucleotide sequences from *Culex* and *Lutzia* mosquitoes, presenting the subgenera *Oculeomyia, Culiciomyia,* and *Eumelanomyia*. At specific branches, the number of * indicates the tree topological support revealed by the different phylogenetic reconstruction methods, assuming relevant bootstrap and aLRT values above 75% and posterior probability values above 0.80. The sequences obtained in this work are indicated by the "EM" code and those with associated genitalia are indicated by horizontal blue arrows. The sequences downloaded from GenBank and Boldsystems are indicated by their respective accession numbers and access codes (respectively); the symbol "." after the code of our sequence indicates that said sequence was not identified by Boldsystems. Their country of origin [South Africa (ZA), China (CN), Japan (JP), Malawi (MW), Mozambique (MZ), Pakistan (PK), Kenya (KE), Vietnam (VN), Uganda (UG)] are also indicated. The vertical lines mark the subgenera of *Culex*; the collapsed branches are the species of the genus *Lutzia* and informal groups of the genus *Culex*.

Sequences derived from *Lutzia (Metalutzia) tigripes* (de Grandpre & de Charmoy, 1901) from Angola and South Africa (File S-II Figure S45) were pooled with conspecific ones from other African countries (Figure 13). When analyzing the relationship of the genus *Lutzia* with the other genera studied in this work, it grouped within a strongly supported clade that combined it with species of the subgenera *Culex, Oculeomyia,* and *Culiciomyia* (Figure 14).

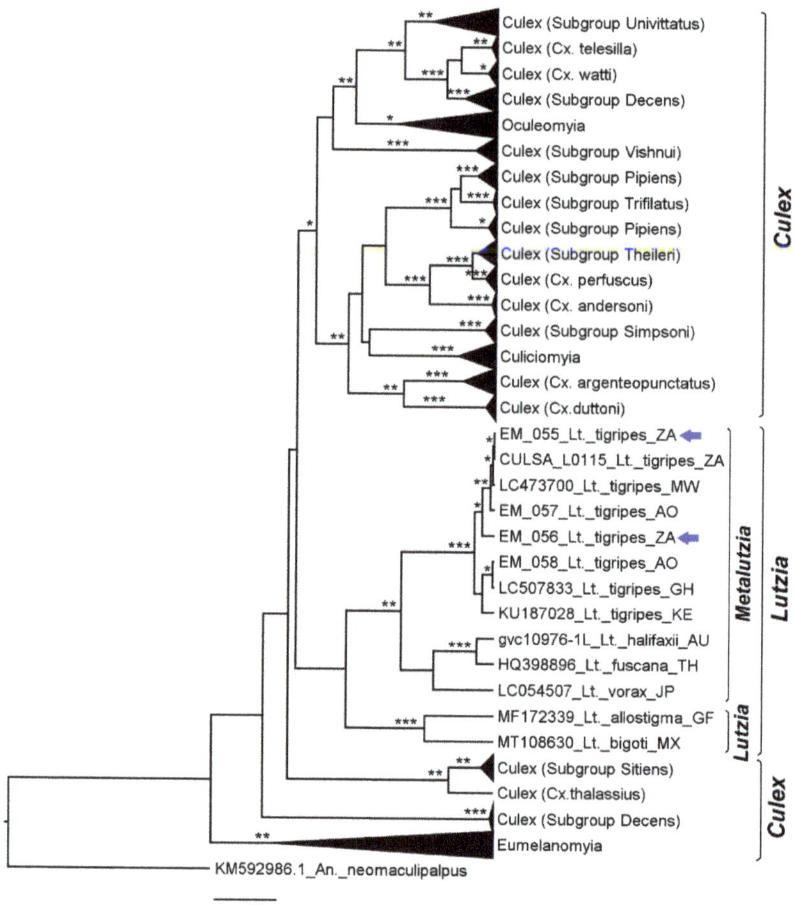

Figure 13. Phylogenetic analysis of 170 partial *coxI* nucleotide sequences from mosquitoes of the *Culex* and *Lutzia* genera, presenting the *Lutzia* genus. At specific branches, the number of * indicates the tree topological support revealed by the different phylogenetic reconstruction methods, assuming relevant bootstrap and aLRT values above 75% and posterior probability values above 0.80. The sequences obtained in this work are indicated by the "EM" code and those with associated genitalia are indicated by horizontal blue arrows. The sequences downloaded from GenBank and Boldsystems are indicated by their respective accession numbers and access codes (respectively). Their country of origin [South Africa (ZA), Angola (AO), Australia (AU), Ghana (GH), French Guiana (GF), Japan (JP), Malawi (MW), Mexico (MX), Kenya (KE), Thailand (TH)] are also indicated. The vertical lines mark the *Culex* genus and the *Lutzia* genus and its subgenera.

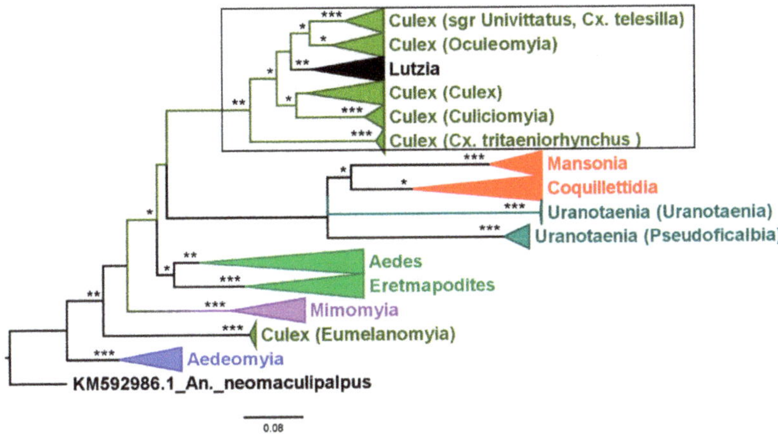

Figure 14. Phylogenetic analysis of 179 partial *coxI* nucleotide sequences from mosquitoes of the genera *Aedeomyia, Aedes, Coquillettidia, Culex, Lutzia, Mimomyia,* and *Uranotaenia*. At specific branches, the number of * indicates the tree topological support revealed by the different phylogenetic reconstruction methods, assuming relevant bootstrap and aLRT values above 75% and posterior probability values above 0.80. The collapsed branches indicate the different genera; "sgr" indicates the subgroup.

3.2.5. Genera *Ficalbia* and *Mimomyia*

The *CoxI* sequence obtained from *Fi. uniformis* (File S-II Figure S46) clustered inside the *Mimomyia* radiation, distant from the *Fi. minima* clade (Figure 15). Sequences of *Mi. (Mimomyia) mimomyiaformis* (Newstead, 1907) (File S-II Figure S47) and *Mi. (Mim) hispida* (File S-II Figure S48) clustered in a large clade, in which the former was organized into two strongly supported paraphyletic clades, with an overall intraspecific variation of 0.9% ± 0.3 (File S-IV, Table S11). Networks and PCOORD analyses revealed an identical pattern (File S-IV, Figure S17).

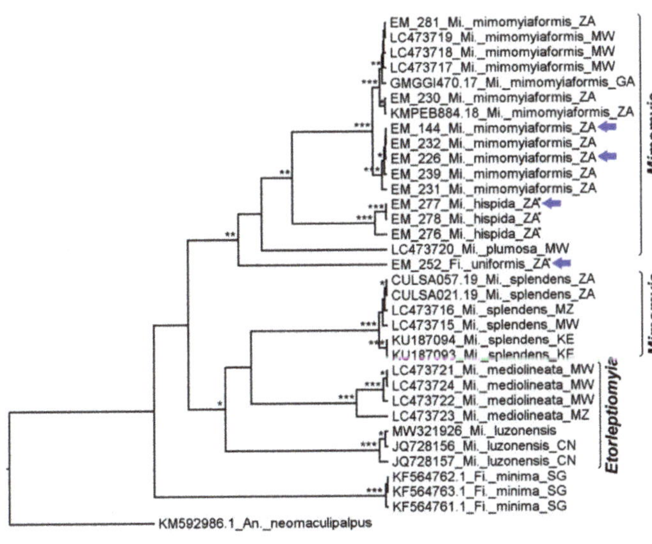

Figure 15. Phylogenetic analysis of 33 partial *coxI* nucleotide sequences from mosquitoes of the Ficalbia and Mimomyia genera. At specific branches, the number of * indicates the tree topological

support revealed by the different phylogenetic reconstruction methods, assuming relevant bootstrap and aLRT values above 75% and posterior probability values above 0.80. The sequences obtained in this work are indicated by the "EM" code and those with associated genitalia are indicated by horizontal blue arrows. The sequences downloaded from GenBank and Boldsystems are indicated by their respective accession numbers and access codes (respectively); the symbol "." after the code of our sequence indicates that said sequence was not identified by Boldsystems. Their country of origin [South Africa (ZA), China (CN), Gabon (GA), Malawi (MW), Mozambique (MZ), Kenya (KE)] are also indicated. The vertical lines mark the subgenera of Mimomyia and Etorleptiomyia.

3.2.6. Genus *Coquillettidia*

Sequences from the South African *Cq.* (*Coquillettidia*) *chrysosoma* (Edwards, 1915) specimens (File S-II Figure S51) grouped with *Cq.* (*Coq.*) *fuscopennata* (Theobald, 1903), *Cq.* (*Coq.*) *aurites* (Theobald, 1901), and *Cq. chrysosoma* sequences from Kenya, with an intra-clade variation of 0.2% ± 0.1 (Figure 16). Sequences of *Cq. fuscopennata* from South Africa (File S-II Figure S49) clustered with a sequence from Malawi in a monophyletic clade with an intraspecific variation of 0.5% ± 0.2, while another clade clustered GenBank sequences from *Cq. fuscopennata*, *Cq.* (*Coq.*) *versicolor* (Edwards, 1913) and *Cq.* (*Coq.*) *microannulata* (Theobald, 1911). The sequence of *Cq.* (*Coq.*) *metallica* (Theobald, 1901) from Mozambique (File S-II Figure S50) clustered in a monophyletic clade with an intraspecific variation of 1.1% ± 0.3. This was confirmed by network and PCOORD analyses (File S-IV, Figure S18).

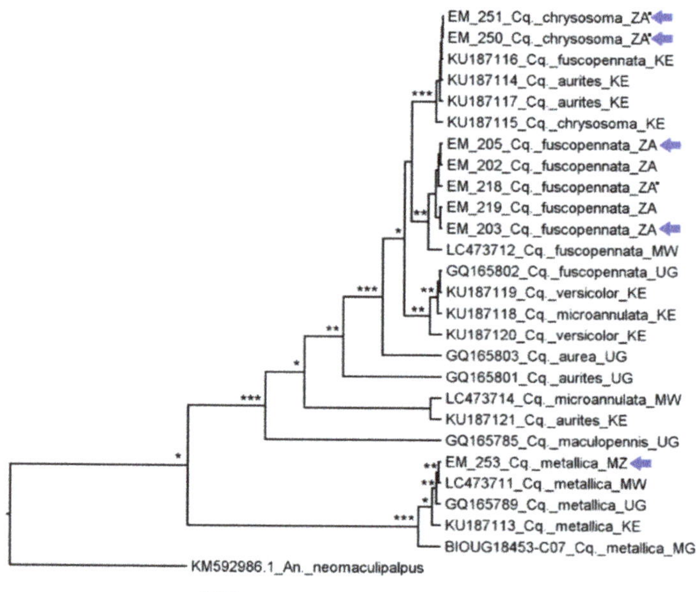

Figure 16. Phylogenetic analysis of 26 partial *coxI* nucleotide sequences from mosquitoes of the genus Coquillettidia. At specific branches, the number of * indicates the tree topological support revealed by the different phylogenetic reconstruction methods, assuming relevant bootstrap and aLRT values above 75% and posterior probability values above 0.80. The sequences obtained in this work are indicated by the "EM" code and those with associated genitalia are indicated by horizontal blue arrows. The sequences downloaded from GenBank and Boldsystems are indicated by their respective accession numbers and access codes (respectively); the symbol "." after the code of our sequence indicates that said sequence was not identified by Boldsystems. Their country of origin [South Africa (ZA), Madagascar (MG), Malawi (MW), Mozambique (MZ), Kenya (KE), Uganda (UG)] are also indicated.

3.2.7. Genus *Mansonia*

Mansonia (Mansonioides) africana (Theobald, 1901) and *Ma. (Mnd.) uniformis* (Theobald, 1901) were identified in this study based on female and male genitalia structures (File S-II Figures S52 and S53). Sequences of *Ma. uniformis*, from the Afrotropical and Indomalayan regions, were placed in two sister clades (Figure 17) with low intra-clade variation (ranging from 0.4 to 0.7%) but diverging from one another by 4.1% ± 0.9. *Mansonia africana* joined conspecific sequences from various African origins, with a divergence from *Ma. uniformis* ≥9%. These results were congruent with the network and PCOORD analyses (File S-IV, Figure S19).

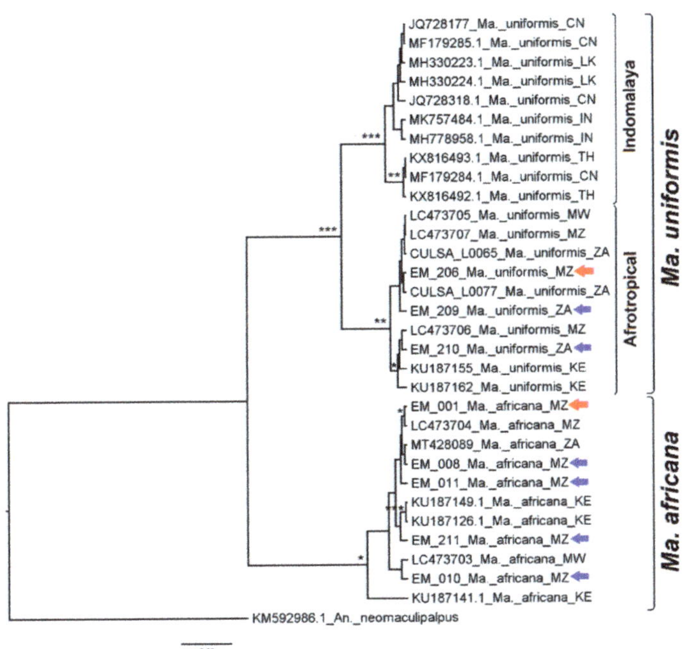

Figure 17. Phylogenetic analysis of 31 partial *coxI* nucleotide sequences from mosquitoes of the genus Mansonia. At specific branches, the number of * indicates the tree topological support revealed by the different phylogenetic reconstruction methods, assuming relevant bootstrap and aLRT values above 75% and posterior probability values above 0.80. The sequences obtained in this work are indicated by the "EM" code and those with associated genitalia are indicated by horizontal arrows (blue = males, red = females). The sequences downloaded from GenBank and Boldsystems are indicated by their respective accession numbers and access codes (respectively). Their country of origin [South Africa (ZA), China (CN), India (IN), Malawi (MW), Mozambique (MZ), Kenya (KE), Sri Lanka (LK), Thailand (TH)] are also indicated.

3.2.8. Genus *Uranotaenia*

Uranotaenia alboabdominalis (File S-II Figure S55) formed a strongly supported monophyletic clade (Figure 18) with an intraspecific variation of 0.2% ± 0.1. The sequences from *Ur. mashonaensis* (File S-II Figure S54) clustered into a monophyletic clade with strong support; however, the intraspecific variation was 4.1% ± 0.8, with a divergence of 5.6% ± 1.1 between the two branches.

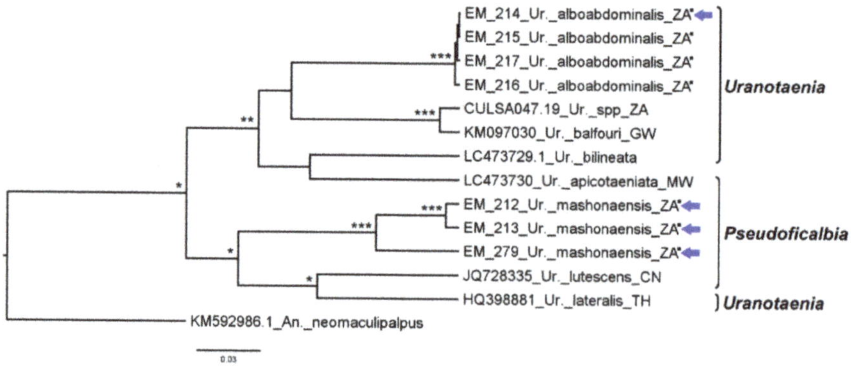

Figure 18. Phylogenetic analysis of 13 partial *coxI* nucleotide sequences from mosquitoes of the genus Uranotaenia. At specific branches, the number of * indicates the tree topological support revealed by the different phylogenetic reconstruction methods, assuming relevant bootstrap and aLRT values above 75% and posterior probability values above 0.80. The sequences obtained in this work are indicated by the "EM" code and those with associated genitalia are indicated by horizontal blue arrows. The sequences downloaded from GenBank and Boldsystems are indicated by their respective accession numbers and access codes (respectively); the symbol "." after the code of our sequence indicates that said sequence was not identified by Boldsystems. Their country of origin [South Africa (ZA), China (CN), Guinea-Bissau (GW), Malawi (MW), Thailand (TH)] are also indicated. The vertical lines mark the subgenera Uranotaenia and Pseudoficalbia.

4. Discussion

The genitalia of 663 mosquitoes (both male and female) were dissected and 55 species were identified; 247 partial sequences of the *CoxI* gene from 65 species were obtained and analyzed using complementary approaches, yielding a total of 67 species from 10 genera, identified by either method. This corresponded to *circa* 40% (60/150) of Culicinae mosquito fauna from South Africa and 34% (31/91) from Mozambique. Eleven of these partial *CoxI* sequences are, to the best of our knowledge, here published for the first time, with corresponding morphologic confirmation. Curiously, a considerable proportion of sequences that were generated failed to be identified either using the BOLD taxonomy tool (36%) or BLASTn (47%). In these cases, formal species assignment was carried out based on a fine morphological confirmation (genitalia) and/or by phylogenetic reconstruction.

Interspecific congeneric distances ranged between 1% and 20%, with mean values between 7% and 15% (Fiel S-IV Table S12). These values are mostly within the range observed for divergence in congeneric species, 2.3–21.8%, although the majority of cases are in the 4–11% interval (Ashfaq et al., 2014; Wang et al., 2012; refs. [21,27]). Low divergence values in congeneric taxa, such as 0.6–2.0%, can be interpreted as species of recent divergence [47]. Divergence between conspecific specimens typically ranges between 0% and 2.4% [21,27] or as high as 3% or 5.4% [18,48].

In some instances, divergence values > 2% in conspecific sequences were observed; hence, they were greater than expected in members of the same taxa; conversely, divergence values < 2% between taxa of different species were also obtained, revealing a failure of the *CoxI* marker to separate such taxa.

Higher than expected genetic diversity was observed in *Ad. africana,* where sequences from Malawi and Mozambique on the one hand and Kenya on the other formed separate clades with a divergence of 6.9%. *Aedes cumminsii* from South Africa fell within the Dentatus group as expected, jointly with *Ae. dentatus* from South Africa, a vector of RVFV and the Middelburg virus (MIDV) [2,3], while sequences from Kenya, Guinea, and Senegal formed a separate clade, diverging from the former by >7%. *Aedes unilineatus*, a monotypic member of the Unilineatus group [5] was considered a potential vector of ZIKV [49], with a very wide distribution in sub-Saharan Africa, the Middle East, and Asia [1], in which South African sequences and those originally from Pakistan had an interclade variation of 5.3%. *Aedes metallicus*, the monotypic member of the Metallicus group [5], an important vector of sylvatic YFV and potential vector of ZIKV in Africa [2], had high sequence divergence of 7.4% between South Africa and Mozambique versus that from Kenyan specimens. *Aedes furcifer* from South Africa, Kenya, and Senegal, had an intraspecific variation of 8.1%. *Aedes caballus* from South Africa clustered separately from those originating in Iran, with a divergence of $\geq$3.6% between them. *Mansonia uniformis* formed separate clades, according to African or Indomalayan-Asian origin, with a divergence of 4% between them. *Uranotaenia mashonaensis* from South Africa had a considerable variation of 4.1%.

Such divergence may be explained, in some cases, by comparing sequences of naturally different conspecific populations collected far apart geographically, such as in this study with *Ae. unilineatus*, *Ae. caballus*, and *Ma. uniformis*, which compared sequences of specimens originating from South Africa to those from the Asian region. However, others such as *Ad. africana*, *Ae. cumminsii*, *Ae. metallicus*, *Ae. furcifer*, and *Ur. mashonaensis* exhibited a large genetical divergence in sequences between specimens originating from a span of regional context in the African continent, e.g., South Africa, Mozambique, and Kenya.

It is not surprising that *Aedes cumminsii*, a vector of MIDV, Spondweni virus (SPOV), and RVFV [2], had considerable within-species sequence variations, as it is likely a sibling species complex. Throughout its broad savanna- and forest-dwelling distribution in Africa, many morphological variations have been noted (AJC personal communication and [4]). In addition, *Ae. cumminsii* has undergone some taxonomic confusion since it was originally described as a now designated valid subspecies, ssp *mesostictus* [1], which was originally named ssp *mediopunctatus* (Theobald, 1909) and later placed synonymously and elevated to a subspecies of *Ae. cumminsii* [50]. This subspecies was originally described from specimens collected in Ghana and differs from the typical form of *Ae. cumminsii* by the presence of small basal median whitish spots on the abdominal tergites in both sexes [4,5]; however, McIntosh [46] suggested that this subspecies occurs only in southern Africa. We identified *Ae. cumminsii* with the typical features of the subspecies *mesostictus* in northeastern South Africa, such as Guarido et al. [12], with a divergence of >7% from specimens from Kenya [15].

A lack of CoxI sequence separation of taxa of different species was found between *Ad. madagascarica* and *Ad. africana*; *Ae. mcintoshi*, *Ae. circumluteolus*, and *Ae. unidentatus*; *Ae. simpsoni* and *Ae. bromeliae*; *Ae. scatophagoides* and *Ae. sudanensis*; *Cx. quinquefasciatus* and *Cx. pipiens*; *Cx. simpsoni* and *Cx. sinaiticus*; and *Cx. bitaeniorhynchus* and *Cx. infula*. In these cases, they could not be correctly segregated into species-specific clusters either by traditional phylogenetic reconstruction, networks, or PCOORD analyses. Fortunately, although being morphologically similar species, they can still be differentiated by fine morphological details or male genitalia. Furthermore, genetic distance analysis disclosed overlapping intra- and inter-specific values, circa <2%, showing a limitation in their resolution capacity. Such overlap has been responsible for misidentifications and impossibilities of delimiting species based on pairwise distances [17,51]. One such example includes the segregation of *Ad. africana* from Mozambique (EM_245, LC662529) and Malawi (LC473725) with *Ad. madagascarica* (MK033247.1). Although *Ad. madagascarica* has only been described in Madagascar, the genetic divergence between this species and Ad. africana was only 0.2%. In a contrasting situation, *Ad. africana CoxI* sequences from neighboring countries were separated as aforementioned, raising the need for further clarification of the significance of both the similarity between sequences of *Ad. madagascarica* and *Ad. africana* from Mozambique and Malawi and the divergence of *Ad. africana* from Mozambique and Malawi versus Kenya.

Among the *Aedes*, the subgenus *Neomelaniconion* includes potential vectors of arboviruses, such as *Ae. mcintoshi*, a major vector of RVFV, and *Ae. circumluteolus* and *Ae. unidentatus* as potential RVFV vectors [2], but also potential vectors of the Shuni virus (SHUV) [31]. These three taxa could not be differentiated through phylogenetic reconstruction, in agreement with previous findings [12,13]. Evidence from Kenya suggests that *Ae. mcintoshi* forms a complex of morphologically indistinguishable species, with discordant results between *CoxI* and *ITS* markers [38], particularly *ITS2*, thereby failing to resolve species and species complexes in the subgenus *Neomelaniconion* in Madagascar [52].

Culex pipiens and *Cx. quinquefasciatus*, which are members of the *Culex pipiens* complex, not only display wide geographic distribution but are also highly relevant in the transmission of various pathogens, including arboviruses such as USUV, WNV, and SINV [1]. The female specimens of the two species are morphologically similar but differ in their vectorial efficiency and may occur sympatrically; additionally, hybridization has been reported in some locations [53–55], but not in South Africa [53]. In this study, the absence of *pipiens-quinquefasciatus* hybrids in southern Africa was also noted. Hybrids of the *molestus* and *pipiens* ecotypes, which, so far, have only been reported in the United States [56], Southern Europe [55], and North Africa [57], have also been described in this work as male *Cx. pipiens* specimens from South Africa. Curiously, while *CoxI* analyses could not resolve the closely convergent *Cx. quinquefasciatus* and *Cx. pipiens* into species-specific clusters, the *molestus* ecotype sequences clustered out, diverging 2.2–2.8% in relation to *pipiens* ecotype, *pipiens-molestus* hybrids, and *Cx. quinquefasciatus*. Other studies have reported an intraspecific variation in *Cx. pipiens* (3%), larger than the interspecific distance with *Cx. quinquefasciatus* (1.6%) and their lack of separation [18,24]. The *CoxI* gene has been successful in differentiating members of the *Cx. pipiens* complex, though with low variability; therefore, it may not be the better marker to infer the evolutionary relationship of such close taxa, and more polymorphic markers or a multilocus analysis may be more informative [58]. A lack of differences in *CoxI* between different species can also be explained by possible introgression of mitochondrial DNA after several interspecific crosses, which was proven for *Culex* species using several DNA markers [58].

Culex bitaeniorhynchus and *Cx. infula* belong to the *Cx. bitaeniorhynchus* complex of the subgenus *Oculeomyia*; *Cx. bitaeniorhynchus* has a wide distribution, being present in tropical and subtropical areas of the Afrotropical, Southern Palearctic, and Indomalayan regions, and on the mainland and islands of Southeast Asia and Australasia [1], and can be involved in the transmission of arboviruses [34]. Its status, as well as that of *Cx. infula* and *Cx. ethiopicus* Edwards, 1941, has been the subject of controversy [34,36]. The morphological characteristics of the specimens identified as *Cx. bitaeniorhynchus* (synonymous *Cx. ethiopicus*) corresponded to those described by Edwards [4] and Jupp [5], while the specimen of *Cx. infula* from Mozambique only allowed us to ascertain it as a *Cx.* sp., confirmed by the analysis of the genitalia. Malawian mosquitoes identified as *Cx. ethiopicus* were found to differ in the shape of the wing scales and diverge in the *CoxI* gene > 2% with *Cx. bitaeniorhynchus* from Asia [13]. In our study, the genetic divergence ranged from 2.0% (±0.6) to 2.7% (±0.7) with sequences from Asia, and the tree topology, networks, and PCOORD analysis suggest that *CoxI* does not have discriminating power for separating *Cx. bitaeniorhynchus* from *Cx. infula*. The *Cx. infula CoxI* sequence from Mozambique also deviated from *Cx. infula* from Asia by a similar range. So far, *Cx. infula* has only been described in Asia [1]; however, we were able to confirm its previous identification by Ribeiro in Angola of five male specimens which he designated as *Cx. bitaeniorhynchus* [59]. Furthermore, specimens collected in Africa continue to be classified as *Cx. bitaeniorhynchus* and *Cx. ethiopicus*, according to Edwards' [4] nomenclature [13,29]. Altogether, based on the evidence presented, we believe that most of the specimens identified as *Cx. bitaeniorhynchus* in Africa actually are *Cx. infula*, a situation that Harbach [34] had already anticipated.

Aedes aegypti Linnaeus, 1762 is one of the most important vectors of several global health impact arboviruses, such as DENV, ZIKV, CHIKV, and YFV, not only in Africa but globally [1,2,60]. In the phylogenetic analysis, all sequences of *Ae. aegypti* grouped into a single clade with strong branch support. Although this clade was divided into two branches, albeit only one with reasonable support, and both with a very small distance, there were no data to support the notion that these may correspond to either subspecies *Ae. aegypti aegypti* or *Ae. aegypti formosus* Walker, 1848. Our samples and data set were not adequate for such a separation as no clear morphological differentiation was noticed in our specimens. A small set of sequences (N = 10) was analyzed, and for low variability data, as in this case, the phylogenetic study performed was not that indicated, and haplotype analyses of markers such as mtDNA ND4 [61] or microsatellites [62] were necessary.

As to the arrangement of genera, on the whole, *Aedes* sequences formed clusters that mainly corresponded to the subgenera, in agreement with the morphology-based taxonomy, including the informal groups proposed by McIntosh for the species of Subgenus *Aedimorphus* [5,46]. Subgenera *Neomelaniconion*, *Stegomyia*, *Catageiomyia*, *Fredwardsius*, *Ochlerotatus*, and *Mucidus* formed monophyletic clades, but *Diceromyia*, *Aedimorphus*, and *Albuginosus* did not. Subgenus *Diceromyia* was represented by *Ae. furcifer* and *Ae. fascipalpis*, which, although well separated, yielded a paraphyletic arrangement. Genus *Aedes* is a highly complex entity, the taxonomy of which is in dire need of clarification [35], and that mitogenome evolutionary analysis has shown to be paraphyletic [63].

Genus *Culex* segregated into a highly paraphyletic topology, where most species of subgenus *Culex* coincided with the informal groups and subgroups proposed by Harbach [64]; however, mitogenome phylogenetics has found genus *Culex* to be monophyletic [63]. To the best of our knowledge, this is the first phylogenetic study including African members of the subgenus *Culiciomyia*. While sequences of only two species were included in this study, the monophyly of *Culiciomyia* was well supported, and the *CoxI* marker performed well in the discrimination of *Cx. nebulosus* and *Cx. cinereus*. Multiple species of *Culiciomyia* occur in Africa and all have identical female and male external morphologies, relying solely on male genitalia structures for species identification [65]. Separation of species by *CoxI* gene sequences may be fruitful in the case of *Culiciomyia*.

Lutzia tigripes is the only representative of the genus *Lutzia* in the Afrotropical region [1], clustering as a monophyletic assemblage within the *Culex* radiation, as previously found with *CoxI* and *ITS2* [20,22,66]. Morphological data from adults and larvae support different patterns of relationships between *Lutzia* and *Culex* [67], while a recent analysis of the complete mitochondrial genome concluded that *Lutzia* forms a monophyletic group with genus status [68], emphasizing the limitations of phylogenetic studies with a single marker. However, the classification controversy is not limited to the genus *Lutzia*. Support for the monophyly of Culicini generic-level groups is granted for all except subgenera *Culex*, *Eumelanomyia*, and *Neoculex* [67]. Our analysis could not confirm subgenus *Culex* as a monophyletic group, while *Eumelanomyia* formed a clade distant from the remaining *Culex*, in agreement with previous works [23,28,69]. Nevertheless, the study of all genera together yielded some interesting results; except for subgenus *Eumelanomyia* of *Culex*, there was support for tribes Aedini, Culicini, Ficalbiini, and Mansoniini, in agreement with the monophyly of genera *Mansonia*, *Coquillettidia*, and *Culex*, through mitochondrial phylogenomics [63].

In most of these cases, representative studies involving more taxa, a higher number of specimens per taxa sampled over a wider geographic range, and merging morphological and molecular characterization are needed to unravel the specific status of different populations and characterize species complexes in Africa and their relationship with their members elsewhere and/or the monophyly/paraphyly of some subgenera or genera. The systematics within the Culicini tribe cannot be resolved with morphological data alone [67], stressing the relevance of obtaining new molecular data.

Circa 40% (36–47%) of the sequences obtained in this study could not be correctly identified using BOLD and BLASTn as identification tools; this was because (i) the sequence was obtained for the first time, (ii) they had been obtained from members of species complexes, or (iii) there was an incorrect assignment, including at the genus level, such as sequences from *Cx. inconspicuosus* and *Ae. durbanensis* that were identified in BOLD as *Ae. argenteopuntatus* and *Cx. tritaeniorhynchus*, respectively, with >99% probability. Although studies that associate morphological and *CoxI* barcode-based molecular identifications are increasing, few include a definite diagnostic identification [20,24,48]. This absence is a potential source of error, as many species are only distinguishable by subtle morphological differences in the male genitalia. In fact, we have detected sequences from the barcode fragment of the *CoxI* gene from GenBank and Boldsystems which, given their phylogenetic signal, suggested the possibility of misassignment to another species. Examples of such are the *Coquillettidia* heterospecific sequences that clustered with our sequence of *Cq. chrysosoma* and the sequence ascribed to *Er. Silvestris*, which had complete identity with our sequence of *Er. quinquevittatus*, given that in both cases we had the morphology-based identification to the level of male genitalia for our sequences. This type of error has already been identified in other studies based on the *CoxI* marker [25,26] and *ITS* [33]. Such species assignment errors are perpetuated and amplified when authors consider only genetic similarities with previous GenBank entries. *CoxI*-based barcoding should complement morphologically-based identification [20], rather than species identification being based only on genetic similarities with existing sequences in the GenBank database [15].

Incorrect assignments also cause irregular situations in the BINs (barcode index numbers) assigned to what the BOLD system defines as operational taxonomic units, ideally corresponding to different species. For example, as in the case of the *Aedeomyia* species from the Afrotropical region, where three BINs were identified, the first for *Ad. furfurea* from Malawi (BOLD:AEH5592), the second BIN was shared between *Ad. madagascarica* and *Ad. africana* (BOLD:ADV5603) and the third was shared between *Ad. africana* and *Ad. furfurea* (BOLD:ACK8488). In the third case, there may have been an incorrect assignment of certain sequence(s) to the species *Ad. furfurea*, a situation that phylogenetic analysis was able to resolve. In other cases, more than one species clustering into one BIN have been registered, and another species has been split into more than one BIN [26].

5. Conclusions

Our study has contributed to the barcode library of Afrotropical mosquitoes, some of which are known potential vectors of arboviruses [2,3] or have recently been found to be so, or carriers of insect specific flavivirus [30–32,70]. This was achieved by associating careful morphologically identified referenced voucher specimens to specific molecular marker *CoxI* partial sequences. However, partial *CoxI* sequences have been shown to fail in unambiguously discriminating some proximal species or members of species complexes in addition to overestimating the diversity of *Culex* spp. [17]. Hence, it will be necessary to use alternative molecular markers, including nuclear, such as *Ace2* [11], microsatellites [54–56], or mitochondrial, such as *ITS*, to molecularly delineate species. However, that may prove to not always be sufficient [52] and other markers such as 16S [26], ND4 [71], or the complete mitochondrial genome [63,68] may be required.

Supplementary Materials: The following supporting information can be downloaded at: https://www.mdpi.com/article/10.3390/d14110940/s1, File S-I—Table SI-1; File S-II—Photos of mounted genitalia; File S-III—Original phylogenetic trees + likelihood mapping; File S-IV—Extra figures + tables.

Author Contributions: A.P.G.d.A., R.P.—study conceptualization; A.P.G.d.A., R.P., M.V., L.B., A.J.C., A.P.A., J.F.—study design and methodology; A.P.G.d.A., A.P.A., M.M.G., A.K., A.J.C., L.B., P.N.T., C.A.S.—field surveys and sample collections; E.M.-S., A.P.A., M.M.G., A.K., A.P.G.d.A., A.J.C., V.V., M.T.N., R.P.—sample processing and analysis; E.M.-S., R.P., A.P.G.d.A.—phylogenetics and statistical analysis; E.M.-S., A.P.G.d.A., R.P.—manuscript drafting. All authors have read and agreed to the published version of the manuscript.

Funding: Eddyson Montalvo-Sabino was recipient of a grant from "Programa Nacional de Becas y Crédito Educativo" (PRONABEC), 2019—Beca Generacion del Bicentenario, from the "Ministerio de Educación" of Peru. A.P. Abilio was a recipient of a grant from Wellcome Trust (Grant WT087546MA) through SACIDS RVF and NPHI-Phase-II from the National Institute for Health of Mozambique through a cooperative agreement number [5NU14GH001237-03-00]. Marietjie Venter was a recipient of a sub-award from the Global Disease Detection Program, US-CDC award 5U19GH000571-02 with the NICD and University of Pretoria that funded vector surveillance in South Africa (2012–2015) and by the Cooperative Agreement Number (5 NU2GGH001874-02-00) with the University of Pretoria (2014–2017). Milehna M. Guarido received a studentship through this grant. A.P.G. Almeida has been a recipient of the Visiting Professor Programme by the University of Pretoria for the work in South

Africa. This work received financial support from the Global Health and Tropical Medicine Center (GHTM | IHMT | NOVA), which is funded through FCT contract UID/Multi/04413/2013, Portugal. The findings and conclusions expressed in this manuscript are those of the author(s) and do not necessarily represent the official position of the funding agencies.

Institutional Review Board Statement: Collection of mosquitoes in South Africa was cleared under Section 20 approval by the Department of Agriculture Land Reform and Rural Development. Informed consent was obtained from the head of the household or property owners in Mozambique and South Africa.

Data Availability Statement: The slides with the mounted dissected genitalia of the mosquitoes in this study are deposited in the Institute of Hygiene and Tropical Medicine | NOVA University Lisbon (IHMT | NOVA) Insect Collection, Lisbon, Portugal.

Acknowledgments: We are grateful to Lapalala Wilderness, Marataba Conservation, South African National Parks, and private farm owners for logistical assistance and permission to collect mosquitoes on their properties. Carla A. Sousa acknowledges Filomeno Fortes, "Diretor nacional do programa de combate às grandes endemias", and his team, responsible for the "Missão de apoio técnico especializado no âmbito do combate ao surto epidémico de febre-amarela, 2016", under which mosquito collections in Angola took place.

Conflicts of Interest: The authors declare no conflict of interest. The funders had no role in the design of the study; collection, analysis, or interpretation of data; writing of the manuscript; or decision to publish the results.

References

1. Wilkerson, R.; Linton, Y.-M.; Strickman, D. *Mosquitoes of the World*; Johns Hopkins University Press: Baltimore, MA, USA, 2021; Volume 1.
2. Braack, L.; De Almeida, A.P.G.; Cornel, A.J.; Swanepoel, R.; De Jager, C. Mosquito-Borne Arboviruses of African Origin: Review of Key Viruses and Vectors. *Parasites Vectors* **2018**, *11*, 29. [CrossRef] [PubMed]
3. Venter, M. Assessing the Zoonotic Potential of Arboviruses of African Origin. *Curr. Opin. Virol.* **2018**, *28*, 74–84. [CrossRef] [PubMed]
4. Edwards, F.W. *Mosquitoes of the Ethiopian Region. III.-Culicine Adults and PupAe*; British Museum (N.H.): London, UK, 1941.
5. Jupp, P.G. *Mosquitoes of Southern Africa: Culicinae and Toxorhynchitinae*; Ekogilde Publishers: New York, NY, USA, 1996.
6. Huang, Y. Aedes (Stegomyia) Simpsoni Complex in the Ethiopian Region with Lectotype Designation for Simpsoni (Theobald) (Diptera: Culicidae). *Mosq. Syst.* **1979**, *11*, 221–234.
7. Huang, Y. Aedes (Stegomyia) Bromeliae (Diptera: Culicidae), the Yellow Fever Virus Vector in East Africa. *J. Med. Entomol.* **1986**, *23*, 196–200. [CrossRef] [PubMed]
8. Huang, Y. Notes on the Aedes (Diceromyia) Furcifer Group, with a Description of a New Species (Diptera: Culicidae). *Proc. Entomol. Soc. Washingt.* **1986**, *88*, 634–649.
9. Harbach, R.E.; Knight, K.L. *Taxonomists' Glossary of Mosquito Anatomy*; Plexus Publishing, Inc.: Marlton, NJ, USA, 1980; ISBN 9780937548004.
10. Walton, C.; Sharpe, R.G.; Pritchard, S.J.; Thelwell, N.J.; Butlin, R.K. Molecular Identification of Mosquito Species. *Biol. J. Linn. Soc.* **1999**, *68*, 241–256. [CrossRef]
11. Smith, J.L.; Fonseca, D.M. Rapid Assays for Identification of Members of the Culex (Culex) Pipiens Complex, Their Hybrids, and Other Sibling Species (Diptera: Culicidae). *Am. J. Trop. Med. Hyg.* **2004**, *70*, 339–345. [CrossRef]
12. Guarido, M.M.; Riddin, M.A.; Johnson, T.; Braack, L.E.O.; Schrama, M.; Gorsich, E.E.; Brooke, B.D.; Almeida, A.P.G.; Venter, M. Aedes Species (Diptera: Culicidae) Ecological and Host Feeding Patterns in the North-Eastern Parts of South Africa, 2014–2018. *Parasites Vectors* **2021**, *14*, 339. [CrossRef]
13. Maekawa, Y.; Pemba, D.; Kumala, J.; Gowelo, S.; Higa, Y.; Futami, K.; Sawabe, K.; Tsuda, Y. DNA Barcoding of Mosquitoes Collected through a Nationwide Survey in 2011 and 2012 in Malawi, Southeast Africa. *Acta Trop.* **2021**, *213*, 105742. [CrossRef]
14. Mixão, V.; Bravo Barriga, D.; Parreira, R.; Novo, M.T.; Sousa, C.A.; Frontera, E.; Venter, M.; Braack, L.; Almeida, A.P.G. Comparative Morphological and Molecular Analysis Confirms the Presence of the West Nile Virus Mosquito Vector, Culex Univittatus, in the Iberian Peninsula. *Parasites Vectors* **2016**, *9*, 601. [CrossRef]
15. Makanda, M.; Kemunto, G.; Wamuyu, L.; Bargul, J.; Muema, J.; Mutunga, J. Diversity and Molecular Characterization of Mosquitoes (Diptera: Culicidae) in Selected Ecological Regions in Kenya. *F1000Research* **2018**, *7*, 262. [CrossRef]
16. Ajamma, Y.U.; Villinger, J.; Omondi, D.; Salifu, D.; Onchuru, T.O.; Njoroge, L.; Muigai, A.W.T.; Masiga, D.K. Composition and Genetic Diversity of Mosquitoes (Diptera: Culicidae) on Islands and Mainland Shores of Kenya's Lakes Victoria and Baringo. *J. Med. Entomol.* **2016**, *53*, 1348–1363. [CrossRef] [PubMed]
17. Hemmerter, S.; Šlapeta, J.; Beebe, N.W. Resolving Genetic Diversity in Australasian Culex Mosquitoes: Incongruence between the Mitochondrial Cytochrome c Oxidase I and Nuclear Acetylcholine Esterase 2. *Mol. Phylogenet. Evol.* **2009**, *50*, 317–325. [CrossRef]

18. Hernández-Triana, L.M.; Brugman, V.A.; Nikolova, N.I.; Ruiz-Arrondo, I.; Barrero, E.; Thorne, L.; de Marco, M.F.; Krüger, A.; Lumley, S.; Johnson, N.; et al. DNA Barcoding of British Mosquitoes (Diptera, Culicidae) to Support Species Identification, Discovery of Cryptic Genetic Diversity and Monitoring Invasive Species. *Zookeys* **2019**, *2019*, 57–76. [CrossRef]
19. Awedoba, A.K. West Africa. *Afr. Yearb.* **2020**, *16*, 32–47. [CrossRef]
20. Chan, A.; Chiang, L.P.; Hapuarachchi, H.C.; Tan, C.H.; Pang, S.C.; Lee, R.; Lee, K.S.; Ng, L.C.; Lam-Phua, S.G. DNA Barcoding: Complementing Morphological Identification of Mosquito Species in Singapore. *Parasites Vectors* **2014**, *7*, 569. [CrossRef]
21. Ashfaq, M.; Hebert, P.D.N.; Mirza, J.H.; Khan, A.M.; Zafar, Y.; Mirza, M.S. Analyzing Mosquito (Diptera: Culicidae) Diversity in Pakistan by DNA Barcoding. *PLoS ONE* **2014**, *9*, e97268. [CrossRef] [PubMed]
22. Demari-Silva, B.; Vesgueiro, F.T.; Sallum, M.A.M.; Marrelli, M.T. Taxonomic and Phylogenetic Relationships between Species of the Genus Culex (Diptera: Culicidae) from Brazil Inferred from the Cytochrome c Oxidase I Mitochondrial Gene. *J. Med. Entomol.* **2011**, *48*, 272–279. [CrossRef]
23. Kumar, N.P.; Rajavel, A.R.; Natarajan, R.; Jambulingam, P. DNA Barcodes Can Distinguish Species of Indian Mosquitoes (Diptera: Culicidae). *J. Med. Entomol.* **2007**, *44*, 1–7. [CrossRef]
24. Laurito, M.; De Oliveira, T.M.P.; Almiron, W.R.; Sallum, M.A.M. COI Barcode versus Morphological Identification of Culex (Culex) (Diptera: Culicidae) Species: A Case Study Using Samples from Argentina and Brazil. *Mem. Inst. Oswaldo Cruz* **2013**, *108*, 110–122. [CrossRef]
25. Pagac, B.B.; Spring, A.R.; Stawicki, J.R.; Dinh, T.L.; Lura, T.; Kavanaugh, M.D.; Pecor, D.B.; Justi, S.A.; Linton, Y.M. Incursion and Establishment of the Old World Arbovirus Vector Aedes (Fredwardsius) Vittatus (Bigot, 1861) in the Americas. *Acta Trop.* **2021**, *213*, 105739. [CrossRef] [PubMed]
26. Talaga, S.; Leroy, C.; Guidez, A.; Dusfour, I.; Girod, R.; Dejean, A.; Murienne, J. DNA Reference Libraries of French Guianese Mosquitoes for Barcoding and Metabarcoding. *PLoS ONE* **2017**, *12*, e0176993. [CrossRef] [PubMed]
27. Wang, G.; Li, C.; Guo, X.; Xing, D.; Dong, Y.; Wang, Z.; Zhang, Y.; Liu, M.; Zheng, Z.; Zhang, H.; et al. Identifying the Main Mosquito Species in China Based on DNA Barcoding. *PLoS ONE* **2012**, *7*, e47051. [CrossRef] [PubMed]
28. Weeraratne, T.C.; Surendran, S.N.; Karunaratne, S.H.P.P. DNA Barcoding of Morphologically Characterized Mosquitoes Belonging to the Subfamily Culicinae from Sri Lanka. *Parasites Vectors* **2018**, *11*, 266. [CrossRef] [PubMed]
29. Johnson, T.; Braack, L.; Guarido, M.; Venter, M.; Gouveia Almeida, A.P. Mosquito Community Composition and Abundance at Contrasting Sites in Northern South Africa, 2014–2017. *J. Vector Ecol.* **2020**, *45*, 104–117. [CrossRef] [PubMed]
30. Abílio, A.P.; Silva, M.; Kampango, A.; Narciso, I.; Gudo, E.S.; Das Neves, L.C.B.; Sidat, M.; Fafetine, J.M.; De Almeida, A.P.G.; Parreira, R. A Survey of RNA Viruses in Mosquitoes from Mozambique Reveals Novel Genetic Lineages of Flaviviruses and Phenuiviruses, as Well as Frequent Flavivirus-like Viral DNA Forms in Mansonia. *BMC Microbiol.* **2020**, *20*, 225. [CrossRef]
31. Guarido, M.M.; Motlou, T.; Riddin, M.A.; MacIntyre, C.; Manyana, S.C.; Johnson, T.; Schrama, M.; Gorsich, E.E.; Brooke, B.D.; Almeida, A.G.P.; et al. Potential Mosquito Vectors for Shuni Virus, South Africa, 2014–2018. *Emerg. Infect. Dis.* **2021**, *27*, 3142–3146. [CrossRef]
32. Guarido, M.M.; Govender, K.; Riddin, M.A.; Schrama, M.; Gorsich, E.E.; Brooke, B.D.; Almeida, A.P.G.; Venter, M. Detection of Insect-Specific Flaviviruses in Mosquitoes (Diptera: Culicidae) in Northeastern Regions of South Africa. *Viruses* **2021**, *13*, 2148. [CrossRef]
33. Marrelli, M.T.; Sallum, M.A.M.; Marinotti, O. The Second Internal Transcribed Spacer of Nuclear Ribosomal DNA as a Tool for Latin American Anopheline Taxonomy—A Critical Review. *Mem. Inst. Oswaldo Cruz* **2006**, *101*, 817–832. [CrossRef]
34. Harbach, R.E. The Mosquitoes of the Subgenus Culex in Southwestern Asia and Egypt (Diptera: Culicidae). *Contrib. Am. Entomol. Inst.* **1988**, *24*, 1.
35. Wilkerson, R.C.; Linton, Y.M.; Fonseca, D.M.; Schultz, T.R.; Price, D.C.; Strickman, D.A. Making Mosquito Taxonomy Useful: A Stable Classification of Tribe Aedini That Balances Utility with Current Knowledge of Evolutionary Relationships. *PLoS ONE* **2015**, *10*, e0133602. [CrossRef] [PubMed]
36. Sirivanakarn, S. Medical Entomology Studies—A Revision of the Subgenus Culex in the Oriental Region (Diptera: Culicidae). *Contrib. Am. Entomol. Inst.* **1976**, *12*, 1–272.
37. Folmer, O.; Black, M.; Hoeh, W.; Lutz, R.; Vrijenhoek, R. DNA Primers for Amplification of Mitochondrial Cytochrome c Oxidase Subunit I from Diverse Metazoan Invertebrates. *Mol. Mar. Biol. Biotechnol.* **1994**, *3*, 294–299. [PubMed]
38. Tchouassi, D.P.; Bastos, A.D.S.; Sole, C.L.; Diallo, M.; Lutomiah, J.; Mutisya, J.; Mulwa, F.; Borgemeister, C.; Sang, R.; Torto, B. Population Genetics of Two Key Mosquito Vectors of Rift Valley Fever Virus Reveals New Insights into the Changing Disease Outbreak Patterns in Kenya. *PLoS Negl. Trop. Dis.* **2014**, *8*, e3364. [CrossRef]
39. Bahnck, C.M.; Fonseca, D.M. Rapid Assay to Identify the Two Genetic Forms of Culex (Culex) Pipiens L. (Diptera: Culicidae) and Hybrid Populations. *Am. J. Trop. Med. Hyg.* **2006**, *75*, 251–255. [CrossRef]
40. Schmidt, H.A.; von Haeseler, A. Maximum-Likelihood Analysis Using TREE-PUZZLE. *Curr. Protoc. Bioinform.* **2007**, *17*, 6.6.1–6.6.23. [CrossRef]
41. Trifinopoulos, J.; Nguyen, L.T.; von Haeseler, A.; Minh, B.Q. W-IQ-TREE: A Fast Online Phylogenetic Tool for Maximum Likelihood Analysis. *Nucleic Acids Res.* **2016**, *44*, W232–W235. [CrossRef]
42. Suchard, M.A.; Lemey, P.; Baele, G.; Ayres, D.L.; Drummond, A.J.; Rambaut, A. Bayesian Phylogenetic and Phylodynamic Data Integration Using BEAST 1.10. *Virus Evol.* **2018**, *4*, vey016. [CrossRef]

43. Huson, D.H.; Bryant, D. Application of Phylogenetic Networks in Evolutionary Studies. *Mol. Biol. Evol.* **2006**, *23*, 254–267. [CrossRef]
44. Bryant, D.; Moulton, V. Neighbor-Net: An Agglomerative Method for the Construction of Phylogenetic Networks. *Mol. Biol. Evol.* **2004**, *21*, 255–265. [CrossRef]
45. Dress, A.W.M.; Huson, D.H. Constructing Splits Graphs. *IEEE/ACM Trans. Comput. Biol. Bioinf.* **2004**, *1*, 109–115. [CrossRef] [PubMed]
46. McIntosh, B.M. A Taxonomic Revision of Certain Aedes Species (Diptera: Culicidae) of the Subgenus Aedimorphus in Southern Africa. *J. Entomol. Soc. S. Afr.* **1975**, *38*, 251–287.
47. Hebert, P.D.N.; Cywinska, A.; Ball, S.L.; DeWaard, J.R. Biological Identifications through DNA Barcodes. *Proc. R. Soc. B Biol. Sci.* **2003**, *270*, 313–321. [CrossRef]
48. Torres-Gutierrez, C.; Bergo, E.S.; Emerson, K.J.; de Oliveira, T.M.P.; Greni, S.; Sallum, M.A.M. Mitochondrial COI Gene as a Tool in the Taxonomy of Mosquitoes Culex Subgenus Melanoconion. *Acta Trop.* **2016**, *164*, 137–149. [CrossRef]
49. Diagne, C.T.; Diallo, D.; Faye, O.; Ba, Y.; Faye, O.; Gaye, A.; Dia, I.; Faye, O.; Weaver, S.C.; Sall, A.A.; et al. Potential of Selected Senegalese Aedes Spp. Mosquitoes (Diptera: Culicidae) to Transmit Zika Virus. *BMC Infect. Dis.* **2015**, *15*, 2–7. [CrossRef] [PubMed]
50. Harbach, R.E.; Howard, T.M. Corrections in the Status and Rank of Names Used to Denote Varietal Forms of Mosquitoes (Diptera: Culicidae). *Zootaxa* **2007**, *48*, 35–48. [CrossRef]
51. Meier, R.; Shiyang, K.; Vaidya, G.; Ng, P.K.L. DNA Barcoding and Taxonomy in Diptera: A Tale of High Intraspecific Variability and Low Identification Success. *Syst. Biol.* **2006**, *55*, 715–728. [CrossRef] [PubMed]
52. Kengne, P.; Goff, G.L.; Fontenille, D. Molecular Genetic Investigation of Morphological Species of Members of the Neomelaniconion Subgenus (Diptera: Culicidae: Aedini) from Madagascar Using Ribosomal Internal Transcribed Spacer 2. *J. Med. Entomol.* **2009**, *46*, 403–407. [CrossRef]
53. Cornel, A.J.; Mcabee, R.D.; Rasgon, J.; Stanich, M.A.; Scott, T.W.; Coetzee, M. Differences in Extent of Genetic Introgression between Sympatric Culex Pipiens and Culex Quinquefasciatus (Diptera: Culicidae) in California and South Africa. *J. Med. Entomol.* **2003**, *40*, 36–51. [CrossRef]
54. Gomes, B.; Alves, J.; Sousa, C.A.; Santa-Ana, M.; Vieira, I.; Silva, T.L.; Almeida, A.P.G.; Donnelly, M.J.; Pinto, J. Hybridization and Population Structure of the Culex Pipiens Complex in the Islands of Macaronesia. *Ecol. Evol.* **2012**, *2*, 1889–1902. [CrossRef]
55. Gomes, B.; Sousa, C.A.; Novo, M.T.; Freitas, F.B.; Alves, R.; Côrte-Real, A.R.; Salgueiro, P.; Donnelly, M.J.; De Almeida, A.P.G.; Pinto, J. Asymmetric Introgression between Sympatric Molestus and Pipiens Forms of Culex Pipiens (Diptera: Culicidae) in the Comporta Region, Portugal. *BMC Evol. Biol.* **2009**, *9*, 262. [CrossRef] [PubMed]
56. Huang, S.; Molaei, G.; Andreadis, T.G. Genetic Insights into the Population Structure of Culex Pipiens (Diptera: Culicidae) in the Northeastern United States by Using Microsatellite Analysis. *Am. J. Trop. Med. Hyg.* **2008**, *79*, 518–527. [CrossRef] [PubMed]
57. Amraoui, F.; Tijane, M.; Sarih, M.; Failloux, A.B. Molecular Evidence of Culex Pipiens Form Molestus and Hybrids Pipiens/Molestus in Morocco, North Africa. *Parasites Vectors* **2012**, *5*, 2010–2013. [CrossRef] [PubMed]
58. Shaikevich, E.V.; Vinogradova, E.B.; Bouattour, A.; De Almeida, A.P.G. Genetic Diversity of Culex Pipiens Mosquitoes in Distinct Populations from Europe: Contribution of Cx. Quinquefasciatus in Mediterranean Populations. *Parasites Vectors* **2016**, *9*, 47. [CrossRef]
59. Ribeiro, H. Research on the Mosquitoes of Angola (Diptera: Culicidae): II–Some New Culicine Records. *An. Do Inst. Med. Trop.* **1966**, *23*, 163–166.
60. Weetman, D.; Kamgang, B.; Badolo, A.; Moyes, C.L.; Shearer, F.M.; Coulibaly, M.; Pinto, J.; Lambrechts, L.; McCall, P.J. Aedes Mosquitoes and Aedes-Borne Arboviruses in Africa: Current and Future Threats. *Int. J. Environ. Res. Public Health* **2018**, *15*, 220. [CrossRef]
61. Moore, M.; Sylla, M.; Goss, L.; Burugu, M.W.; Sang, R.; Kamau, L.W.; Kenya, E.U.; Bosio, C.; de Munoz, M.L.; Sharakova, M.; et al. Dual African Origins of Global Aedes Aegypti s.l. Populations Revealed by Mitochondrial DNA. *PLoS Negl. Trop. Dis.* **2013**, *7*, e2175. [CrossRef]
62. Gloria-Soria, A.; Ayala, D.; Bheecarry, A.; Calderon-Arguedas, O.; Chadee, D.D.; Chiappero, M.; Coetzee, M.; Elahee, K.B.; Fernandez-Salas, I.; Kamal, H.A.; et al. Global Genetic Diversity of Aedes Aegypti. *Mol. Ecol.* **2016**, *25*, 5377–5395. [CrossRef]
63. da Silva, A.F.; Machado, L.C.; de Paula, M.B.; da Silva, P.V.C.J.; de Morais, B.R.V.; de Melo, S.M.A.V.; Wallau, G.L. Culicidae Evolutionary History Focusing on the Culicinae Subfamily Based on Mitochondrial Phylogenomics. *Sci. Rep.* **2020**, *10*, 18823. [CrossRef]
64. Harbach, R.E. Classification within the Cosmopolitan Genus Culex (Diptera: Culicidae): The Foundation for Molecular Systematics and Phylogenetic Research. *Acta Trop.* **2011**, *120*, 1–14. [CrossRef]
65. Cornel, A.J.; Mayi, M.P.A.; Kowo, C.; Foncha, D.; Andongma, E.; Anong, D.N.; Elad, M.; Djomo, C.; Tchuinkam, T.; Brisco, K.K.; et al. New Species of Culex (Culiciomyia) (Diptera: Culicidae) from Talangaye Forest in Cameroon and Descriptions and Identification Keys for Males of the Afrotropical Species of the Subgenus. *Zootaxa* **2020**, *4858*, zootaxa-4858. [CrossRef]
66. Vesgueiro, F.T.; Demari-Silva, B.; dos Santos, M.R.; Sallum, M.A.M.; Marrelli, M.T. Intragenomic Variation in the Second Internal Transcribed Spacer of the Ribosomal DNA of Species of the Genera Culex and Lutzia (Diptera: Culicidae). *Mem. Inst. Oswaldo Cruz* **2011**, *106*, 1–8. [CrossRef] [PubMed]

67. Kitching, I.J.; Culverwell, C.L.; Harbach, R.E. The Phylogenetic Conundrum of Lutzia (Diptera: Culicidae: Culicini): A Cautionary Account of Conflict and Support. *Insect Syst. Evol.* **2015**, *46*, 269–290. [CrossRef]
68. Sun, L.; Li, T.J.; Fu, W.B.; Yan, Z.T.; Si, F.L.; Zhang, Y.J.; Mao, Q.M.; Demari-Silva, B.; Chen, B. The Complete Mt Genomes of Lutzia Halifaxia, Lt. Fuscanus and Culex Pallidothorax (Diptera: Culicidae) and Comparative Analysis of 16 Culex and Lutzia Mt Genome Sequences. *Parasites Vectors* **2019**, *12*, 368. [CrossRef] [PubMed]
69. Harbach, R.E.; Kitching, I.J.; Culverwell, C.L.; Dubois, J.; Linton, Y.M. Phylogeny of Mosquitoes of Tribe Culicini (Diptera: Culicidae) Based on Morphological Diversity. *Zool. Scr.* **2012**, *41*, 499–514. [CrossRef]
70. van den Bergh, C.; Thompson, P.N.; Swanepoel, R.; Almeida, A.P.G.; Paweska, J.T.; van Vuren, P.J.; Wilson, W.C.; Kemp, A.; Venter, E.H. Detection of Rift Valley Fever Virus in Aedes (Aedimorphus) Durbanensis, South Africa. *Pathogens* **2022**, *11*, 125. [CrossRef]
71. Andreeva, Y.V.; Khrabrova, N.V.; Alekseeva, S.S.; Abylkassymova, G.M.; Simakova, A.V.; Sibataev, A.K. First Record of the Invasive Mosquito Species Aedes Koreicus (Diptera, Culicidae) in the Republic of Kazakhstan. *Parasite* **2021**, *28*, 52. [CrossRef]

diversity

MDPI

Communication

Seasonal and Nocturnal Activity of *Culicoides* spp. (Diptera: Ceratopogonidae) Adapted to Different Environments in the Balearic Islands

Carlos Barceló *, Ricardo del Río and Miguel A. Miranda

Applied Zoology and Animal Conservation Group, University of the Balearic Islands (UIB), Cra. Valldemossa Km. 7.5, 07122 Palma, Mallorca, Balearic Islands, Spain; ricardo.delrio@uib.es (R.d.R.); ma.miranda@uib.es (M.A.M.)

* Correspondence: carlos.barcelo@uib.es

Abstract: Several pathogens are known to be transmitted by arthropods. One of the most relevant, in economic terms, affecting animals is bluetongue virus. Its known vectors are several species of *Culicoides* midges. In Europe, the considered main vector species are *Culicoides imicola* and *Culicoides obsoletus*, though other species may be implicated at different levels. In the present work, the activity of these vector species between sunset and sunrise from May to November is analyzed according to their captures in a rotator bottle collector adapted to a light trap. Additionally, the *Culicoides* populations of two areas in Mallorca (Spain) with different characteristics (rural vs. urban) are compared. The results indicated that June is the month of higher abundance of *C. obsoletus* in our climatic conditions—being active during the first three hours after sunset. Conversely, *C. imicola* reached the maximum level of captures during October, and captures were more evenly distributed during the night. Collections from the two areas revealed that *Culicoides* populations were composed by the same species; however, abundance and sex ratio presented marked differences. These results add valuable insight into the ecology of *Culicoides* and may be used to design more accurate strategies to control diseases associated with *Culicoides*.

Keywords: Bluetongue; bottle rotator; *Culicoides imicola*; *Culicoides obsoletus*; sex ratio; Spain

Citation: Barceló, C.; del Río, R.; Miranda, M.A. Seasonal and Nocturnal Activity of *Culicoides* spp. (Diptera: Ceratopogonidae) Adapted to Different Environments in the Balearic Islands. *Diversity* **2023**, *15*, 690. https://doi.org/10.3390/d15050690

Academic Editor: Olivier A.E. Sparagano

Received: 14 March 2023
Revised: 7 May 2023
Accepted: 17 May 2023
Published: 20 May 2023

1. Introduction

Culicoides Latreille (Diptera: Ceratopogonidae) midges are small (1–3 mm) Nematocera insects known to act as vectors of several pathogens. The most important, from an economical point of view, are bluetongue (BT) and African horse sickness [1]. However, a significant number of other pathogens –ranging from nematodes to several types of virus—may be transmitted to animals and humans by the bites of some species of *Culicoides* [2–4]. In Europe, the Southern species *Culicoides imicola* Kieffer 1913 and the Northern species from the Obsoletus complex ((mainly *C. obsoletus* (Meigen 1818) and *Culicoides scoticus* Downes and Kettle 1952)) are considered to play a major role in the transmission of BT disease [5–9]. An large number of BT outbreaks have occurred in Europe since 2000, when it was detected for the first time in the Balearic Islands (Spain) [10,11]. Nowadays, restrictive and prophylactic measures have succeeded in reducing the incidence of BT disease in Europe [12,13]. However, a certain degree of entomological and serological surveillance should always be maintained since one of the main entry pathways of the disease to Europe is considered to be via infected *Culicoides* transported by warm air currents from North Africa to southern Europe [14–16].

Due to their small size and, as a result, the difficulty in rearing the vector species of *Culicoides* in laboratory conditions, knowledge of the ecological and behavioral aspects of this genus is still quite limited. Flight patterns and optimal environmental conditions for the different *Culicoides* species have been determined in previous studies [17–20];

however, sub-populations of the same species may exhibit different behavioral patterns depending on their suitable environment. The most important environmental drivers for *Culicoides* vector species are related to temperature and precipitation [21]. In Spain, the species *C. obsoletus/C. scoticus* are active almost the whole year in Northern provinces, while *C. imicola* is absent in the North, with peaks of abundance between September and November in Southeastern areas [22]. The activity period of adult *C. obsoletus/C. scoticus* is longest in low elevation sites with warmer springs and high livestock abundance; meanwhile, *C. imicola* adult females prefer not only lower elevation regions but also sites with broad-leaved vegetation [23].

For the present study, our aim was to improve the knowledge of the above mentioned ecological aspects of *Culicoides* midges, focusing on the vector species *C. imicola* and *C. obsoletus/C. scoticus*. Seasonal and nocturnal flight activity patterns were determined for several *Culicoides* in two different habitats (rural vs. urban), and the composition and abundance of midges between the two study areas were compared.

2. Materials and Methods

Culicoides midges for this trial were collected from two different areas in Mallorca (Balearic Islands, Spain). The areas separated 31.4 km from each other. One of them was a rural farm named Ca's Boter, located 6 km away from the closest urban area (Felanitx; 39°31′27″ N; 3°7′36″ E), dedicated to cattle raising and milking (≈30 dairy cattle and 20 veal calves), and 18.4 km from the nearest city (Manacor). The farm had an area of 7413 m^2, of which 5620 m^2 (75.8%) was intended for the animals and surrounded by several areas of different sizes dedicated mainly to agricultural or farming practices (sheep flocks of different sizes were scattered in some of the surrounding areas). In this area, no light sources were available during the night. The other was a non-farming property in a residential area named Rialema (s'Aranjassa; 39°32′22″ N, 2°47′28″ E) and located 10 km away from the capital city of Mallorca island (Palma). The area covers 7530 m^2 of Mediterranean scrub and vegetation (pine grove area). The area has streetlights, and the property was surrounded by other residences with similar sizes and characteristics. Animals commonly found in the area mainly include pet dogs and cats.

Insects were collected from the two areas during 2012 using a collection bottle rotator (model 1512, John W. Hock Company, Gainesville, FL, USA) adapted to a downdraft blacklight (UV) trap (CDC miniature light trap model 912, John W. Hock Company, Gainesville, FL, USA). The rotator traps were set at 1.8 m from the ground. In Ca's Boter, the trap was placed 1 m close to the livestock premises, while in Rialema, it was set in the backyard. Rotator traps operated during the main *Culicoides* activity period in the area (May–November). A number of nights were sampled per month (Table 1), avoiding rain and strong winds, and bottles were rotated every 1:15 h to 2 h depending on the scotoperiod. Trappings were conducted during the same nights between both places. Eight bottles were rotated during the sampling nights, the first one starting soon after sunset (between 19:00 h to 21:00 h depending on the month) and the last one collecting insects before sunrise (between 7:00 h to 8:00 h) (GTM + 1).

Collected midges were taken to the laboratory of zoology of the University of the Balearic Islands to identify their wing pattern according to the method described by Rawlings [24], and were separated by sex and gonotrophic condition according to Dyce [25]. Composition and activity patterns of the *Culicoides* populations from the two areas (rustic vs. urban) were compared. Flight activity of vector species *C. imicola* and *C. obsoletus/C. scoticus* were analyzed. Two flight activity patterns were observed: the activity pattern (from May to November) and the nocturnal activity pattern (from sunset to sunrise). Temperatures during the trial were obtained from a nearby climatic station of AEMET [26].

Species from Obsoletus complex were sent to CIRAD (Centre International de Recherche de l'Agriculture et du Dévéloppement) and identified via PCR assay according to the procedures of Nolan et al. [8].

Table 1. Collected *C. obsoletus/C. scoticus* and *C. imicola* species from two areas in Mallorca with different environmental conditions—Ca's Boter (rural) and Rialema (urban). Letters denote significant differences between months (ANOVA test, $p < 0.011$ for Obsoletus complex and $p < 0.006$ for *C. imicola*).

Species	Month	Total Captures	Ca's Boter Nights Sampled	Captures/Night	Total Captures	Rialema Nights Sampled	Captures/Night
C. obsoletus/ C. scoticus	May	1993	3	664.3 ± 11.2^{a}			
	June	5514	7	853.2 ± 9.2^{a}	6	1	6.0 ± 0.0
	July	272	3	90.7 ± 4.7^{a}	12	5	2.4 ± 0.3
	August	47	3	15.7 ± 2.0^{b}	2	2	1.0 ± 0.0
	September	28	6	4.7 ± 1.3^{b}	6	4	1.5 ± 0.7
	October	43	10	4.3 ± 0.6^{b}	6	4	1.5 ± 1.0
	November	7	3	2.3 ± 1.0^{b}	2	5	0.4 ± 0.4
	Total	**7904**	**35**	**225.8 ± 21.9**	**34**	**21**	**1.6 ± 0.3**
C. imicola	May	7	3	2.3 ± 0.2^{b}			
	June	36	7	5.1 ± 0.9^{b}	0	1	0.0
	July	9	3	3.0 ± 1.2^{b}	2	5	0.4 ± 0.4
	August	7	3	2.3 ± 1.2^{b}	1	2	0.5 ± 0.7
	September	56	6	9.7 ± 1.9^{b}	3	4	0.8 ± 0.9
	October	259	10	28.8 ± 1.3^{a}	5	4	1.3 ± 0,4
	November	41	3	13.7 ± 3.7^{b}	1	5	0.2 ± 0,4
	Total	**415**	**35**	**9.3 ± 0.5**	**12**	**21**	**1.0 ± 0.6**

A Shapiro–Wilk test for checking normal distribution of the data followed by an Analysis of Variance (ANOVA) was performed with the STATGRAPHICS plus V 3.0 program to determine the statistical differences between *Culicoides* species abundance among sites, months, and sex ratio. Differences were considered significant at $p < 0.05$.

3. Results

3.1. Species Composition, Sex Ratio, and Gonothrophic Condition

During the complete trial, 12,376 *Culicoides* midges were collected—12,050 from Ca's Boter and 326 from Rialema. At least 10 species were collected from Ca's Boter, namely the following: Obsoletus complex (7904 specimens; 66.1%), *C. imicola* (415 specimens; 3.5%), *Culicoides cataneii* Clastrier 1957 (494 specimens; 4.1%), *Culicoides circumscriptus* Kieffer 1918 (1683 specimens; 14.1%), *Culicoides jumineri* (Callot and Kremer 1969) (345 specimens; 2.9%), *Culicoides longipennis* (Khalaf 1957) (26 specimens; 0.2%), *Culicoides maritimus* (Kieffer 1924) (36 specimens; 0.3%), *Culicoides newsteadi* (Austen 1921) (527 specimens; 4.4%), *Culicoides paolae* (Boorman 1996) (436 specimens; 3.6%), *Culicoides puncticollis* (Becker 1903) (86 specimens; 0.7%). The remaining 98 specimens had missing wings or were unidentifiable. A molecular analysis of the Obsoletus complex population revealed that 0.6% of the samples belonged to *C. scoticus*, while the remaining (99.4%) belonged to *C. obsoletus*.

Nine out of the ten species encountered in Ca's Boter were also present in Rialema. However, the predominant species in the latter area were *C. circumscriptus* and *C. paolae* (instead of *C. obsoletus*), which represented more than 45% of the *Culicoides* collected. The species composition of the *Culicoides* collected from Rialema was as follows: Obsoletus complex (34 specimens; 10.4%), *C. imicola* (12 specimens; 3.7%), *C. cataneii* (47 specimens; 14.4%), *C. circumscriptus* (88 specimens; 27%), *C. jumineri* (9 specimens; 2.8%), *C. maritimus* (20 specimens; 6.1%), *C. newsteadi* (48 specimens; 14.7), *C. paolae* (67 specimens; 20.6%), and *C. puncticollis* (1 specimen; 0.3%).

Regarding the sex and gonotrophic condition of the *Culicoides* midges collected from Ca's Boter, 3.7% were males, while 95.3% were females, which included 34.6% nulliparous, 23.7% parous, 35.3% gravid, 1.3% blood-fed, and 0.5% specimens without an abdomen. The remaining 1% belonged to intersexed specimens of *C. circumscriptus*. For the *Culicoides* collected from Rialema, 48.2% were males, while the remaining females included 20.2% nulliparous, 17.5% parous, 13.2% gravid, and 0.9% blood-fed specimens. Differences in the abundance and sex ratio between the sampled areas were highly significant ($p < 0.016$).

3.2. Seasonal and Nocturnal Activity Pattern

The *Culicoides* collected from Ca's Boter farm indicated that June was the month in which the population of the *C. obsoletus*/*C. scoticus* reached its peak (Av. ± SE = 853.2 ± 9.2 *Culicoides*/night) (Table 1)—when the average temperature was 24.4 °C (max: 39.7 °C, min: 13.2 °C) (Table 2). Differences in the number of captures were not significant between June and May–July ($p > 0.011$), but they were with the rest of the months ($p < 0.011$). Regarding *C. imicola*, the highest number of captures were obtained also in Ca's Boter in October (Av. ± SE = 28.8 ± 1.3 *Culicoides*/night), recording an average temperature of 18.8 °C (max: 30.1 °C, min: 3.9 °C). Significant differences were observed between the collections of this species during this month and the rest of the season ($p < 0.006$). Populations of these two species reached their highest levels in Rialema during the same months.

Table 2. Average temperatures during the sampling months.

Site	Month	Av. T (C°)	Max. T (C°)	Min. T (C°)
Ca's Boter	May	18.8	33.1	6.0
	June	24.4	39.7	13.2
	July	24.9	36.4	14.0
	August	27.0	39.4	16.8
	September	22.1	33.3	12.0
	October	18.8	30.1	3.9
	November	14.1	23.3	3.8
Rialema	May	18.9	32.0	7.4
	June	24.2	36.7	13.5
	July	25.3	37.5	14.9
	August	27.1	38.5	17.5
	September	22.5	32.6	12.2
	October	19.6	30.4	5.2
	November	15.0	23.6	4.5

The rates of collection regarding *Culicoides* midges from the Rialema area were too low to display a proper nocturnal activity pattern; hence, the results presented here are those obtained from Ca's Boter farm. More than 50% of the midges were collected during the first three hours after sunset, both for the *C. obsoletus*/*C. scoticus* and for the total population, while *C. imicola* was collected more evenly throughout the night (Figure 1). During the months wherein there was a higher abundance of *C. obsoletus*/*C. scoticus* (May, June, and July), the monthly distribution pattern of collection was well defined and in accordance with the seasonal pattern (Figure 2). However, the distribution pattern changed during the months wherein there was a lower abundance of these species (August, September, October, and November), and the rates of collection of individuals of these species were more uniform during the night or even before the sunset (Figure 3). In contrast, the nocturnal activity of *C. imicola* showed a different pattern, and collections of this species were obtained more evenly throughout the night, with peaks pre-sunset in October and November, as seen in Figure 4.

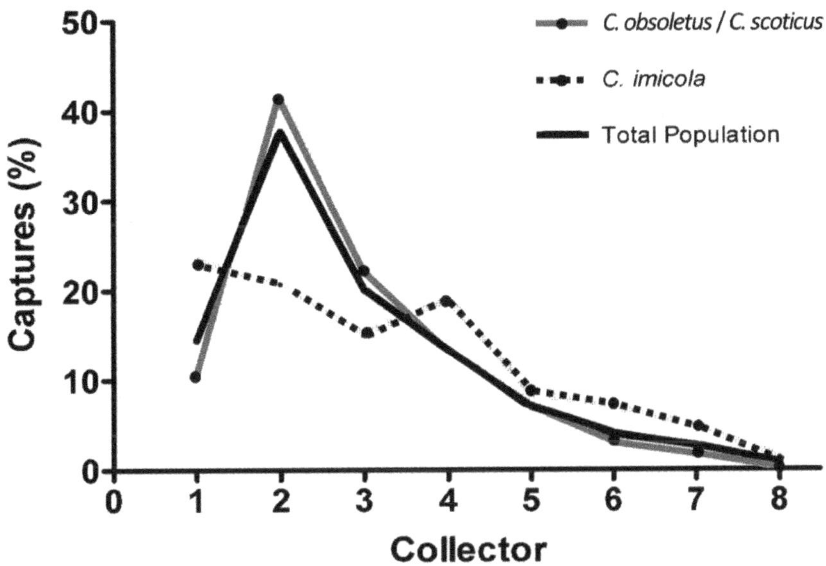

Figure 1. Percentage of different *Culicoides* species collected with a bottle rotator operated between sunset (collector 1) and sunrise (collector 8) in 2012 in Ca's Boter farm.

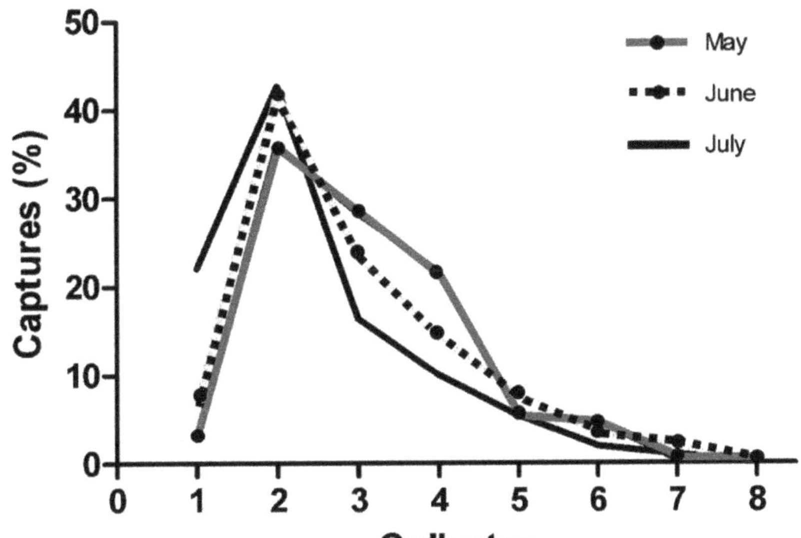

Figure 2. Percentage of *C. obsoletus/C. scoticus* collected with a bottle rotator operated between sunset (collector 1) and sunrise (8) during the months of highest abundance with regards to this species during 2012 in Ca's Boter farm.

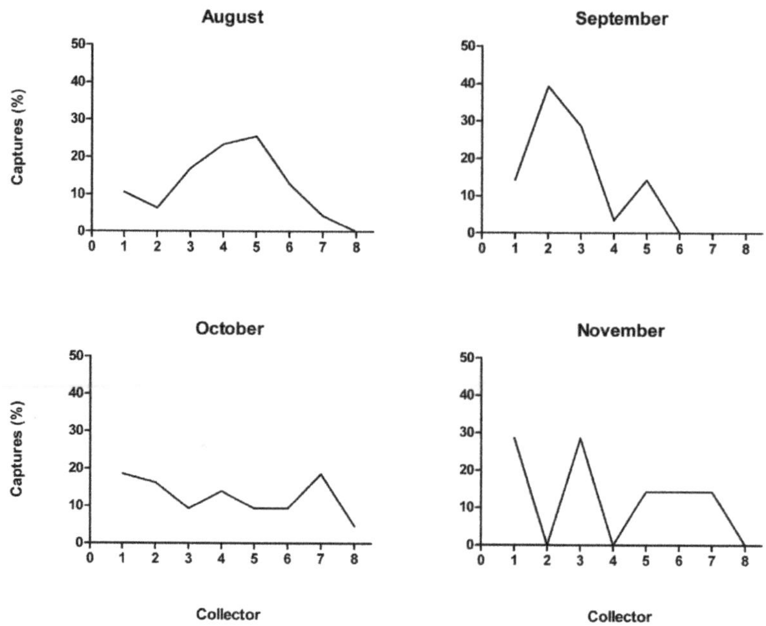

Figure 3. Percentage of *C. obsoletus/C. scoticus* collected with a bottle rotator operated between sunset (collector 1) and sunrise (collector 8) during the months of lowest abundance with regards to this species during 2012 in Ca's Boter farm.

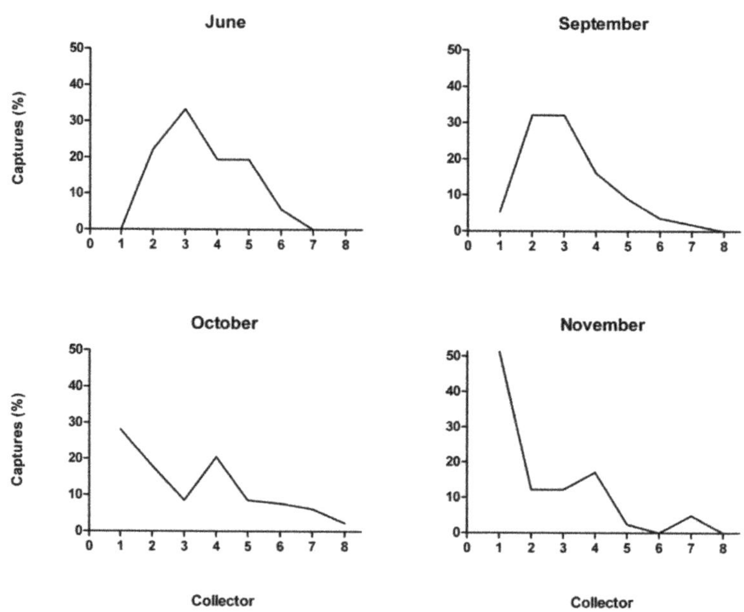

Figure 4. Percentage of *Culicoides imicola* collected with a bottle rotator operated between sunset (collector 1) and sunrise (collector 8) during the months of highest abundance with regards to this species during 2012 in Ca's Boter farm.

4. Discussion

The collection of *Culicoides* from areas with different environmental conditions revealed that urban areas host a significantly lower abundance of *Culicoides* than rural areas. However, results showed that in spite of the lower rates of capture in Rialema, the species composition was mostly the same in both areas (nine out of ten) with significant differences in sex ratio (F/M ratio from Ca's Boter = 23.9 times higher than F/M ratio from Rialema). It has been suggested that males do not disperse as far as females, and the collection of high numbers of males are probably indicative of a proximal breeding site and larval habitat [27]. No identification of larval habitats was observed in Rialema, but the breeding conditions in that area did not seem optimal (i.e., absence of humid or standing water to lay eggs and lack of host animals to parasitically infect). Previous reports indicated that some species, such as *C. circumscriptus* or *C. puncticollis*, bred near the collection site of Ca's Boter [28], and the proportion of males collected in the last were extremely low compared to those obtained from Rialema. This fact suggests that most of the males of the different *Culicoides* species do not linger near the place wherein they spent time as larvae, as previously mentioned and supported by the results of Braverman [29]. Conversely, streetlights in the residential area of Rialema could interfere with the results. Further studies, including more locations and habitats (e.g., peri-urban sites) or even records of the moon phase [30], should be performed to compare *Culicoides* populations regarding breeding sites, available hosts, and anthropic presence.

The results obtained in the present trial indicated that, in our climatic conditions, June is the month in which the abundance of *C. obsoletus*/*C. scoticus* is at its peak (avg. temp = 24.4 °C), while, during October, environmental conditions are optimal for the development of *C. imicola* populations (avg. temp = 18.9 °C). These results are in accordance with those presented by Miranda et al. [31], indicating that, in our climatic conditions, the higher numbers of *C. obsoletus* are able to be observed and obtained between June–July, while, during October, the abundance of *C. imicola* reaches its peak. Hence, particular attention should be paid to the preventive measures against biting midges during these months due to the higher epidemiological risk associated with vector abundance. In fact, the BT outbreaks that occurred in Mallorca during 2000, 2004, and 2021 were detected during these months [12].

When *C. obsoletus*/*C. scoticus* populations are at their higher level, the species was most commonly captured during the first 3 h after sunset. However, *C. imicola* seemed to be active before sunset, with the collection of this species being more evenly distributed throughout the night. Peak nocturnal activity in *C. obsoletus* and *C. scoticus* during the first hours after sunset was also observed in mainland conditions by Viennet et al. [32]. Information concerning the most frequent hours of activity among *C. obsoletus* and *C. scoticus* indicates the period of time whereby the risk of disease transmission associated with this vector species is maximal; hence, the proper preventive measures (such as stabling during peak hours of vector activity) should be implemented during these time periods. However, as our results indicate, the activity of *C. imicola* is highest during the majority of the night; therefore, preventive measures should be maintained throughout the night during the period of maximal abundance of this vector species (October, in our conditions). The same *Culicoides* species, but in different numbers, were found in areas with very different environmental conditions. It should be considered that, in regions with temperate and cool climates, some *Culicoides* adult species enter into diapause or shift their diel activity patterns when the number of daylight hours or the temperature declines [33,34]. Hence, the lack of any daytime data, especially as overnight temperatures cool later in the year, could leave some uncertainty about *Culicoides* activity patterns and potential transmission dynamics.

The activity of *Culicoides* midges was measured following their capture in UV light traps. However, the composition of *Culicoides* populations may differ broadly depending on the collection method [35–37]. Traps with CO_2 and/or animal baited traps should be the preferable method to more accurately determine the epidemiologic risk depending on the hour of the night and the month of the season, and the results obtained from that trial could be compared with the *Culicoides* obtained with light traps to assess their validity. Additionally, environmental conditions may vary from one year to the next, and parameters such as temperature, rain, and relative humidity should be taken into consideration when conducting risks assessments for the area.

Future trials assessing the biting rate of *Culicoides* vector species among farm animals depending on the hour of the night and months of the season are encouraged. Light traps are easier to handle and less time-consuming than animal-baited traps; hence, if results derived from the use of UV light traps and animal baiting are found to be related, a gold standard to assess the risk of infection in a determined area could be adopted. This approach would provide information to more accurately determine the disease risk periods associated with *Culicoides* vectors and hence contribute to the optimization of available resources against potential disease outbreaks.

Author Contributions: Conceptualization, R.d.R. and M.A.M.; methodology, C.B. and R.d.R.; software, R.d.R.; validation, C.B., R.d.R., and M.A.M.; formal analysis, C.B. and R.d.R.; investigation, C.B. and R.d.R.; resources, R.d.R. and M.A.M.; data curation, R.d.R.; writing—original draft preparation, R.d.R.; writing—review and editing, C.B.; visualization, C.B.; supervision, M.A.M.; project administration, M.A.M.; funding acquisition, M.A.M. All authors have read and agreed to the published version of the manuscript.

Funding: This research was funded by the Ministry of Research and Innovation of Spain, grant number AGL 2009-12640 and the EDENext project: biology and control of vector-borne diseases in Europe from the 7th framework program for research of the European Union, contract n° 261504.

Institutional Review Board Statement: Not applicable.

Data Availability Statement: Not applicable.

Acknowledgments: The authors would like to thank the landowners J. Capó and M. López and their families for their selfless help and assistance during the execution of the present trial.

Conflicts of Interest: The authors declare no conflict of interest.

References

1. Du Toit, R.M. The transmission of blue-tongue and horse-Sickness by *Culicoides*. *Onderst. J. Vet. Sci. An. Indust.* **1944**, *19*, 7–16.
2. Linley, J.R. Biting midges (Diptera: Ceratopogonidae) as vectors of non-viral animal pathogens. *J. Med. Entomol.* **1985**, *22*, 589–599. [CrossRef] [PubMed]
3. Meiswinkel, R.; Venter, G.J.; Nevill, E.M. Vectors: *Culicoides* spp. In *Infectious Diseases of Livestock*; Coetzer, J.A.W., Tustin, R.C., Eds.; Oxford University Press: Cape Town, South Africa, 2004; pp. 93–136.
4. Gibbens, N. Schmallenberg virus: A novel viral disease in northern Europe. *Vet. Rec.* **2012**, *170*, 58. [CrossRef]
5. Calvete, C.; Miranda, M.A.; Estrada, R.; Borràs, D.; Sarto i Monteys, V.; Collantes, F.; García de Francisco, J.M.; Moreno, N.; Lucientes, J. Spatial distribution of *Culicoides imicola*, the main vector of Bluetongue virus, in Spain. *Vet. Rec.* **2006**, *158*, 130–131. [CrossRef] [PubMed]
6. Barros, S.C.; Ramos, F.; Luís, T.M.; Vaz, A.; Duarte, M.; Henriques, M.; Cruz, B.; Fevereiro, M. Molecular epidemiology of Bluetongue virus in Portugal during 2004-2006 outbreak. *Vet. Microbiol.* **2007**, *124*, 25–34. [CrossRef]
7. Thiry, E.; Saegerman, C.; Guyot, H.; Kirten, P.; Losson, B.; Rollin, F.; Bodmer, M.; Czaplicki, G.; Toussaint, J.F.; De Clercq, K. Bluetongue in northern Europe. *Vet. Rec.* **2006**, *159*, 327. [CrossRef]
8. Nolan, D.V.; Carpenter, S.; Barber, J.; Mellor, P.S.; Dallas, J.F.; Mordue, J.A.; Piertney, S.B. Rapid diagnostic PCR assays for members of the *Culicoides obsoletus* and *Culicoides pulicaris* species complexes, implicated vectors of bluetongue virus in Europe. *Vet. Microbiol.* **2007**, *24*, 82–94. [CrossRef]
9. Versteirt, V.; Balenghien, T.; Tack, W.; Wint, W. A first estimation of *Culicoides imicola* and *Culicoides obsoletus/Culicoides scoticus* seasonality and abundance in Europe. *EFSA Support. Publ.* **2017**, *14*, 1182E. [CrossRef]
10. Gómez-Tejedor, C. Brief overview of the Bluetongue situation in Mediterranean Europe, 1998–2004. *Vet. Ital.* **2004**, *40*, 57–60.
11. Pérez de Diego, A.C.; Sánchez-Cordón, P.J.; Sánchez-Vizcaíno, J.M. Bluetongue in Spain: From the first outbreak to 2012. *Transbound. Emerg. Dis.* **2014**, *61*, e1–e11. [CrossRef]
12. RASVE (Red de Alerta Sanitaria Veterinaria). Programa nacional de erradicación de lengua azul. *Dirección General de Sanidad de la Producción Agraria. Ministerio de Medio Ambiente y Medio Rural y Marino.* Available online: https://www.mapa.gob.es/es/ganaderia/temas/sanidad-animal-higiene-ganadera/sanidad-animal/ (accessed on 12 December 2018).
13. WOAH (World Organization for Animal Health). *Terrestrial Animal Health Code*, 28th ed.; OIE: Paris, France, 2019; Volumes 1 and 2, 518p.

14. Sellers, R.F.; Pedgley, D.E.; Tucker, M.R. Possible windborne spread of Bluetongue to Portugal, June–July 1956. *J. Hyg.* **1978**, *81*, 189–196. [CrossRef]
15. Sellers, R.F.; Gibbs, E.P.J.; Herniman, K.A.J.; Pedgeley, D.E.; Tucker, M.R. Possible origin of the bluetongue epidemic in Cyprus, August 1977. *J. Hyg.* **1979**, *83*, 547–555. [CrossRef]
16. Hendricks, G.; Gilbert, M.; Staubach, C.; Elbers, A.; Mintiens, K.; Gerbier, G.; Ducheyne, E. A wind density model to quantify the airborne spread of *Culicoides* species during north-western Europe bluetongue epidemic, 2006. *Prev. Vet. Med.* **2008**, *87*, 162–181. [CrossRef] [PubMed]
17. Nelson, R.L.; Bellamy, R.E. Patterns of flight activity of *Culicoides variipennis* (Coquillett) (Diptera: Ceratopogonidae). *J. Med. Entomol.* **1971**, *8*, 283–291. [CrossRef] [PubMed]
18. Barnard, D.R.; Jones, R.H. Diel and seasonal patterns of flight activity of Ceratopogonidae in northeast Colorado: *Culicoides*. *Environ. Entomol.* **1980**, *9*, 446–451. [CrossRef]
19. Sanders, C.J.; Shortall, C.R.; Gubbins, S.; Burgin, L.; Gloster, J.; Harrington, R.; Reynolds, D.R.; Mellor, P.S.; Carpenter, S. Influence of season and meteorological parameters on flight activity of *Culicoides* biting midges. *J. Appl. Ecol.* **2011**, *48*, 1355–1364. [CrossRef]
20. Purse, B.V.; Carpenter, S.; Venter, G.J.; Bellis, G.; Mullens, B.A. Bionomics of temperate and tropical *Culicoides* midges: Knowledge gaps and consequences for transmission of *Culicoides*-borne viruses. *Annu. Rev. Entomol.* **2015**, *60*, 373–392. [CrossRef]
21. Cuéllar, A.C.; Jung Kjær, L.; Baum, A.; Stockmarr, A.; Skovgard, H.; Nielsen, S.A.; Andersson, M.G.; Lindström, A.; Chirico, J.; Lühken, R.; et al. Monthly variation in the probability of presence of adult *Culicoides* populations in nine European countries and the implications for targeted surveillance. *Parasites Vectors* **2018**, *11*, 608. [CrossRef]
22. Barceló, C.; Estrada, R.; Lucientes, J.; Miranda, M.A. A Mondrian matrix of seasonal patterns of *Culicoides* nulliparous and parous females at different latitudes in Spain. *Res. Vet. Sci.* **2020**, *129*, 154–163. [CrossRef]
23. Barceló, C.; Purse, B.V.; Estrada, R.; Lucientes, J.; Miranda, M.A.; Searle, K.R. Environmental drivers of adult seasonality and abundance of biting midges *Culicoides* (Diptera: Ceratopogonidae), bluetongue vector species in Spain. *J. Med. Entomol.* **2021**, *58*, 350–364. [CrossRef]
24. Rawlings, P. A key, based on wing patterns of biting midges (genus *Culicoides* Latreille-Diptera: Ceratopogonidae) in the Iberian Peninsula, for use in epidemiological studies. *Graellsia* **1996**, *52*, 57–71. [CrossRef]
25. Dyce, A.L. The recognition of nulliparous and parous *Culicoides* (Diptera: Ceratopogonidae) without dissection. *Aust. J. Entomol.* **1969**, *8*, 11–15. [CrossRef]
26. AEMET (Agencia Estatal de Meteorología) Ministry for the Ecological Transition and the Demographic Challenge. Available online: https://eportal.mapa.gob.es/websiar/SeleccionParametrosMap.aspx?dst=1 (accessed on 5 May 2023).
27. Kramer, W.L.; Greiner, E.C.; Gibbs, E.P.J. Seasonal variations in population size, fecundity, and parity rates of *Culicoides insignis* (Diptera: Ceratopogonidae) in Florida, USA. *J. Med. Entomol.* **1985**, *22*, 163–169. [CrossRef] [PubMed]
28. Del Río, R.; Barceló, C.; Lucientes, J.; Miranda, M.A. Detrimental effect of cypermethrin treated nets on *Culicoides* populations (Diptera; Ceratopogonidae) and non-targeted fauna in livestock farms. *Vet. Parasitol.* **2014**, *199*, 230–234. [CrossRef]
29. Braverman, Y. Characteristics of *Culicoides* (Diptera, Ceratopogonidae) breeding places near Salisbury, Rhodesia. *Ecol. Entomol.* **1978**, *3*, 163–170. [CrossRef]
30. Bishop, A.L.; McKenzie, H.J.; Barchia, I.M.; Spohr, L.J. Moon phase and other factors affecting light-trap catches of Culicoides brevitarsis Kieffer (Diptera: Ceratopogonidae). *Aust. J. Entomol.* **2000**, *39*, 29–32. [CrossRef]
31. Miranda, M.A.; Borràs, D.; Rincón, C. Seasonal abundance of *Culicoides imicola* and *C. obsoletus* group in the Balearic Islands (Spain). *Vet. Ital.* **2004**, *40*, 292–295.
32. Viennet, E.; Garros, C.; Rakotoarivony, I.; Allene, X.; Gardes, L.; Lhoir, J.; Fuentes, I.; Venail, R.; Crochet, D.; Lancelot, R.; et al. Host-seeking activity of bluetongue virus vectors: Endo/exophagy and circadian rhythm of *Culicoides* in Western Europe. *PLoS ONE* **2012**, *7*, e48120. [CrossRef]
33. Lysyk, T.J.; Danyk, T. Effect of temperature on life history parameters of adult *Culicoides sonorensis* (Diptera: Ceratopogonidae) in relation to geographic origin and vectorial capacity for Bluetongue virus. *J. Med. Entomol.* **2007**, *44*, 741–751. [CrossRef]
34. Rawlings, P.; Mellor, P.S. African horse sickness and the overwintering of *Culicoides* spp. in the Iberian peninsula. *Rev. Sci. Tech. Off. Int. Epiz.* **1994**, *13*, 753–761. [CrossRef]
35. Anderson, J.R.; Linhares, A.X. Comparison of several different trapping methods for *Culicoides variipennis* (Diptera: Ceratopogonidae). *J. Am. Mosq. Control. Assoc.* **1989**, *5*, 325–334. [PubMed]
36. Carpenter, S.; Szmaragd, C.; Barber, J.; Labuschagne, K.; Gubbins, S.; Mellor, P. An assessment of *Culicoides* surveillance techniques in northern Europe: Have we underestimated a potential bluetongue virus vector? *J. Appl. Ecol.* **2008**, *45*, 1237–1245. [CrossRef]
37. Viennet, E.; Garros, C.; Lancelot, R.; Allène, X.; Gardès, L.; Rakotoarivony, I.; Crochet, D.; Delécolle, J.C.; Moulia, C.; Baldet, T.; et al. Assessment of vector/host contact: Comparison of animal-baited traps and UV-light/suction trap for collecting *Culicoides* biting midges (Diptera: Ceratopogonidae), vectors of Orbiviruses. *Parasites Vectors* **2011**, *4*, 119. [CrossRef] [PubMed]

diversity

MDPI

Article

Development of Conventional Multiplex PCR Assays for the Identification of 21 West Palaearctic Biting Midge Taxa (Diptera: Ceratopogonidae) Belonging to the *Culicoides* Subgenus *Culicoides*, including Recently Discovered Species and Genetic Variants

Oliver Dähn [1,*], Doreen Werner [2], Bruno Mathieu [3] and Helge Kampen [1]

1 Friedrich-Loeffler-Institut (FLI), Federal Research Institute for Animal Health, Südufer 10, 17493 Greifswald, Germany
2 Leibniz Centre for Agricultural Landscape Research (ZALF), Eberswalder Str. 84, 15374 Müncheberg, Germany
3 Institute of Parasitology and Tropical Pathology, Université de Strasbourg, DIHP UR 7292, 67000 Strasbourg, France
* Correspondence: daehno@gmx.de

Abstract: In 2006 and 2011, two biting-midge-borne arboviruses of high veterinary importance emerged for the first time in Central Europe: bluetongue virus (BTV) and Schmallenberg virus (SBV). Members of the native Obsoletus and Pulicaris Groups were soon identified as the potential vectors. However, despite several years of extensive taxonomic research on these groups, correct species identification and differentiation from closely related species are still challenging due to isomorphic features, the existence of cryptic species and obsolete PCR identification assays. At present, 17 valid West Palaearctic biting midge species of the *Culicoides* subgenus *Culicoides*, including the Pulicaris Group, are known, and additional genetic variants have been described. For many of them, no identification tests are available, and their roles in disease transmission have remained unknown. In this study, 465 GenBank DNA sequence entries of the mitochondrial cytochrome c oxidase subunit I (COI) gene were used to design PCR primers as specific genetic markers for 21 West Palaearctic biting midge taxa of the *Culicoides* subgenus *Culicoides*. During their validation with DNA from field-collected biting midges and synthetic DNA from biting midge genotypes not available from the field, all primers detected their target taxa, while few showed cross-reactions. Our results indicate the great potential of the new primers in PCR assays and clearly demonstrate the suitability of the COI gene as an excellent marker for the identification of different biting midge species and genetic variants of the *Culicoides* subgenus *Culicoides*.

Keywords: *Culicoides*; Pulicaris Group; West Palaearctic; vectors; polymerase chain reaction (PCR); mitochondrial cytochrome c oxidase subunit I (COI)

Citation: Dähn, O.; Werner, D.; Mathieu, B.; Kampen, H. Development of Conventional Multiplex PCR Assays for the Identification of 21 West Palaearctic Biting Midge Taxa (Diptera: Ceratopogonidae) Belonging to the *Culicoides* Subgenus *Culicoides*, including Recently Discovered Species and Genetic Variants. *Diversity* **2023**, *15*, 699. https://doi.org/10.3390/d15060699

Academic Editor: Luc Legal

Received: 21 April 2023
Revised: 20 May 2023
Accepted: 22 May 2023
Published: 24 May 2023

1. Introduction

Biting midges of the genus *Culicoides* Latreille (Diptera: Ceratopogonidae) are considered the smallest hematophagous dipterans on our planet [1], capable of transmitting a variety of nematodes, protozoan parasites and viruses to domestic animals, an unknown number of wild animal species and—in rare cases—humans [2,3]. Despite their high abundance in Europe, the early isolation of African horse sickness virus (AHSV) from mixed pools of the widely distributed Obsoletus and Pulicaris Groups [4,5] and the detection of bluetongue virus (BTV) in field-collected *C. obsoletus* specimens [4], little attention had been given to indigenous biting midge species for many years. In 2006, the unprecedented outbreak of bluetongue disease (BT) in European areas where the Afro-Asian vector species

C. imicola was absent changed this attitude and highlighted the importance of Palaearctic biting midges in pathogen transmission. Subsequent entomological surveillance confirmed the long-suspected involvement of species of both the Obsoletus Group (genus *Culicoides*, subgenus *Avaritia* Fox, 1955) and the Pulicaris Group (genus *Culicoides*, subgenus *Culicoides* Latreille, 1809) in the transmission of BTV [6–14] and, a few years later, also associated them with the newly emerged Schmallenberg virus (SBV) [15–26]. However, for most species, vector competence has so far been suggested rather than experimentally proven, and correct species identification is challenging, especially in the heterogeneous subgenus *Culicoides*, to which the Pulicaris Group belongs.

To date, 17 valid species of the subgenus *Culicoides* have been described for the western Palaearctic: *C. almeidae* Cambournac, 1970; *C. boyi* Nielsen, Kristensen and Pape, 2015; *C. bysta* Sarvašová and Mathieu, 2017; *C. cryptipulicaris* Talavera, Muñoz-Muñoz, Verdún and Pagès, 2017; *C. delta* Edwards, 1939/*C. lupicaris* Downes and Kettle, 1952 (considered synonymous); *C. fagineus* Edwards, 1939; *C. flavipulicaris* Dzhafarov, 1964; *C. grisescens* Edwards, 1939/*C. remmi* Damien-Georgescu, 1972 (considered synonymous); *C. impunctatus* Goetghebuer, 1920; *C. kalix* Nielsen, Kristensen and Pape, 2015; *C. newsteadi* Austen, 1921; *C. paradoxalis* Ramilo and Delécolle, 2013; *C. pulicaris* (Linnaeus, 1758); *C. punctatus* (Meigen, 1804); *C. quasipulicaris* Talavera, Muñoz-Muñoz, Verdún and Pagès, 2017; *C. selandicus* Nielsen, Kristensen and Pape, 2015; and *C. subfagineus* Delécolle and Ortega, 1998 [27]. However, the taxonomy of the subgenus *Culicoides* is apparently far more complex, and the elucidation of biting midge phylogeny must be seen as an ongoing process [2].

Confusingly, a variety of synonyms are commonly used for one and the same species in the subgenus *Culicoides*: e.g., *C. pulicaris* is also known as *C. setosinervis* Kieffer, 1913; *C. pullatus* Kieffer, 1915; *C. stephensi* Carter, 1916; *C. cinerellus* Kieffer, 1919; *C. quinquepunctatus* Goetghebuer, 1921; *C. flaviplumus* Kieffer, 1924; and *C. sawamotoi* Kono and Takahasi, 1940 [27]. In some cases, it has been further discussed whether synonyms, for example, *C. delta*/*C. lupicaris* and *C. grisescens*/*C. remmi*, should be considered separate species [2,28–31].

The identification of biting midges is classically based on morphological features, particularly the wing pigmentation of adult insects, which allows a quick separation into the Obsoletus Group, Pulicaris Group and other *Culicoides* spp. [32,33]. This method proves to be difficult for the identification to the species level of females of closely related species that have very similar or identical (isomorphic) features and is even more challenging for juvenile specimens (larvae, pupae) [34–36], for which either identification keys have not yet been developed or distinguishing features are not yet even established for corresponding adults. Furthermore, morphological species identification cannot be used for the determination of phylogenetic distances [37] and may require time-consuming analyses of slide-mounted microscopical insect preparations to visualize fine structures [38], a lot of practical experience and fresh material with distinct coloration.

The development and implementation of molecular tools such as species-specific PCR tests and DNA barcoding have improved the knowledge of phylogenetic relationships and revolutionized the species identification of biting midges. For these genetic techniques, various molecular markers have been used, such as mitochondrial and nuclear genes, including ribosomal markers [39]. The mitochondrial cytochrome oxidase c subunit I (COI or COX1) gene has been by far the most widely utilized marker for phylogenetic studies and identification purposes in culicoid biting midges, as it is a sufficiently long high-copy gene that is composed of both conserved and variable regions [40–42].

In the past two decades, comprehensive studies of the COI region have revealed a considerable number of genetic variants in the subgenus *Culicoides*: Pagès et al. [43] described a new haplotype of *C. pulicaris* (*C. pulicaris* haplotype P3) from Spain, which was described later as *C. cryptipulicaris* [31], a previously unknown haplotype of *C. fagineus* (referred to as *C. fagineus* haplotype F1), and three new genetic variants of *C. newsteadi* (haplotypes N1, N2 and N3). These findings were supported by COI analyses of biting midges collected in Denmark and Sweden [44]. Similar heterogeneity in the mitochondrial gene sequence was found in *C. grisescens* specimens from Switzerland (haplotypes G1 and G2) [45] and in *C. lupicaris* biting midges from various European countries, including Spain, Denmark, Czech Republic, France, Slovakia, Sweden, Turkey and Austria (*C. lupicaris* sensu stricto (s.s.), *C. lupicaris* haplotypes L2 and L3) [30,43,44,46–49]. Additionally, another genotype that is morphologically similar to *C. fagineus* haplotype F1—*Culicoides* WBS—was recently reported from the Black Sea region of Turkey [48], and it is expected that further genetic variants will be discovered in the future [43,46,47].

Several PCR tests have been designed to distinguish between common West Palaearctic species of the subgenus *Culicoides* and some of their genetic variants known at that time [34,43,45,50]. However, these PCR tests are incapable of differentiating newly discovered haplotypes and were developed more than ten years ago using a small and spatially restricted gene pool, which limits their applicability. Commonly used COI barcoding is not an appropriate alternative for species identification, as it cannot be applied to pooled samples due to the risk of mixed taxa/sequences and the consequent detection of the more abundant species. Moreover, previous analyses of engorged females led to the unintended identification of the blood-donor species and failed to characterize individuals stored in ethanol for extended time periods. Thus, a revision of published PCR tests and the development of new diagnostic assays are urgently necessary.

In the present study, a huge dataset of West Palaearctic subgenus *Culicoides* COI gene sequences from GenBank was analyzed with the aim to develop easy-to-use multiplex PCR assays for the differentiation of their species and genetic variants. The COI gene features both variable and conserved regions and is represented in GenBank by a number of entries sufficient to provide comprehensive and reliable information on DNA variations and homologies between species and haplotypes. Some authors, however, have already addressed the issue of wrong entries in such data repositories [46,51,52], which are mainly regarded as attributable to the preceding incorrect morphological identification of specimens. Since classical taxonomists are becoming progressively scarce all the while cryptic taxa are being detected, reliable alternative techniques have to be developed, thus providing the basis for the improved identification of potential vector species and a better understanding of the *Culicoides* biting midge distribution and ecology.

2. Materials and Methods

2.1. Biting Midge Collection

Culicoid biting midges were collected with BG-Sentinel UV-light suction traps (Biogents, Regensburg, Germany) operated once a week for 24 h during various German monitoring activities. The individual specimens analyzed originated from samplings in other European countries. Biting midges were morphologically pre-identified under a stereomicroscope to the group or species level using commonly used identification keys [38,53–55]. Pre-sorted biting midges were kept in 75% EtOH for subsequent molecular analysis.

2.2. Genetic Identification of Field-Collected Biting Midges

After discarding the ethanol and evaporating the remaining fixative for 1 min at room temperature, three steel beads with a diameter of 3 mm (TIS GmbH, Gauting, Germany) were added to morphologically pre-identified, single specimens of the subgenus *Culicoides*. The samples were supplemented with either 180 µL of buffer ATL and 20 µL of Proteinase K (Qiagen, Hilden, Germany) or 350 µL of in-house ZB5d medium (Eagle's minimal essential medium with Earle's and Hank's salts plus non-essential amino acids) containing 3.5 µL of penicillin–streptomycin (100 U/mL) and 0.7 µL of gentamycin–amphotericin (0.01 mg/mL, 0.25 µg/mL) (Thermo Fisher Scientific, Dreieich, Germany). Samples were homogenized for 3 min at 30 Hz with a TissueLyser II (Qiagen), and total DNA was isolated using the QIAamp DNA Mini Kit (Qiagen) or the NucleoMag VET Kit (Macherey-Nagel, Düren, Germany) according to the manufacturers' instructions, with a final elution volume of 50 µL of AE-buffer (QIAamp DNA Mini Kit) or 100 µL of VEL-buffer (NucleoMag VET Kit), respectively.

DNA extracts were used to generate amplicons of the COI gene with the species-specific PCR described in Nolan et al. [50], the universal primers PanCuli-COX1-211F and PanCuli-COX1-727R according to the authors' protocol [56] or the self-designed generic primer PanCuli-COX1-025F (5′-ACTTTATATTTTATTTTGGAGYWTGRGC-3′) in combination with PanCuli-COX1-727R using an adapted protocol (54 °C annealing temperature) from Lehmann et al. [56]. PCR products with expected lengths were excised and extracted with the QIAquick Gel Extraction Kit (Qiagen). For sequencing, DNA fragments were cycled with the PCR primers using the BigDye Termintor v1.1 Cycle Sequencing Kit (Thermo Fisher Scientific). The resulting PCR products were purified with the Bioanalysis NucleoSEQ Kit (Macherey-Nagel), and 15 µL of the eluates were mixed with the same volume of Hi-Di formamide (Thermo Fisher Scientific). Each sample was sequenced on a 3500 Genetic Analyzer (Applied Biosystems/Hitachi, Darmstadt, Germany), followed by sequence editing with Geneious Prime software version 2021.0.1 (Biomatters, Auckland, New Zealand). Edited sequences were deposited in GenBank.

2.3. COI Data Analysis and Primer Design

All available GenBank entries of West Palaearctic taxa of the subgenus *Culicoides* were collected and checked for plausibility: COI sequences (Table S1) were compared with sequences of the first description to find incorrect entries. Dubious sequences were re-analyzed with the NCBI nucleotide BLAST tool (https://blast.ncbi.nlm.nih.gov/Blast.cgi (accessed on 28 October 2022)), assigned to fitting species or—if no sequence match could be found—excluded from further analysis. Remaining sequences were used for the generation of consensus sequences with Geneious Prime software (Biomatters), which were finally compared in a Geneious multiple alignment using initial settings. In this context, it is to be noted that no sequences were found in GenBank under the species name '*C. delta*'. Instead, sequences of that taxon had been deposited using the name '*C. deltus*', which had been used for '*C. delta*' until renaming in 2015.

Inter- and intraspecific variances in the DNA sequence were used to design specific forward primers, including wobble sites, according to common guidelines for primer design [57–59]. Promising primer candidates were checked regarding melting temperature, GC content, self-dimerization and primer-dimer formation with the Oligo Analysis Tool (https://eurofinsgenomics.eu/en/ecom/tools/oligo-analysis/ (accessed on 14 December 2021)) and analyzed with the NCBI BLAST tool for repetitive sequences before ordering. Primers were checked for functionality, specificity and the capability of multiplexing and finally validated with genetically identified biting midge material from the field or—in case no field-collected material of the respective taxon was available—synthetic COI gene DNA (Table S2), produced by GenExpress (Berlin, Germany).

2.4. Multiplex PCRs

The newly designed specific forward primers were applied in combination with the published universal reverse primer PanCuli-COX1-727R [56] in several multiplex PCRs (mPCRs), which can be carried out in parallel or successively. For easier handling, the approach was based on a universal annealing temperature instead of primer-specific annealing temperatures, although this increased the risk of reduced primer sensitivity and specificity. As far as possible, the primers were combined in the various mPCRs according to morphological similarities of the species or haplotypes they were meant to detect, but this was not possible in all cases.

The master mixes were composed of 10 µL of 2× QuantiTect Multiplex PCR NoROX reagent (Qiagen), 0.5 µM of each primer and 2 µL of DNA template and replenished with water to give a total volume of 20 µL. DNA amplification was performed using the following thermoprofile: 15 min at 95 °C (activation of Taq polymerase), followed by 42 cycles of 30 s at 95 °C (denaturation), 45 s at 63 °C (primer annealing) and 45 s at 72 °C (primer elongation), and a final elongation step for 5 min at 72 °C. The complete PCR reaction mixture was supplemented with 2.5 µL of 6× DNA Loading Dye (Thermo Fisher

Scientific) and applied to 1.5% agarose gels, pre-mixed with 5 mg/mL ethidium bromide solution. After running for 50 min at 100 V, gels were visualized with a ChemiDoc MP Imaging System (Bio-Rad, Feldkirchen, Germany).

3. Results

During the analysis of the comprehensive GenBank dataset of COI sequences from 21 different countries, a significant number of subgenus *Culicoides* sequences was found to be incorrectly deposited, especially those of specimens with similar morphology (e.g., *C. newsteadi* group). However, implausible entries could be identified, and a total number of 465 COI sequences could accurately be assigned to the respective species and haplotypes (Table S1).

For each species and haplotype, specific consensus sequences with lengths between 412 and 1535 base pairs were generated, which revealed high intraspecific pairwise identity between 98.3% and 99.8% (average: 99.4%, Table S1). Multiple alignment comparison of the consensus sequences displayed interspecific differences between 6.0% and 19.7% (Figure 1). Interestingly, genetic distances between synonymous *C. delta*/*C. lupicaris* taxa and recently described haplotypes of the same species, especially *C. newsteadi*, were comparatively high (10.0% to 18.0%), questioning their taxonomic status. Thus, these taxa were considered separate taxonomic entities during PCR development.

Inter	pul	cry	del	lup L1	pun	imp	kal	boy	bys	new N3	lup L2	fla	new N1	gri G1	sel	fag F1	fag F2	sub	new N2	gri G2	new
pul		89.1	86.1	85.9	85.5	85.0	84.7	84.5	84.4	84.2	84.2	84.0	83.8	83.1	82.9	82.8	82.7	82.7	82.6	82.1	81.8
cry	10.9		84.7	85.0	85.6	83.4	83.1	80.3	80.8	82.5	85.4	82.2	83.8	81.2	82.8	82.8	83.2	81.5	83.0	80.3	81.4
del	13.9	15.3		86.7	85.1	84.9	87.0	85.3	84.1	85.3	84.9	83.5	83.3	84.2	85.6	82.5	84.1	82.8	83.9	82.9	84.6
lup L1	14.1	15.0	13.3		84.2	85.7	84.9	85.8	85.1	83.6	85.1	83.6	84.3	82.1	85.0	81.8	83.6	84.0	84.3	83.0	84.8
pun	14.5	14.4	14.9	15.8		84.8	89.0	84.0	83.8	82.3	86.7	85.3	85.5	85.3	85.9	80.8	82.6	83.7	84.3	83.8	84.6
imp	15.0	16.6	15.1	14.3	15.2		86.2	85.9	84.8	86.7	87.1	83.4	83.4	83.8	83.5	81.3	83.6	82.5	84.1	81.7	84.7
kal	15.3	16.9	13.0	15.1	11.0	13.8		85.1	84.2	84.6	85.7	84.1	84.5	85.4	93.9	83.0	84.3	85.1	84.1	81.8	85.5
boy	15.5	19.7	14.7	14.2	16.0	14.1	14.9		94.0	85.3	84.5	83.8	84.4	82.8	84.3	84.7	82.6	83.3	82.4	84.6	86.2
bys	15.6	19.2	15.9	14.9	16.2	15.2	15.8	6.0		83.8	85.1	83.4	84.7	82.6	83.4	82.8	83.4	81.3	81.7	83.2	85.4
new N3	15.8	17.5	14.7	16.4	17.7	13.3	15.4	14.7	16.2		86.9	83.8	82.3	82.5	83.8	83.3	84.6	84.1	84.5	82.2	85.7
lup L2	15.8	14.6	15.1	14.9	13.3	12.9	14.3	15.5	14.9	13.1		83.7	83.7	84.1	84.5	81.8	84.1	82.8	84.3	82.0	84.9
fla	16.0	17.8	16.5	16.4	14.7	16.6	15.9	16.2	16.6	16.2	16.3		85.7	85.9	85.0	88.3	89.1	87.0	81.7	86.1	85.4
new N1	16.2	16.2	16.7	15.7	14.5	16.6	15.5	15.6	15.3	17.7	16.3	14.3		84.3	84.8	81.1	83.7	82.7	82.0	83.4	85.6
gri G1	16.9	18.8	15.8	17.9	14.7	16.2	14.6	17.2	17.4	17.5	15.9	14.1	15.7		84.1	83.7	85.3	85.2	81.1	90.0	82.1
sel	17.1	17.2	14.4	15.0	14.1	16.5	6.1	15.7	16.6	16.2	15.5	15.0	15.6	15.9		82.5	83.8	83.5	82.8	82.5	85.4
fag F1	17.2	17.2	17.5	18.2	19.2	18.7	17.0	15.3	17.2	16.7	18.2	11.7	18.9	16.3	17.5		89.8	89.9	80.8	85.0	82.3
fag F2	17.3	16.8	15.9	16.4	17.4	16.4	15.7	17.4	16.6	15.4	15.9	10.9	16.3	14.7	16.2	10.2		89.1	84.1	85.4	84.8
sub	17.3	18.5	17.2	16.0	16.3	17.5	14.9	16.7	18.7	15.9	17.2	13.0	17.3	14.8	16.5	10.1	10.9		81.8	84.8	83.3
new N2	17.4	17.0	16.1	15.7	15.7	15.9	15.9	17.6	18.3	15.5	15.7	18.3	18.0	18.9	17.2	19.2	15.9	18.2		81.5	83.2
gri G2	17.9	19.7	17.1	17.0	16.2	18.3	18.2	15.4	16.8	17.8	18.0	13.9	16.6	10.0	17.5	15.0	14.6	15.2	18.5		83.5
new	18.2	18.6	15.4	15.2	15.4	15.3	14.5	13.8	14.6	14.3	15.1	14.6	14.4	17.9	14.6	17.7	15.2	16.7	16.8	16.5	
Intra	0.3	0.2	0.4	0.5	0.7	0.3	0.6	0.2	1.7	0.3	0.2	0.2	0.9	0.2	0.7	0.3	0.9	0.7	1.0	1.1	0.3
n	82	4	14	31	105	30	14	13	21	19	9	7	16	42	7	3	14	6	7	6	15

Figure 1. Inter- and intraspecific pairwise genetic comparison of COI gene DNA sequences between tested West Palaearctic taxa of the subgenus *Culicoides*: genetic distances are displayed in the left-bottom half of the matrix and highlighted with graded colors from red (low distance) through yellow (medium distance) to green (high distance). Genetic similarities are presented in the right-upper half of the matrix. Values (in %) were calculated through the comparison of species- and haplotype-specific consensus sequences of respective GenBank entries (n). *C. pulicaris* (pul), *C. cryptipulicaris* (cry), *C. delta* (del), *C. lupicaris* haplotype L1 (lup L1), *C. punctatus* (pun), *C. impunctatus* (imp), *C. kalix* (kal), *C. boyi* (boy), *C. bysta* (bys), *C. newsteadi* haplotype N3 (new N3), *C. lupicaris* haplotype L2 (lup L2), *C. flavipulicaris* (fla), *C. newsteadi* haplotype N1 (new N1), *C. grisescens* haplotype G1 (gri G1), *C. selandicus* (sel), *C. fagineus* haplotype F1 (fag F1), *C. fagineus* haplotype F2 (fag F2), *C. subfagineus* (sub), *C. newsteadi* haplotype N2 (new N2), *C. grisescens* haplotype G2 (gri G2) and *C. newsteadi* s.s. (new).

Genetic differences were subsequently used to develop specific forward primers for 21 West Palaearctic biting midge taxa of the *Culicoides* subgenus *Culicoides* according to the PCR concept published in Lehmann et al. [56] (Table 1). On average, 14 primers per species (290 in total, Table S3) were tested, and in many cases, pre-testing revealed cross-reactivity with other subgenus *Culicoides* taxa. However, the targeted insertion of wobbles and mismatch bases significantly reduced unspecific detection. The best-performing forward primers (Table 1), in combination with PanCuli-COX1-727R as a reverse primer [56], were put together in various single-tube mPCRs. Merely the number of species and haplotypes to be considered and the limited length of generated consensus sequences made it necessary to subdivide the PCR approach into four reactions (mPCRs A–D). The mPCRs were pre-tested with DNA extracts of single specimens or with an equivalent of 10^6 copies of synthetic COI gene DNA (calculated based on serial dilution quantitative real-time PCR) of subgenus *Culicoides* taxa to verify multiplexability (Figure 2A–D). Each multiplex PCR showed the expected amplicons of the target species between 139 bp and 491 bp, whereas no amplification was observed for no-template negative controls (Figure 2, lanes 2, 9, 15 and 21).

Table 1. Newly designed forward primers specific for 21 taxa of the *Culicoides* subgenus *Culicoides* to be used in combination with the universal reverse primer PanCuli-COX1-727R.

mPCR	Species/Haplotype	Primer Code	Primer Sequence (5′-3′)	Modification (Position)	Amplicon (bp)
A	*C. bysta*	bys-COI-158F	AATCTTACTTCTCTTATCTCTRC	R-wobble (2)	158
	C. punctatus	pun-COI-227F	TCATATGCGATCAAACGGG	A > C (18)	227
	C. boyi	boy-COI-275F	AGCTATTTCATCAATTCTTGGA	G > C (20)	275
	C. grisescens G2	gri2-COI-346F	CCACACCTTTCTGCAAACA	C > A (15)	346
	C. kalix	kal-COI-419F	CCACCCTTCTCTAACATTGC	C > A (18)	419
	C. grisescens G1	gri1-COI-463F	GATATAGCTTTCACACGAATG	C > A (9)	463
B	*C. fagineus* F2	fag2-COI-151F [3]	TTGCATCTTTCCCTCCCTGTA	T > A (17)	151
	C. flavipulicaris	fla-COI-215F [3]	CAATCGTATTACTTTTGATCGT	G > C (18)	215
	C. subfagineus	sub-COI-318F [3]	CTGTRGCTTCTGTAGATC	R-wobble (14), G > T (15)	318
	C. fagineus F1	fag1-COI-420F	TTCCTCCATCTCTTTCCCTAT	C > T (17)	420
	C. impunctatus	imp-COI-491F	ATTGGTTCCATTAATACTCGGA	none	491
C	*C. delta*	del-COI-161F	TGCTATATTACTTCTTTTGTCAC	T > A (17)	161
	C. lupicaris L1 [1]	lup1-COI-214F	AATGGAATGTCATTCGACCGT	T > G (13)	214
	C. pulicaris s.s.	pul-COI-313F [2,3]	GCATCCGTAGACTTGGCC	none	313
	C. cryptipulicaris	cry-COI-405F	CGTTACTCTTATTGAGCAGAT	none	405
	C. lupicaris L2	lup2-COI-467F	TCCTGATATAGCTTTTCCC	none	467
D	*C. newsteadi* N2	new2-COI-139F [3]	CTCCCAGTTCTTGCTGGT	none	139
	C. newsteadi s.s.	new-COI-231F	TTATTAATATGCGATCCGCC	none	231
	C. newsteadi N3	new3-COI-296F [3]	CATCTTCTCCCTACACCTG	none	296
	C. newsteadi N1	new1-COI-351F	TATATCCGCCTCTTTCAAGA	none	351
	C. selandicus	sel-COI-403F	TGACTATTATTAAGTAGCTTGGTA	T > G (23)	403

[1] For easier demarcation from *C. lupicaris* haplotype L2, *C. lupicaris* s.s. is designated as *C. lupicaris* haplotype L1 according to the nomenclature of Ander et al. [46]. [2] Modified primer from Nolan et al. [50]. [3] Modified primer from Pagès et al. [43].

To validate the specificity and sensitivity of the designed forward primers, the mPCRs were further tested with the DNA material of 41 genetically pre-identified specimens or the synthetic DNA of various taxa of the subgenus *Culicoides* (Table 2). All forward primers reliably detected their specific DNA, with the exception of the primer sel-COI-403F (mPCR D), which only generated specific PCR amplicons for two of three *C. selandicus* DNA samples, resulting in a total sensitivity of all mPCRs of 97.6%.

In terms of specificity, mPCRs A and B showed no unspecific annealing of the forward primers to non target subgenus *Culicoides* taxa at all (100% specificity). However, three forward primers of mPCRs C and D showed weak signals with other subgenus *Culicoides* taxa: while using mPCR C, unspecific reaction signals were observed for del-COI-161F with the only *C. impunctatus* sample, for cry-COI-405F with one out of five tested *C. punctatus* samples and with *C. grisescens* haplotype G2, and for lup2-COI-467F with the synthetic DNA of *C. flavipulicaris*. In the case of mPCR D, the forward primer new3-COI-296F incorrectly reacted with the only *C. impunctatus* sample and with four of five DNA samples pre-identified as *C. lupicaris* haplotype L2. Additionally, one of five tested *C. punctatus* samples were identified as *C. newsteadi* haplotype N1 with the primer new1-COI-351F, and a 403 bp fragment of the genomic DNA of *C. grisescens* haplotype G2 (one sample tested) was amplified with the primer sel-COI-403F.

Finally, the mPCRs were tested with the genomic DNA of single biting midge specimens not belonging to the subgenus *Culicoides* (n = 21) but to other subgenera of the genus *Culicoides* to check whether the pre-sorting of biting midges to the group level is necessary before using the new PCRs. The agarose gel analyses summarized in Table 3 show no unspecific detection of tested *Culicoides* species with mPCR B. In the case of mPCR D, three unspecific DNA fragments with lengths of 120 bp, 550 bp and 900 bp were amplified when using the only *C. dewulfi* and *C. sanguisuga* DNAs as templates, but no unspecific signals occurred with the other 19 *Culicoides* taxa. mPCR A showed no unspecific amplicons at all, but the primer pun-COI-227F incorrectly detected *C. griseidorsum* and *C. pictipennis* as *C. punctatus*. Most cross-reactivity was observed for mPCR C: the primer lup1-COI-214F detected *C. riethi*, and the primer pul-COI-313F *C. poperinghensis* and the forward primer lup2-COI-467F amplified a 467 bp fragment with the genomic DNA of *C. festivipennis*, *C. kibunensis*, *C. obsoletus* clade O1 and *C. sanguisuga*. Additionally, one or more forward primers of mPCR C generated unspecific PCR amplicons with lengths of approximately 600 bp if *C. festivipennis* DNA was tested.

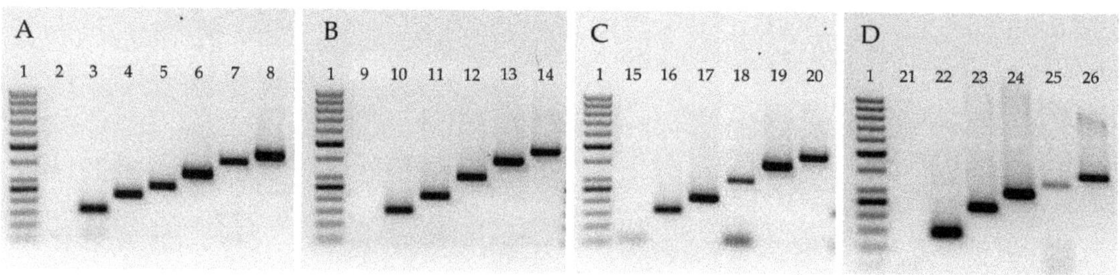

Figure 2. Validation of the different multiplex PCR tests (mPCRs (**A–D**)) for the subgenus *Culicoides* taxa. The specific primers used were as follows: bys-COI-158F, pun-COI-227F, boy-COI-275F, gri2-COI-346F, kal-COI-419F and gri1-COI-463F (lanes 2–8) for mPCR (**A**); fag2-COI-151F, fla-COI-215F, sub-COI-318F, fag1-COI-420F and imp-COI-491F (lanes 9–14) for mPCR (**B**); del-COI-161F, lup1-COI-214F, pul-COI-313F, cry-COI-405F and lup2-COI-467F (lanes 15–20) for mPCR (**C**); new2-COI-139F, new-COI-231F, new3-COI-296F, new1-COI-351F and sel-COI-403F (lanes 21–26) for mPCR (**D**). In all multiplex PCRs, forward primers were used in combination with the universal reverse primer PanCuli-COX1-727R. DNA samples used for PCR validation contained 10^6 synthetic COI gene copies or DNA extracts of single *Culicoides* specimens. Lane 1: 50 bp ladder (Gene Ruler, 50–1000 bp); lanes 2, 9, 15 and 21: no-template control; lane 3: *C. bysta*; lane 4: *C. punctatus*; lane 5: *C. boyi*; lane 6: *C. grisescens* haplotype G2; lane 7: *C. kalix*; lane 8: *C. grisescens* haplotype G1; lane 10: *C. fagineus* haplotype F2 (=*C. fagineus* s.s.); lane 11: *C. flavipulicaris*; lane 12: *C. subfagineus*; lane 13: *C. fagineus* haplotype F1; lane 14: *C. impunctatus*; lane 16: *C. delta*; lane 17: *C. lupicaris* haplotype L1; lane 18: *C. pulicaris* s.s.; lane 19: *C. cryptipulicaris*; lane 20: *C. lupicaris* haplotype L2; lane 22: *C. newsteadi* haplotype N2; lane 23: *C. newsteadi* s.s.; lane 24: *C. newsteadi* haplotype N3; lane 25: *C. newsteadi* haplotype N1; lane 26: *C. selandicus*. Primer-dimer formation was observed in the case of mPCR C (lanes 15 and 18).

Table 2. Results of specificity tests within the subgenus *Culicoides*. A total of 41 samples belonging to 21 taxa of the subgenus *Culicoides* were tested with the new multiplex PCRs, either with genomic DNA of single specimens or with 10^6 copies of specific synthetic COI gene DNA.

Species/Haplotype DNA	No. Samples	GenBank Accession No.	Multiplex PCR A						Multiplex PCR B					Multiplex PCR C					Multiplex PCR D				
			bys	pun	boy	G2	kal	G1	F2	fla	sub	F1	imp	del	L1	pul	cry	L2	N2	new	N3	N1	sel
C. bysta [1]	1	n.a	+ (1)	-	-	-	-	-	-	-	-	-	-	-	-	-	-	-	-	-	-	-	-
C. punctatus [2]	5	OQ789061-65	-	+ (5)	-	-	-	-	-	-	-	-	-	-	-	-	+ (1)	-	-	-	-	+ (1)	-
C. boyi [1]	1	n.a	-	-	+ (1)	-	-	-	-	-	-	-	-	-	-	-	-	-	-	-	-	-	-
C. grisescens G2 [2]	1	OQ789038	-	-	-	+ (1)	-	-	-	-	-	-	-	-	-	-	+ (1)	-	-	-	-	-	+ (1)
C. kalix [1]	1	n.a.	-	-	-	-	+ (1)	-	-	-	-	-	-	-	-	-	-	-	-	-	-	-	-
C. grisescens G1 [2]	1	OQ789037	-	-	-	-	-	+ (1)	-	-	-	-	-	-	-	-	-	-	-	-	-	-	-
C. fagineus F2 [2]	1	OQ789036	-	-	-	-	-	-	+ (1)	-	-	-	-	-	-	-	-	-	-	-	-	-	-
C. flavipulicaris [1]	1	n.a.	-	-	-	-	-	-	-	+ (1)	-	-	-	-	-	-	-	+ (1)	-	-	-	-	-
C. subfagineus [1]	1	n.a.	-	-	-	-	-	-	-	-	+ (1)	-	-	-	-	-	-	-	-	-	-	-	-
C. fagineus F1 [1]	1	n.a.	-	-	-	-	-	-	-	-	-	+ (1)	-	-	-	-	-	-	-	-	-	-	-
C. impunctatus [2]	1	OQ789034	-	-	-	-	-	-	-	-	-	-	+ (1)	+ (1)	-	-	-	-	-	-	+ (1)	-	-
C. delta [2]	1	OQ789035	-	-	-	-	-	-	-	-	-	-	-	+ (1)	-	-	-	-	-	-	-	-	-
C. lupicaris L1 [2]	1	OQ789039	-	-	-	-	-	-	-	-	-	-	-	-	+ (1)	-	-	-	-	-	-	-	-
C. pulicaris s.s. [2]	6	OQ789055-60	-	-	-	-	-	-	-	-	-	-	-	-	-	+ (6)	-	-	-	-	-	-	-
C. cryptipulicaris [1]	1	n.a.	-	-	-	-	-	-	-	-	-	-	-	-	-	-	+ (1)	-	-	-	-	-	-
C. lupicaris L2 [2]	5	OQ789040-44	-	-	-	-	-	-	-	-	-	-	-	-	-	-	-	+ (5)	-	-	+ (4)	-	-
C. newsteadi N2 [1]	1	n.a.	-	-	-	-	-	-	-	-	-	-	-	-	-	-	-	-	+ (1)	-	-	-	-
C. newsteadi s.s. [1]	1	n.a.	-	-	-	-	-	-	-	-	-	-	-	-	-	-	-	-	-	+ (1)	-	-	-
C. newsteadi N3 [2]	4	OQ789048-51	-	-	-	-	-	-	-	-	-	-	-	-	-	-	-	-	-	-	+ (4)	-	-
C. newsteadi N1 [2]	3	OQ789045-47	-	-	-	-	-	-	-	-	-	-	-	-	-	-	-	-	-	-	-	+ (3)	-
C. selandicus [2]	3	OQ789052-54	-	-	-	-	-	-	-	-	-	-	-	-	-	-	-	-	-	-	-	-	+ (2)

+: Amplification; -: no amplification; (n): number of samples with amplification; n.a.: not applicable. [1] Synthetic COI gene DNA. [2] Genomic DNA of single specimens.

Table 3. Results of specificity tests with the four different multiplex PCRs with 21 species of the genus *Culicoides* not belonging to the subgenus *Culicoides*. One specimen of each species or haplotype was tested.

Species/Haplotype DNA	GenBank Accession No.	Multiplex PCR A						Multiplex PCR B						Multiplex PCR C				Multiplex PCR D				
		bys	pun	boy	G2	kal	G1	F2	fla	sub	F1	imp	del	L1	pul	cry	L2	N2	new	N3	N1	sel
C. achrayi	OQ789066	-	-	-	-	-	-	-	-	-	-	-	-	-	-	-	-	-	-	-	-	-
C. alazanicus	OQ789067	-	-	-	-	-	-	-	-	-	-	-	-	-	-	-	-	-	-	-	-	-
C. chiopterus	OQ789068	-	-	-	-	-	-	-	-	-	-	-	-	-	-	-	-	-	-	-	-	-
C. dewulfi	OQ789069	-	-	-	-	-	-	-	-	-	-	-	-	-	-	-	-	+[2,3]	+[2,3]	+[2,3]	+[2,3]	+[2,3]
C. festivipennis	OQ789070	-	-	-	-	-	-	-	-	-	-	-	+[1]	+[1]	+[1]	+[1]	+[1]	-	-	-	-	-
C. griseidorsum	OQ789071	-	+	-	-	-	-	-	-	-	-	-	-	-	-	-	-	-	-	-	-	-
C. imicola	OQ789072	-	-	-	-	-	-	-	-	-	-	-	-	-	-	-	-	-	-	-	-	-
C. kibunensis	OQ789073	-	-	-	-	-	-	-	-	-	-	-	-	-	-	-	+	-	-	-	-	-
C. montanus	OQ789074	-	-	-	-	-	-	-	-	-	-	-	-	-	-	-	-	-	-	-	-	-
C. obsoletus clade O1	OQ789075	-	-	-	-	-	-	-	-	-	-	-	-	-	-	-	+	-	-	-	-	-
C. obsoletus clade O2	OQ789076	-	-	-	-	-	-	-	-	-	-	-	-	-	-	-	-	-	-	-	-	-
C. obsoletus clade O3	OQ789077	-	-	-	-	-	-	-	-	-	-	-	-	-	-	-	-	-	-	-	-	-
C. pallidicornis	OQ789078	-	-	-	-	-	-	-	-	-	-	-	-	-	-	-	-	-	-	-	-	-
C. pictipennis	OQ789079	-	+	-	-	-	-	-	-	-	-	-	-	-	-	-	-	-	-	-	-	-
C. poperinghensis	OQ789080	-	-	-	-	-	-	-	-	-	-	-	-	-	+	-	-	-	-	-	-	-
C. riethi	OQ789081	-	-	-	-	-	-	-	-	-	-	-	-	+	-	-	-	-	-	-	-	-
C. riouxi	OQ789082	-	-	-	-	-	-	-	-	-	-	-	-	-	-	-	-	-	-	-	-	-
C. salinarius	OQ789083	-	-	-	-	-	-	-	-	-	-	-	-	-	-	-	-	-	-	-	-	-
C. sanguisuga	MK760238	-	-	-	-	-	-	-	-	-	-	-	-	-	-	-	+	+[4]	+[4]	+[4]	+[4]	+[4]
C. scoticus clade 2	OQ789084	-	-	-	-	-	-	-	-	-	-	-	-	-	-	-	-	-	-	-	-	-
C. sinanoensis	MK760244	-	-	-	-	-	-	-	-	-	-	-	-	-	-	-	-	-	-	-	-	-

+: Amplification; -: no amplification. Unspecific amplicons with lengths of approx. [1] 600 bp, [2] 550 bp, [3] 900 bp and [4] 120 bp.

4. Discussion

In the past, several PCR tests were developed using either the mitochondrial COI gene [34,43,45,50,56,60] or nuclear-encoded ribosomal ITS1 [61–65] and ITS2 genes [65–67] to identify the species of the Obsoletus and Pulicaris Groups, which contain the putative biting midge vectors of BTV and SBV. GenBank DNA sequence analyses performed in the framework of the present study demonstrated that only the COI gene provides a sufficient number of entries for species and haplotypes of the subgenus *Culicoides* (COI: 495 sequences; ITS1: 15 sequences; ITS2: 39 sequences; as of September 2022), allowing the design of specific primers.

While most existent PCR assays for indigenous biting midge species are based on a limited number of COI sequences from field-collected specimens caught in a certain country or region, the present approach made use of all available sequence data for 21 subgenus *Culicoides* taxa deposited in GenBank. However, searching for suitable DNA sequences was not trivial with regard to the species variety within the subgenus, the wide use of synonyms for one and the same species [27] and misidentifications, which increase the chance of incorrect entries in databases. In this study, a significant number of GenBank COI sequences of subgenus *Culicoides* taxa were found to be incorrectly assigned to species (Table S1), which confirmed the results of previous studies [46,51,52] demonstrating the limitations of biting midge classification based on genetic data alone. Instead, it highlights the importance of the morphological definition of species and taxa whose sequences are to be entered into databases by experienced *Culicoides* taxonomists. It should also be discussed whether taxonomic changes have to be updated in such data repositories.

Despite these problems and the general issue of the genetic delimitation of species [68], a procedure was found to generate consensus sequences for the subgenus *Culicoides* taxa with high intraspecific homology. Further comparison of consensus sequences revealed variations between taxa, which suggests that some described cryptic taxa, especially those of *C. newsteadi* and the synonymous *C. delta* and *C. lupicaris*, should be regarded as separate species rather than genetic variants. This observation confirms the results obtained by Yildirim et al. [48]. The genetic variations were deployed for the development of specific primers, using identical cycling conditions for simultaneous sample analysis. The concept is based on the use of specific forward primers in combination with a universal reverse primer (PanCuli-COX1-727R) in a cost-effective and easy-to-use single-tube (multiplex) approach, generating one characteristic band for each taxon after the gel electrophoresis of PCR products. Another advantage of the idea of using one and the same reverse primer for all possible target DNAs is the efficient use of the COI gene fragment available in GenBank [50,56,69]. Its limited length of usually less than 500 base pairs reduces the options for primer positioning and makes the development of conventional multiplex PCR tests extremely difficult.

Despite many advantages, the application of a universal reverse primer simultaneously implies a great challenge: the specificity of the PCR is exclusively provided by the forward primer. Thus, in most cases, initial experiments resulted in the cross-reactive binding of potential primers to other taxa of the subgenus *Culicoides* and required more intensive testing. This issue could be solved by inserting single mismatch bases into the conserved regions of the primer sequence, assuming that mismatch base pairing would be less detrimental to the detection of the target taxa, according to the higher binding strength of the primer, than to non-target taxa. There was an attempt to apply the 'general hierarchy of mismatch impact' described in the literature [70–76]. However, except for the observation that incorrect base pairing at the 3′-terminal part of the primer should be avoided, it seemed to be more trial-and-error to find the best working mismatch in this study, which is not unexpected since many factors can influence the mismatch behavior [74,76]. After a considerable optimization process, primer specificity was adapted in a way that allowed the different taxa of the subgenus *Culicoides* to be distinguished, although some of the primers containing mismatches showed weaker binding strength than others. Thus, we elongated the affected primers despite the risk of reduced specificity, resulting in four functional mPCRs detecting five to six species or haplotypes, each in a single reaction mixture. The sensitivity and specificity of these multiplex PCRs were checked as extensively as possible, although sample materials of species or haplotypes from the field were restricted. Among other things, this dilemma is caused by the comparatively low abundance of biting midges of the subgenus *Culicoides* as compared to those of the subgenus *Avaritia* (e.g., Obsoletus Group) in general and the over-representation of *C. pulicaris* and *C. punctatus* specimens within the subgenus *Culicoides* in particular in field collections from Germany [77–80]. In addition, in some taxa, only a few specimens have been found so far, probably due to the limited knowledge of their specific ecological niches and difficulties in identifying them. Missing materials for certain taxa were compensated for by using synthetic COI DNA.

In this study, PCR tests on four recently described species of the Pulicaris Group (*C. boyi*, *C. bysta*, *C. kalix* and *C. selandicus*) were developed for the first time, and all 21 newly designed specific forward primers for the subgenus *Culicoides* were able to identify their target taxa. In addition, 15 of the 21 forward primers showed no cross-reactivity with other members of the group if total DNA from single biting midges or equivalent (10^6 copies of synthetic COI gene) were tested, while six primers showed cross-talk within the subgenus *Culicoides* without a comprehensible explanation for how these primers were able to anneal to unspecific targets, especially if the number of mismatches (c.f. Table S4), the high annealing temperature and the use of hot-start Taq polymerase to reduce unspecific annealing are considered [57,58]. In one case, the forward primer for *C. delta* (del-COI-161F, mPCR C) produced a weak signal with the genomic DNA of the only *C. impunctatus* specimen available, although the primer sequence differed in 8 of the 23 bases, basically at the 3′-terminal part of the primer, from the sequence of the tested sample. Usually, such mismatch values prevent primer binding to unintended targets, and mismatches toward the 3′-end particularly hamper primer annealing [59]. Similar implausible results were obtained with the cry-COI-405F primer (mPCR C), incorrectly detecting one out of five *C. punctatus* samples (seven-base difference between primer and template sequence, with all tested samples having identical sequences) and the primer new3-COI-296F (mPCR D), incorrectly detecting four out of five *C. lupicaris* haplotype L2 samples (again, all with the same primer binding site sequence) despite mismatch pairing at seven positions. In another case, the primer (sel-COI-403F) only detected two out of three *C. selandicus* DNA samples despite the 100% sequence identity of all three specimens in the primer binding region. This false-negative result was attributed to the low DNA quantity of the non-identified sample. The DNA extraction of that one was performed with the NucleoMag VET Kit and a 100 μL elution volume, whereas the other two samples were processed with the QIAamp DNA Mini Kit and eluted in 50 μL of buffer. In the case of the only available *C. impunctatus* sample, which was non-specifically detected with the *C. delta*- and *C. newsteadi* haplotype N3-specific forward primers, a DNA extract from a previous study was used, which cannot be excluded without a doubt to have originated from a pool of biting midges. In order to circumvent such uncertainties, each DNA sample was generally sequenced before use; however, it might be possible that only the more abundant species within a mixed pool was determined. With respect to the other cross-reactions observed within the subgenus *Culicoides*, a plausible explanation cannot be found yet. Despite all preventive measures to avoid contamination and methodological measures to avoid unspecific annealing, as well as the application of the 'four eyes' principle during sample preparation, individual mistakes, including the confusion of tubes, cannot be ruled out.

Unspecific binding was also observed in several cases in which biting midge species not belonging to the subgenus *Culicoides* were tested. For instance, although in the primer pun-COI-227F (mPCR A), there are exchanges of two bases compared to *C. griseidorsum* DNA and three bases compared to *C. pictipennis* DNA, all of them exclusively in the middle and at the 5′-terminal part of the primer (Table S4), unspecific binding occurred with the DNA of these species. This is not unusual, as only a few mismatches in the middle or at the 5′-end of the primer do not necessarily lead to the complete loss of primer binding capacity, which is exploited, for example, in site-directed mutagenesis or the insertion of restriction sites [59]. However, the PCRs were meant to differentiate taxa within the subgenus *Culicoides*, and cross-reactivity with taxa not belonging to this subgenus appears to be extremely difficult to avoid, if not impossible, simply because of the huge number of taxa to be considered. These cases in fact demonstrate the importance of the morphological pre-sorting of biting midges to the group level before genetic examination.

Unfortunately, morphological pre-sorting is time-consuming and unsuitable for the high-throughput approaches needed to process the tremendous numbers of biting midges usually obtained from field collections. According to this, and considering that classical taxonomists are becoming an 'extinct species', there is a great need for finding alternative techniques for *Culicoides* classification. A biochemical method for species-specific protein profiling, Matrix-Assisted Laser Desorption/Ionization Time of Flight (MALDI-TOF), was suggested to soon become a reference method for arthropod identification [81], as it is rapid and cost-effective and does not require entomological expertise or special training [81]. Subsequently, several authors successfully used this method for the species identification of adult [82–86] and juvenile *Culicoides* specimens [87]. However, like all methods, MALDI-TOF has some disadvantages: (i) comparatively high costs for initial equipment [88,89], (ii) results do not provide evolutionary information for phylogenetic analyses [39], (iii) incongruent protein spectra, depending on the developmental stage of the insect [81,87] and the selected body parts used for analysis [81], (iv) no public library available with a sufficient number of reference protein profiles of each *Culicoides* species or haplotype and, finally, (v) relatively low resolution and limited sensitivity, insufficient to distinguish very closely related species [81]. Facing these obstacles, scientists need to think of more creative ways for the fast and accurate classification of *Culicoides*. Perhaps, artificial intelligence will help develop new identification methods in the future.

So far, PCR-based approaches seem to be the methods of choice, and in particular, quantitative real-time PCR is becoming increasingly important because of the possibility of analyzing pooled specimens [39]. Since there are indications that variations in the COI gene are insufficient for species delimitation within some subgenera [46,90], which could be confirmed by the present work, the implementation of multi-marker PCR approaches might be a great advantage, as already demonstrated in a recent study using several gene loci for phylogenetic analysis within the subgenus *Avaritia* [91].

Our results clearly demonstrate that developing multiplex PCR tests is a great challenge, merely based on the number of molecules used together in one reaction mixture and the multitude of possible interactions between them. Due to restricted availability, the newly developed tests were evaluated with a limited number of specimens and need further evaluation with additional samples, including pools of subgenus *Culicoides* biting midges, as tested during PCR development for the Obsoletus Group in a previous study [65]. However, the mPCR tests described here enable the parallel identification of almost all taxa of the subgenus *Culicoides* for the first time, among them recently described genetic variants and species not detectable with published PCR tests.

Moreover, it was attempted to keep the PCRs as simple as possible: in contrast to the PCR test of Pagès et al. [43], the utilization of one specific forward primer per species or haplotype, in combination with a universal reverse primer, decreases costs and simplifies the PCR evaluation. As opposed to previous PCRs developed for the Pulicaris Group and its relatives [43,50], the annealing temperature of the new mPCRs was generalized, thus applying a uniform PCR temperature profile and simplifying the experimental protocol. Depending on the aim of the study, each specific forward primer can also be used together with the universal reverse primer in a singleplex approach, which again reduces the cost per reaction. Each of the four mPCRs can be performed with fewer primers than suggested, but specific primers should not be mixed in other combinations in order to reduce primer-dimer formation and avoid the simultaneous production of amplicons indistinguishable by length. Despite the observed unspecific binding of individual primers, the first results with the PCRs are promising and indicate the great potential of our tests to improve the identification of suspected vector species within the subgenus *Culicoides* and the knowledge on biting midge distribution and ecology.

5. Conclusions

The results presented in this study confirm the great potential of the COI marker for species identification within the culicoid subgenus *Culicoides* [28,46]. The aim of the study, the development of PCR tests for the differentiation of species and haplotypes of the subgenus *Culicoides*, and the members of the Pulicaris Group in particular, was achieved through bioinformatic analysis of all COI sequences available from GenBank. This successful approach stresses the importance of such databases and resulted in different multiplex assays now becoming available to identify taxa of the subgenus *Culicoides*. A particular achievement of the assays is the inclusion of recently discovered species and haplotypes, for which no PCR identification tests have been available so far and whose ecologies and vector roles are completely unknown. Nonetheless, further testing with more specimens from field collections has to be performed to confirm the reproducibility and the benefit of the developed tests. Future analysis of the complete mitochondrial genome of *Culicoides* could significantly increase the possibilities of genetic differentiation and help unveil systematic issues.

Supplementary Materials: The following supporting information can be downloaded at https://www.mdpi.com/article/10.3390/d15060699/s1. Table S1: GenBank entries and consensus sequences used for the development of forward primers specific for the different taxa of the subgenus *Culicoides*; Table S2: Sequences of synthetic COI genes of subgenus *Culicoides* taxa used in this study; Table S3: List of all forward primers tested in this study; Table S4: Cross-talk of newly designed primers with the taxa of the subgenus *Culicoides* and of culicoid genera other than *Culicoides*.

Author Contributions: Conceptualization, O.D. and H.K.; methodology, O.D. and H.K.; validation, O.D.; formal analysis, O.D.; investigation, O.D. and H.K.; resources, H.K., D.W. and B.M.; data curation, O.D. and H.K.; writing—original draft preparation, O.D.; writing—review and editing, H.K., D.W. and B.M.; visualization, O.D.; supervision, H.K.; project administration, H.K. and D.W.; funding acquisition, H.K. and D.W. All authors have read and agreed to the published version of the manuscript.

Funding: This study was conducted as part of the CeratoVir project (grant number 281B101816), funded by the German Federal Ministry of Food and Agriculture (BMEL) through the Federal Office for Agriculture and Food (BLE).

Institutional Review Board Statement: Not applicable.

Data Availability Statement: Publicly available datasets were analyzed in this study and are listed in the Supplementary Material (Table S1). Respective data can be found in GenBank at: https://www.ncbi.nlm.nih.gov/genbank/ (accessed on 28 October 2022). New data created in this study are deposited in the same data repository with the following accession numbers: *C. impunctatus*: OQ789034; *C. delta*: OQ789035; *C. fagineus* haplotype F2: OQ789036; *C. grisescens* haplotype G1: OQ789037; *C. grisescens* haplotype G2: OQ789038; *C. lupicaris* haplotype L1: OQ789039; *C. lupicaris* haplotype L2: OQ789040-OQ789044; *C. newsteadi* haplotype N1: OQ789045-OQ789047; *C. newsteadi* haplotype N3: OQ789048-OQ789051; *C. selandicus*: OQ789052-OQ789054; *C. pulicaris*: OQ789055-OQ789060; *C. punctatus*: OQ789061-OQ789065; *C. achrayi*: OQ789066; *C. alazanicus*: OQ789067; *C. chiopterus*: OQ789068; *C. dewulfi*: OQ789069; *C. festivipennis*: OQ789070; *C. griseidorsum*: OQ789071; *C. imicola*: OQ789072; *C. kibunensis*: OQ789073; *C. montanus*: OQ789074; *C. obsoletus* clade O1: OQ789075; *C. obsoletus* clade O2: OQ789076; *C. obsoletus* clade O3: OQ789077; *C. pallidicornis*: OQ789078; *C. pictipennis*: OQ789079; *C. poperinghensis*: OQ789080; *C. riethi*: OQ789081; *C. riouxi*: OQ789082; *C. salinarius*: OQ789083; *C. scoticus* clade 2: OQ789084.

Acknowledgments: We thank our colleagues Sarah Groschupp, Christin Henneberg, Maria Will, Nathalie Richter and Sebastian Schuran from the Leibniz Centre for Agricultural Landscape Research (ZALF) for collecting and pre-sorting biting midges used in this study. We also thank Maxi Uecker and Freja Pfirschke for excellent technical assistance regarding the molecular biological work at FLI.

Conflicts of Interest: The authors declare no conflict of interest.

References

1. Mellor, P.S.; Boorman, J.; Baylis, M. *Culicoides* biting midges: Their role as arbovirus vectors. *Annu. Rev. Entomol.* **2000**, *45*, 307–340. [CrossRef]
2. Meiswinkel, R.; Gomulski, L.M.; Delécolle, J.-C.; Goffredo, M.; Gasperi, G. The taxonomy of *Culicoides* vector complexes-unfinished business. *Vet. Ital.* **2004**, *40*, 151–159.
3. Borkent, A. The biting midges, the Ceratopogonidae (Diptera). In *Biology of Disease Vectors*, 2nd ed.; Marquardt, W., Ed.; Elsevier: Amsterdam, The Netherlands, 2005; pp. 113–126.
4. Mellor, P.S.; Pitzolis, G. Observations on breeding sites and light-trap collections of *Culicoides* during an outbreak of bluetongue in Cyprus. *Bull. Entomol. Res.* **1979**, *69*, 229–234. [CrossRef]
5. Mellor, P.S.; Boned, J.; Hamblin, C.; Graham, S. Isolations of African horse sickness virus from vector insects made during the 1988 epizootic in Spain. *Epidemiol. Infect.* **1990**, *105*, 447–454. [CrossRef]
6. Mehlhorn, H.; Walldorf, V.; Klimpel, S.; Jahn, B.; Jaeger, F.; Eschweiler, J.; Hoffmann, B.; Beer, M. First occurrence of *Culicoides obsoletus*-transmitted Bluetongue virus epidemic in Central Europe. *Parasitol. Res.* **2007**, *101*, 219–228. [CrossRef]
7. Meiswinkel, R.; van Rijn, P.; Leijs, P.; Goffredo, M. Potential new *Culicoides* vector of bluetongue virus in northern Europe. *Vet. Rec.* **2007**, *161*, 564–565. [CrossRef]
8. Dijkstra, E.; van der Ven, I.J.K.; Meiswinkel, R.; Hölzel, D.R.; van Rijn, P.A. *Culicoides chiopterus* as a potential vector of bluetongue virus in Europe. *Vet. Rec.* **2008**, *162*, 422. [CrossRef]
9. Hoffmann, B.; Bauer, B.; Bauer, C.; Bätza, H.J.; Beer, M.; Clausen, P.H.; Geier, M.; Gethmann, J.M.; Kiel, E.; Liebisch, G.; et al. Monitoring of putative vectors of bluetongue virus serotype 8, Germany. *Emerg. Infect. Dis.* **2009**, *15*, 1481–1484. [CrossRef]
10. Vanbinst, T.; Vandenbussche, F.; Vandemeulebroucke, E.; de Leeuw, I.; Deblauwe, I.; de Deken, G.; Madder, M.; Haubruge, E.; Losson, B.; de Clercq, K. Bluetongue virus detection by real-time RT-PCR in *Culicoides* captured during the 2006 epizootic in Belgium and development of an internal control. *Transbound. Emerg. Dis.* **2009**, *56*, 170–177. [CrossRef]
11. Romón, P.; Higuera, M.; Delécolle, J.-C.; Baldet, T.; Aduriz, G.; Goldarazena, A. Phenology and attraction of potential *Culicoides* vectors of bluetongue virus in Basque Country (northern Spain). *Vet. Parasitol.* **2012**, *186*, 415–424. [CrossRef]
12. Goffredo, M.; Catalani, M.; Federici, V.; Portanti, O.; Marini, V.; Mancini, G.; Quaglia, M.; Santilli, A.; Teodori, L.; Savini, G. Vector species of *Culicoides* midges implicated in the 2012–2014 Bluetongue epidemics in Italy. *Vet. Ital.* **2015**, *51*, 131–138. [CrossRef]
13. Foxi, C.; Delrio, G.; Falchi, G.; Marche, M.G.; Satta, G.; Ruiu, L. Role of different *Culicoides* vectors (Diptera: Ceratopogonidae) in bluetongue virus transmission and overwintering in Sardinia (Italy). *Parasit. Vectors* **2016**, *9*, 440. [CrossRef]
14. Foxi, C.; Meloni, G.; Puggioni, G.; Manunta, D.; Rocchigiani, A.; Vento, L.; Cabras, P.; Satta, G. Bluetongue virus detection in new *Culicoides* species in Sardinia, Italy. *Vet. Rec.* **2019**, *184*, 621. [CrossRef]
15. Rasmussen, L.D.; Kristensen, B.; Kirkeby, C.; Rasmussen, T.B.; Belsham, G.J.; Bødker, R.; Bøtner, A. Culicoids as vectors of Schmallenberg virus. *Emerg. Infect. Dis.* **2012**, *18*, 1204–1206. [CrossRef]
16. de Regge, N.; Deblauwe, I.; de Deken, R.; Vantieghem, P.; Madder, M.; Geysen, D.; Smeets, F.; Losson, B.; van den Berg, T.; Cay, A.B. Detection of Schmallenberg virus in different *Culicoides* spp. by real-time RT-PCR. *Transbound. Emerg. Dis.* **2012**, *59*, 471–475. [CrossRef]
17. Elbers, A.R.W.; Meiswinkel, R.; van Weezep, E.; van Sloet Oldruitenborgh-Oosterbaan, M.M.; Kooi, E.A. Schmallenberg virus in *Culicoides* spp. biting midges, the Netherlands, 2011. *Emerg. Infect. Dis.* **2013**, *19*, 106–109. [CrossRef]
18. Goffredo, M.; Monaco, F.; Capelli, G.; Quaglia, M.; Federici, V.; Catalani, M.; Montarsi, F.; Polci, A.; Pinoni, C.; Calistri, P.; et al. Schmallenberg virus in Italy: A retrospective survey in *Culicoides* stored during the bluetongue Italian surveillance program. *Prev. Vet. Med.* **2013**, *111*, 230–236. [CrossRef]
19. Larska, M.; Polak, M.P.; Grochowska, M.; Lechowski, L.; Związek, J.S.; Zmudziński, J.F. First report of Schmallenberg virus infection in cattle and midges in Poland. *Transbound. Emerg. Dis.* **2013**, *60*, 97–101. [CrossRef]
20. Larska, M.; Lechowski, L.; Grochowska, M.; Żmudziński, J.F. Detection of the Schmallenberg virus in nulliparous *Culicoides obsoletus/scoticus* complex and *C. punctatus*—The possibility of transovarial virus transmission in the midge population and of a new vector. *Vet. Microbiol.* **2013**, *166*, 467–473. [CrossRef]
21. Balenghien, T.; Pagès, N.; Goffredo, M.; Carpenter, S.; Augot, D.; Jacquier, E.; Talavera, S.; Monaco, F.; Depaquit, J.; Grillet, C.; et al. The emergence of Schmallenberg virus across *Culicoides* communities and ecosystems in Europe. *Prev. Vet. Med.* **2014**, *116*, 360–369. [CrossRef]
22. Rasmussen, L.D.; Kirkeby, C.; Bødker, R.; Kristensen, B.; Rasmussen, T.B.; Belsham, G.J.; Bøtner, A. Rapid spread of Schmallenberg virus-infected biting midges (*Culicoides* spp.) across Denmark in 2012. *Transbound. Emerg. Dis.* **2014**, *61*, 12–16. [CrossRef]
23. Elbers, A.R.W.; Meiswinkel, R.; van Weezep, E.; Kooi, E.A.; van der Poel, W.H.M. Schmallenberg virus in *Culicoides* biting midges in the Netherlands in 2012. *Transbound. Emerg. Dis.* **2015**, *62*, 339–342. [CrossRef]
24. de Regge, N.; de Deken, R.; Fassotte, C.; Losson, B.; Deblauwe, I.; Madder, M.; Vantieghem, P.; Tomme, M.; Smeets, F.; Cay, A.B. *Culicoides* monitoring in Belgium in 2011: Analysis of spatiotemporal abundance, species diversity and Schmallenberg virus detection. *Med. Vet. Entomol.* **2015**, *29*, 263–275. [CrossRef]
25. Pagès, N.; Talavera, S.; Verdún, M.; Pujol, N.; Valle, M.; Bensaid, A.; Pujols, J. Schmallenberg virus detection in *Culicoides* biting midges in Spain: First laboratory evidence for highly efficient infection of *Culicoides* of the Obsoletus Complex and *Culicoides imicola*. *Transbound. Emerg. Dis.* **2018**, *65*, e1–e6. [CrossRef]

26. Ségard, A.; Gardès, L.; Jacquier, E.; Grillet, C.; Mathieu, B.; Rakotoarivony, I.; Setier-Rio, M.-L.; Chavernac, D.; Cêtre-Sossah, C.; Balenghien, T.; et al. Schmallenberg virus in *Culicoides* Latreille (Diptera: Ceratopogonidae) populations in France during 2011–2012 outbreak. *Transbound. Emerg. Dis.* **2018**, *65*, e94–e103. [CrossRef]
27. Borkent, A.; Dominiak, P. Catalog of the biting midges of the world (Diptera: Ceratopogonidae). *Zootaxa* **2020**, *4787*, 1–377. [CrossRef]
28. Lassen, S.B.; Nielsen, S.A.; Skovgård, H.; Kristensen, M. Molecular differentiation of *Culicoides* biting midges (Diptera: Ceratopogonidae) from the subgenus *Culicoides* Latreille in Denmark. *Parasitol. Res.* **2012**, *110*, 1765–1771. [CrossRef]
29. Ramilo, D.W.; Diaz, S.; Da Pereira Fonseca, I.; Delécolle, J.-C.; Wilson, A.; Meireles, J.; Lucientes, J.; Ribeiro, R.; Boinas, F. First report of 13 species of *Culicoides* (Diptera: Ceratopogonidae) in mainland Portugal and Azores by morphological and molecular characterization. *PLoS ONE* **2012**, *7*, e34896. [CrossRef]
30. Ramilo, D.; Garros, C.; Mathieu, B.; Benedet, C.; Allène, X.; Silva, E.; Alexandre-Pires, G.; Da Fonseca, I.P.; Carpenter, S.; Rádrová, J.; et al. Description of *Culicoides paradoxalis* sp. nov. from France and Portugal (Diptera: Ceratopogonidae). *Zootaxa* **2013**, *3745*, 243–256. [CrossRef]
31. Talavera, S.; Muñoz-Muñoz, F.; Verdún, M.; Pagès, N. Morphology and DNA barcoding reveal three species in one: Description of *Culicoides cryptipulicaris* sp. nov. and *Culicoides quasipulicaris* sp. nov. in the subgenus *Culicoides*. *Med. Vet. Entomol.* **2017**, *31*, 178–191. [CrossRef]
32. Goffredo, M.; Meiswinkel, R. Entomological surveillance of bluetongue in Italy: Methods of capture, catch analysis and identification of *Culicoides* biting midges. *Vet. Ital.* **2004**, *40*, 260–265. [PubMed]
33. Meiswinkel, R.; Baldet, T.; de Deken, R.; Takken, W.; Delécolle, J.-C.; Mellor, P.S. The 2006 outbreak of bluetongue in northern Europe–the entomological perspective. *Prev. Vet. Med.* **2008**, *87*, 55–63. [CrossRef]
34. Schwenkenbecher, J.M.; Mordue, A.J.; Switek, K.; Piertney, S.B. Discrimination of *Culicoides* midge larvae using multiplex polymerase chain reaction assays based on DNA sequence variation at the mitochondrial cytochrome C oxidase I gene. *J. Med. Entomol.* **2009**, *46*, 610–614. [CrossRef]
35. Yanase, T.; Matsumoto, Y.; Matsumori, Y.; Aizawa, M.; Hirata, M.; Kato, T.; Shirafuji, H.; Yamakawa, M.; Tsuda, T.; Noda, H. Molecular identification of field-collected *Culicoides* larvae in the southern part of Japan. *J. Med. Entomol.* **2013**, *50*, 1105–1110. [CrossRef]
36. Jin, T.; Husseneder, C.; Foil, L. Assigning *Culicoides* larvae to species using DNA barcoding of adult females and phylogenetic associations. *Parasit. Vectors* **2022**, *15*, 349. [CrossRef]
37. Schwenkenbecher, J.M.; Mordue, A.J.; Piertney, S.B. Phylogenetic analysis indicates that *Culicoides dewulfi* should not be considered part of the *Culicoides obsoletus* complex. *Bull. Entomol. Res.* **2009**, *99*, 371–375. [CrossRef]
38. Delécolle, J.-C. Nouvelle Contribution à L'étude Systématique et Iconographique des Espèces du Genre *Culicoides* (Diptera: Ceratopogonidae) du Nord-Est de la France. Ph.D. Thesis, University of Strasbourg, Strasbourg, France, 1985.
39. Harrup, L.E.; Bellis, G.A.; Balenghien, T.; Garros, C. *Culicoides* Latreille (Diptera: Ceratopogonidae) taxonomy: Current challenges and future directions. *Infect. Genet. Evol.* **2015**, *30*, 249–266. [CrossRef]
40. Lunt, D.H.; Zhang, D.X.; Szymura, J.M.; Hewitt, G.M. The insect cytochrome oxidase I gene: Evolutionary patterns and conserved primers for phylogenetic studies. *Insect Mol. Biol.* **1996**, *5*, 153–165. [CrossRef]
41. Dobler, S.; Farrell, B.D. Host use evolution in *Chrysochus* milkweed beetles: Evidence from behaviour, population genetics and phylogeny. *Mol. Ecol.* **1999**, *8*, 1297–1307. [CrossRef]
42. Hebert, P.D.N.; Cywinska, A.; Ball, S.L.; deWaard, J.R. Biological identifications through DNA barcodes. *Proc. Biol. Sci.* **2003**, *270*, 313–321. [CrossRef]
43. Pagès, N.; Muñoz-Muñoz, F.; Talavera, S.; Sarto, V.; Lorca, C.; Núñez, J.I. Identification of cryptic species of *Culicoides* (Diptera: Ceratopogonidae) in the subgenus *Culicoides* and development of species-specific PCR assays based on barcode regions. *Vet. Parasitol.* **2009**, *165*, 298–310. [CrossRef] [PubMed]
44. Nielsen, S.A.; Kristensen, M. Delineation of *Culicoides* species by morphology and barcode exemplified by three new species of the subgenus *Culicoides* (Diptera: Ceratopogonidae) from Scandinavia. *Parasit. Vectors* **2015**, *8*, 151. [CrossRef] [PubMed]
45. Wenk, C.E.; Kaufmann, C.; Schaffner, F.; Mathis, A. Molecular characterization of Swiss Ceratopogonidae (Diptera) and evaluation of real-time PCR assays for the identification of *Culicoides* biting midges. *Vet. Parasitol.* **2012**, *184*, 258–266. [CrossRef] [PubMed]
46. Ander, M.; Troell, K.; Chirico, J. Barcoding of biting midges in the genus *Culicoides*: A tool for species determination. *Med. Vet. Entomol.* **2013**, *27*, 323–331. [CrossRef]
47. Sarvašová, A.; Kočišová, A.; Halán, M.; Delécolle, J.-C.; Mathieu, B. Morphological and molecular analysis of the genus *Culicoides* (Diptera: Ceratopogonidae) in Slovakia with five new records. *Zootaxa* **2014**, *3872*, 541–560. [CrossRef]
48. Yildirim, A.; Dik, B.; Duzlu, O.; Onder, Z.; Ciloglu, A.; Yetismis, G.; Inci, A. Genetic diversity of *Culicoides* species within the Pulicaris Complex (Diptera: Ceratopogonidae) in Turkey inferred from mitochondrial COI gene sequences. *Acta Trop.* **2019**, *190*, 380–388. [CrossRef]
49. Zittra, C.; Wöss, G.; van der Vloet, L.; Bakran-Lebl, K.; Shahi Barogh, B.; Sehnal, P.; Fuehrer, H.-P. Barcoding of the genus *Culicoides* (Diptera: Ceratopogonidae) in Austria—An update of the species inventory including the first records of three species in Austria. *Pathogens* **2020**, *9*, 406. [CrossRef]

50. Nolan, D.V.; Carpenter, S.; Barber, J.; Mellor, P.S.; Dallas, J.F.; Mordue Luntz, A.J.; Piertney, S.B. Rapid diagnostic PCR assays for members of the *Culicoides obsoletus* and *Culicoides pulicaris* species complexes, implicated vectors of bluetongue virus in Europe. *Vet. Microbiol.* **2007**, *124*, 82–94. [CrossRef]
51. Bellis, G.; Kim, H.-C.; Kim, M.-S.; Klein, T.A.; Lee, D.-K.; Gopurenko, D. Three species of *Culicoides* Latreille (Diptera: Ceratopogonidae) newly recorded from the Republic of Korea. *Zootaxa* **2013**, *3718*, 171–182. [CrossRef]
52. Rot, A.; Meiswinkel, R.; Fleker, M.; Blum, S.E.; Behar, A. Towards modernizing the taxonomy of Mediterranean *Culicoides* using classical morphology, mtDNA barcoding, and MALDI-TOF MS protein profiling. *Acta Trop.* **2020**, *211*, 105628. [CrossRef]
53. Campbell, J.A.; Pelham-Clinton, E.C. A taxonomic review of the British species of *Culicoides* Latreille (Diptera, Ceratopogonidæ). *Proc. R. Soc. Edinb. Sect. B Biol. Sci.* **1960**, *67*, 181–302. [CrossRef]
54. Glukhova, V.M. *Krovososuščie Mokrecy Rodov Culicoides i Forcipomyia (Ceratopogonidae)*; Nauka: Leningrad, Russia, 1989; ISBN 9785020257603.
55. Mathieu, B.; Cêtre-Sossah, C.; Garros, C.; Chavernac, D.; Balenghien, T.; Carpenter, S.; Setier-Rio, M.L.; Vignes-Lebbe, R.; Ung, V.; Candolfi, E.; et al. Development and validation of IIKC: An interactive identification key for *Culicoides* (Diptera: Ceratopogonidae) females from the Western Palaearctic region. *Parasit. Vectors* **2012**, *5*, 137. [CrossRef] [PubMed]
56. Lehmann, K.; Werner, D.; Hoffmann, B.; Kampen, H. PCR identification of culicoid biting midges (Diptera, Ceratopogonidae) of the Obsoletus Complex including putative vectors of bluetongue and Schmallenberg viruses. *Parasit. Vectors* **2012**, *5*, 213. [CrossRef] [PubMed]
57. Henegariu, O.; Heerema, N.A.; Dlouhy, S.R.; Vance, G.H.; Vogt, P.H. Multiplex PCR: Critical parameters and step-by-step protocol. *Biotechniques* **1997**, *23*, 504–511. [CrossRef]
58. Elnifro, E.M.; Ashshi, A.M.; Cooper, R.J.; Klapper, P.E. Multiplex PCR: Optimization and application in diagnostic virology. *Cli. Microbiol. Rev.* **2000**, *13*, 559–570. [CrossRef]
59. Newton, C.R.; Graham, A. *PCR*; Spektrum Akad. Verl.: Heidelberg, Germany, 1994; ISBN 3-86025-236-4.
60. Pagès, N.; Sarto i Monteys, V. Differentiation of *Culicoides obsoletus* and *Culicoides scoticus* (Diptera: Ceratopogonidae) based on mitochondrial cytochrome oxidase subunit I. *J. Med. Entomol.* **2005**, *42*, 1026–1034. [CrossRef]
61. Cêtre-Sossah, C.; Baldet, T.; Delécolle, J.-C.; Mathieu, B.; Perrin, A.; Grillet, C.; Albina, E. Molecular detection of *Culicoides* spp. and *Culicoides imicola*, the principal vector of bluetongue (BT) and African horse sickness (AHS) in Africa and Europe. *Vet. Res.* **2004**, *35*, 325–337. [CrossRef]
62. Cêtre-Sossah, C.; Mathieu, B.; Setier-Rio, M.-L.; Grillet, C.; Baldet, T.; Delécolle, J.-C.; Albina, E. Development and evaluation of a real-time quantitative PCR assay for *Culicoides imicola*, one of the main vectors of bluetongue (BT) and African horse sickness (AHS) in Africa and Europe. *Res. Vet. Sci.* **2008**, *85*, 372–382. [CrossRef]
63. Mathieu, B.; Perrin, A.; Baldet, T.; Delécolle, J.-C.; Albina, E.; Cêtre-Sossah, C. Molecular identification of western European species of Obsoletus Complex (Diptera: Ceratopogonidae) by an internal transcribed spacer-1 rDNA multiplex polymerase chain reaction assay. *J. Med. Entomol.* **2007**, *44*, 1019–1025. [CrossRef]
64. Stephan, A.; Clausen, P.-H.; Bauer, B.; Steuber, S. PCR identification of *Culicoides dewulfi* midges (Diptera: Ceratopogonidae), potential vectors of bluetongue in Germany. *Parasitol. Res.* **2009**, *105*, 367–371. [CrossRef]
65. Mathieu, B.; Delecolle, J.-C.; Garros, C.; Balenghien, T.; Setier-Rio, M.-L.; Candolfi, E.; Cêtre-Sossah, C. Simultaneous quantification of the relative abundance of species complex members: Application to *Culicoides obsoletus* and *Culicoides scoticus* (Diptera: Ceratopogonidae), potential vectors of bluetongue virus. *Vet. Parasitol.* **2011**, *182*, 297–306. [CrossRef] [PubMed]
66. Gomulski, L.M.; Meiswinkel, R.; Delécolle, J.C.; Goffredo, M.; Gasperi, G. Phylogenetic relationships of the subgenus *Avaritia* Fox, 1955 including *Culicoides obsoletus* (Diptera, Ceratopogonidae) in Italy based on internal transcribed spacer 2 ribosomal DNA sequences. *Syst. Entomol.* **2005**, *30*, 619–631. [CrossRef]
67. Monaco, F.; Benedetto, L.; Di Marcello, V.; Lelli, R.; Goffredo, M. Development and preliminary evaluation of a real-time polymerase chain reaction for the identification of *Culicoides obsoletus* sensu strictu, *C. scoticus* and *C. montanus* in the Obsoletus Complex in Italy. *Vet. Ital.* **2010**, *46*, 215–220.
68. Meyer, C.P.; Paulay, G. DNA barcoding: Error rates based on comprehensive sampling. *PLoS Biol.* **2005**, *3*, e422. [CrossRef]
69. Dallas, J.F.; Cruickshank, R.H.; Linton, Y.-M.; Nolan, D.V.; Patakakis, M.; Braverman, Y.; Capela, R.; Capela, M.; Pena, I.; Meiswinkel, R.; et al. Phylogenetic status and matrilineal structure of the biting midge, *Culicoides imicola*, in Portugal, Rhodes and Israel. *Med. Vet. Entomol.* **2003**, *17*, 379–387. [CrossRef] [PubMed]
70. Kwok, S.; Kellogg, D.E.; McKinney, N.; Spasic, D.; Goda, L.; Levenson, C.; Sninsky, J.J. Effects of primer template mismatches on the polymerase chain reaction: Human immunodeficiency virus type 1 model studies. *Nucleic Acids Res.* **1990**, *18*, 999–1005. [CrossRef] [PubMed]
71. Mendelman, L.V.; Petruska, J.; Goodman, M.F. Base mispair extension kinetics. Comparison of DNA polymerase alpha and reverse transcriptase. *J. Biol. Chem.* **1990**, *265*, 2338–2346. [CrossRef] [PubMed]
72. Creighton, S.; Huang, M.M.; Cai, H.; Arnheim, N.; Goodman, M.F. Base mispair extension kinetics. Binding of avian myeloblastosis reverse transcriptase to matched and mismatched base pair termini. *J. Biol. Chem.* **1992**, *267*, 2633–2639. [CrossRef]
73. Huang, M.M.; Arnheim, N.; Goodman, M.F. Extension of base mispairs by Taq DNA polymerase: Implications for single nucleotide discrimination in PCR. *Nucleic Acids Res.* **1992**, *20*, 4567–4573. [CrossRef]
74. Joyce, C.M.; Sun, X.C.; Grindley, N.D. Reactions at the polymerase active site that contribute to the fidelity of *Escherichia coli* DNA polymerase I (Klenow fragment). *J. Biol. Chem.* **1992**, *267*, 24485–24500. [CrossRef]

75. Day, J.P.; Bergstrom, D.; Hammer, R.P.; Barany, F. Nucleotide analogs facilitate base conversion with 3′ mismatch primers. *Nucleic Acids Res.* **1999**, *27*, 1810–1818. [CrossRef]
76. Stadhouders, R.; Pas, S.D.; Anber, J.; Voermans, J.; Mes, T.H.M.; Schutten, M. The effect of primer-template mismatches on the detection and quantification of nucleic acids using the 5′ nuclease assay. *J. Mol. Diagn.* **2010**, *12*, 109–117. [CrossRef]
77. Kiel, E.; Liebisch, G.; Focke, R.; Liebisch, A.; Werner, D. Monitoring of *Culicoides* at 20 locations in northwest Germany. *Parasitol. Res.* **2009**, *105*, 351–357. [CrossRef] [PubMed]
78. Mehlhorn, H.; Walldorf, V.; Klimpel, S.; Schaub, G.; Kiel, E.; Focke, R.; Liebisch, G.; Liebisch, A.; Werner, D.; Bauer, C.; et al. Bluetongue disease in Germany (2007–2008): Monitoring of entomological aspects. *Parasitol. Res.* **2009**, *105*, 313–319. [CrossRef]
79. Vorsprach, B.; Meiser, C.K.; Werner, D.; Balczun, C.; Schaub, G.A. Monitoring of Ceratopogonidae in southwest Germany. *Parasitol. Res.* **2009**, *105*, 337–344. [CrossRef] [PubMed]
80. Werner, D.; Groschupp, S.; Bauer, C.; Kampen, H. Breeding habitat preferences of major *Culicoides* species (Diptera: Ceratopogonidae) in Germany. *Int. J. Environ. Res. Public Health* **2020**, *17*, 5000. [CrossRef] [PubMed]
81. Yssouf, A.; Almeras, L.; Raoult, D.; Parola, P. Emerging tools for identification of arthropod vectors. *Future Microbiol.* **2016**, *11*, 549–566. [CrossRef]
82. Kaufmann, C.; Ziegler, D.; Schaffner, F.; Carpenter, S.; Pflüger, V.; Mathis, A. Evaluation of matrix-assisted laser desorption/ionization time of flight mass spectrometry for characterization of *Culicoides nubeculosus* biting midges. *Med. Vet. Entomol.* **2011**, *25*, 32–38. [CrossRef]
83. Kaufmann, C.; Schaffner, F.; Ziegler, D.; Pflüger, V.; Mathis, A. Identification of field-caught *Culicoides* biting midges using matrix-assisted laser desorption/ionization time of flight mass spectrometry. *Parasitology* **2012**, *139*, 248–258. [CrossRef]
84. Kaufmann, C.; Steinmann, I.C.; Hegglin, D.; Schaffner, F.; Mathis, A. Spatio-temporal occurrence of *Culicoides* biting midges in the climatic regions of Switzerland, along with large scale species identification by MALDI-TOF mass spectrometry. *Parasit. Vectors* **2012**, *5*, 246. [CrossRef]
85. Sambou, M.; Aubadie-Ladrix, M.; Fenollar, F.; Fall, B.; Bassene, H.; Almeras, L.; Sambe-Ba, B.; Perrot, N.; Chatellier, S.; Faye, N.; et al. Comparison of matrix-assisted laser desorption ionization-time of flight mass spectrometry and molecular biology techniques for identification of *Culicoides* (Diptera: Ceratopogonidae) biting midges in Senegal. *J. Clin. Microbiol.* **2015**, *53*, 410–418. [CrossRef]
86. Uhlmann, K.R.; Gibb, S.; Kalkhof, S.; Arroyo-Abad, U.; Schulz, C.; Hoffmann, B.; Stubbins, F.; Carpenter, S.; Beer, M.; von Bergen, M.; et al. Species determination of *Culicoides* biting midges via peptide profiling using matrix-assisted laser desorption ionization mass spectrometry. *Parasit. Vectors* **2014**, *7*, 392. [CrossRef] [PubMed]
87. Steinmann, I.C.; Pflüger, V.; Schaffner, F.; Mathis, A.; Kaufmann, C. Evaluation of matrix-assisted laser desorption/ionization time of flight mass spectrometry for the identification of ceratopogonid and culicid larvae. *Parasitology* **2013**, *140*, 318–327. [CrossRef] [PubMed]
88. Singhal, N.; Kumar, M.; Kanaujia, P.K.; Virdi, J.S. MALDI-TOF mass spectrometry: An emerging technology for microbial identification and diagnosis. *Front. Microbiol.* **2015**, *6*, 791. [CrossRef] [PubMed]
89. Yssouf, A.; Socolovschi, C.; Flaudrops, C.; Ndiath, M.O.; Sougoufara, S.; Dehecq, J.-S.; Lacour, G.; Berenger, J.-M.; Sokhna, C.S.; Raoult, D.; et al. Matrix-assisted laser desorption ionization—Time of flight mass spectrometry: An emerging tool for the rapid identification of mosquito vectors. *PLoS ONE* **2013**, *8*, e72380. [CrossRef]
90. Bellis, G. Studies on the Taxonomy of Australian Species of *Culicoides* Latreille (Diptera: Ceratopogonidae). Ph.D. Thesis, University of Queensland, St Lucia, Australia, 2013.
91. Mignotte, A.; Garros, C.; Gardès, L.; Balenghien, T.; Duhayon, M.; Rakotoarivony, I.; Tabourin, L.; Poujol, L.; Mathieu, B.; Ibañez-Justicia, A.; et al. The tree that hides the forest: Cryptic diversity and phylogenetic relationships in the Palaearctic vector Obsoletus/Scoticus Complex (Diptera: Ceratopogonidae) at the European level. *Parasit. Vectors* **2020**, *13*, 265. [CrossRef]

diversity

MDPI

Article

North–South Differentiation of Black Flies in the Western Cordillera of North America: A New Species of *Prosimulium* (Diptera: Simuliidae) †

Peter H. Adler [1,*] and Will K. Reeves [2]

1 Department of Plant and Environmental Sciences, Clemson University, Clemson, SC 29634-0310, USA
2 C.P. Gillette Museum of Arthropod Diversity, Fort Collins, CO 80528, USA
* Correspondence: padler@clemson.edu
† urn:lsid:zoobank.org:pub: 7BBA84B5-BD26-4001-8E66-8FF7FE1AEBFA.

Abstract: Glaciation has been a powerful determiner of species distributions and the genetic structure of populations. Contemporary distributions of many organisms in North America's Western Cordillera reflect the influence of Pleistocene glaciation. We identified a pattern of north–south differentiation in the genus *Prosimulium* of western North America, which reflects the separation of northern and southern populations by the North American Ice Sheet during the Pleistocene Epoch. The taxonomic implication is that new species exist within nominal species, requiring formal description or revalidation of names currently in synonymy. We morphologically and cytogenetically examined populations of one nominal species of black fly, *Prosimulium esselbaughi* Sommerman, over its known range from Alaska south to California and Colorado. Chromosomal and morphological evidence supports the presence of two species, *P. esselbaughi* sensu stricto from Alaska to at least southern British Columbia, and a new species, *Prosimulium supernum* in the central Rocky Mountains and high Sierra Nevada range of the United States. The new species is described in all life stages above the egg, along with its polytene chromosomes. The existence of differentiated populations of other nominal species of black flies in northern and southern North America provides a system for investigating possible co-differentiation of vectors and parasites.

Keywords: aquatic insects; cytogenetics; glaciation; Pleistocene; Rocky Mountains; speciation

Citation: Adler, P.H.; Reeves, W.K. North–South Differentiation of Black Flies in the Western Cordillera of North America: A New Species of *Prosimulium* (Diptera: Simuliidae). *Diversity* **2023**, *15*, 212. https://doi.org/10.3390/d15020212

Academic Editor: Michael Wink

Received: 17 December 2022
Revised: 26 January 2023
Accepted: 28 January 2023
Published: 2 February 2023

1. Introduction

Mountains provide rich opportunities for population isolation and, therefore, are often hotspots for speciation, biodiversity, and endemism. North America's vast Western Cordillera, running from the Brook's Range of Alaska southward through the Rockies, Cascades, Coast Ranges, and Sierra Nevada, has an ancient history of tectonic influences [1,2]. Superimposed on the orogenic consequences have been cyclical glacial and interglacial periods, most recently the Last Glacial Period (ca. 115,000–12,000 BP) with its Late Glacial Maximum about 26,500 to 19–20,000 BP [3]. Species distributions and genetic structure have been profoundly influenced by glaciation cycles, particularly during the Quaternary Period of the past 2.6 million years [4–6].

Among the cold-adapted organisms that inhabit the Western Cordillera are members of the dipteran family Simuliidae. The family's oldest extant lineages are in the Holarctic Region, reflecting the hypothesis that these flies evolved in cool, mountainous areas [7]. One of the oldest extant lineages is the Holarctic genus *Prosimulium*, consisting of 80 nominal species, of which 25 inhabit the Western Cordillera, including 23 that are precinctive to this mountain chain [8]. Only two of the 80 species of *Prosimulium* are found in both the Nearctic and Palearctic regions [8]. All species of *Prosimulium* are cold-adapted, particularly the high-elevation inhabitants, and therefore are potential sentinels for monitoring climate change. The female flies of *Prosimulium* are mammalophilic, and those of some species have been incriminated as vectors of onchocercid worms [9,10].

The extensive ranges of many western nominal *Prosimulium* species from northern Alaska southward through the Rocky Mountains and Sierra Nevada, coupled with genetic differences, have suggested the presence of multiple species [11,12]. Populations in California's Sierra Nevada, for example, were tentatively identified as *Prosimulium esselbaughi* Sommerman, with the caveat that they might represent a new species, based on novel sex-linkage of chromosomal rearrangements [11]. Our discovery of chromosomally and morphologically similar populations in the central Rocky Mountains led to an evaluation of the Rocky Mountain and Sierra Nevada populations and the conclusion that they represent a new species. We describe these populations as a new species, based on chromosomal band patterns and morphology of the female, male, pupa, and larva. We highlight the geographic distributions of the new species and its close relative and discuss the general pattern of north–south differentiation of *Prosimulium* species in western North America.

2. Materials and Methods

2.1. Collection Sites

At the type locality of the new species in Wyoming (Figure 1A,B), larvae and pupae were collected in the North Fork drainage basin of the Little Laramie River in Albany County (Table 1). Larvae and pupae were taken from trailing vegetation and stones in 2007 and 2008 and from black plastic bags anchored in the stream on 1 June 2008. Holes in the ice allowed access to the stream. The plastic bags were used as attachment substrates for larvae and particularly for pupae that are typically concealed in sediment. The plastic was checked weekly through the third week of July. Larvae were collected into Carnoy's fixative (3 parts 99% ethanol: 1 part glacial acetic acid) and transferred to fresh fixative 2–4 h later. Pupae were collected by cutting the plastic around them and placing them in Petri dishes with moist filter paper. Adults were allowed to emerge from the pupae in the laboratory (elevation 2184 m) and were dispatched 12–24 h later by freezing. Pupal exuviae and cocoons were associated with each adult.

Larvae in Colorado were collected from the Michigan River and Cache La Poudre drainages in Jackson County (Figure 1C,D) in June 2022 and fixed in Carnoy's fixative, which was refreshed 4–8 h later (Table 1). Plastic bags were placed in the streams in June and retrieved in July 2022, but no larvae or pupae were attached to them.

Table 1. Sites from which type material was collected for *Prosimulium supernum* n. sp. in the Rocky Mountains of Colorado and Wyoming.

Site	Location (Stream Width)	Latitude Longitude	Elevation (m above Sea Level)	Date	Life Stage
1a	WY, Albany County, Snowy Range Pass, Libby Creek (2.0–2.5 m)	41°21′07″ N 106°17′01″ W	3233	11 June 2007	13 larvae (4 chromosome preparations)
1b				12 June 2008	2 larvae (2 chromosome preparations)
1c				7 July 2008	7 pupae, 2 males and 5 females with pupal exuviae
1d				17 July 2008	4 pupae, 2 females with pupal exuviae
2	CO, Jackson County, near Cameron Pass, snowmelt trib. Michigan River (0.5–1.0 m)	40°30′31″ N 105°53′05″ W	3089	11 June 2022	6 larvae (6 chromosome preparations)
3	CO, Jackson County, trib. Michigan River (0.5–1.0 m)	40°30′56″ N 105°53′11″ W	3139	20 June 2022	2 larvae (2 chromosome preparations)
4	CO, Jackson County, Cameron Pass, Michigan Ditch (2.0–2.5 m)	40°31′13″ N 105°53′32″ W	3135	20 June 2022	2 larvae (2 chromosome preparations)

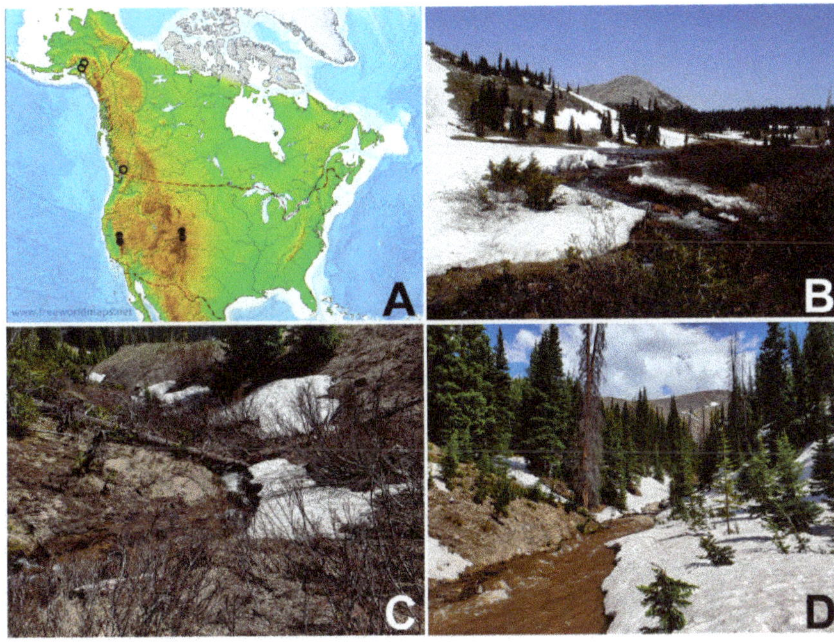

Figure 1. Collecting sites for *Prosimulium supernum* n. sp. (**A**) Map of North America. Solid black circles represent *P. supernum* n. sp. and open circles represent chromosomally confirmed records of its near relative *P. esselbaughi*; map by Daniel Feher, https://www.freeworldmaps.net/about.html (accessed on 20 December 2022). (**B**) Type locality of *P. supernum* n. sp., Libby Creek, Wyoming, 12 June 2008. (**C**) Cameron Pass, snow melt tributary of Michigan River, Colorado, 11 June 2022 (site 2, Table 1). (**D**) Cameron Pass, Michigan Ditch, Colorado, 20 June 2022 (site 4, Table 1).

In addition to material collected in Colorado and Wyoming, we also examined larval material from California (Figure 1A) currently housed, as *Prosimulium esselbaughi*, in the Clemson University Arthropod Collection (Clemson, SC, USA). Because larvae of *P. minifulvum* Adler, Currie and Wood are not morphologically distinguishable from those of so-called *P. esselbaughi* from California, we examined only those larvae that had originally been identified chromosomally.

2.2. Chromosomes

The posterior half of each larval abdomen was removed, and the chromosomes and gonads were Feulgen-stained, dissected from the abdomen, and analyzed under oil immersion [13]. Diagnostic chromosomal sequences were photographed with a Jenoptik ProgRes® SpeedXT Core 5 digital camera mounted on an Olympus BX40 light microscope. The images were imported into Adobe® PhotoShop® Elements 8 to assemble and label the chromosome maps.

The band sequences of the long (L) and short (S) arms of each of the three chromosomes (I, II, and III) were compared with the standard sequence for the genus *Prosimulium* [14,15]. Chromosomal rearrangements previously found in other taxa were named according to their original designations [14]. New rearrangements were assigned unique numbers. Enhanced bands (i.e., heterobands) relative to the standard are identified by the chromosome arm and section number (e.g., IIIS hb80). Fixed inversions in the text and on the maps are italicized; polymorphic rearrangements appear in non-italic type. We also indicate on our maps the following landmarks that are homologous throughout the Simuliidae and that have been used throughout the history of simuliid chromosomal studies [11,14,15]: CI, CII, CIII (centromeres of chromosomes I, II, and III, respectively), NO (nucleolar organizer), Pb (parabalbiani), and RB (ring of Balbiani, a lightly stained area that varies among nuclei and individuals from little or no puffing (e.g., Figure 2C) to well-expressed puffing).

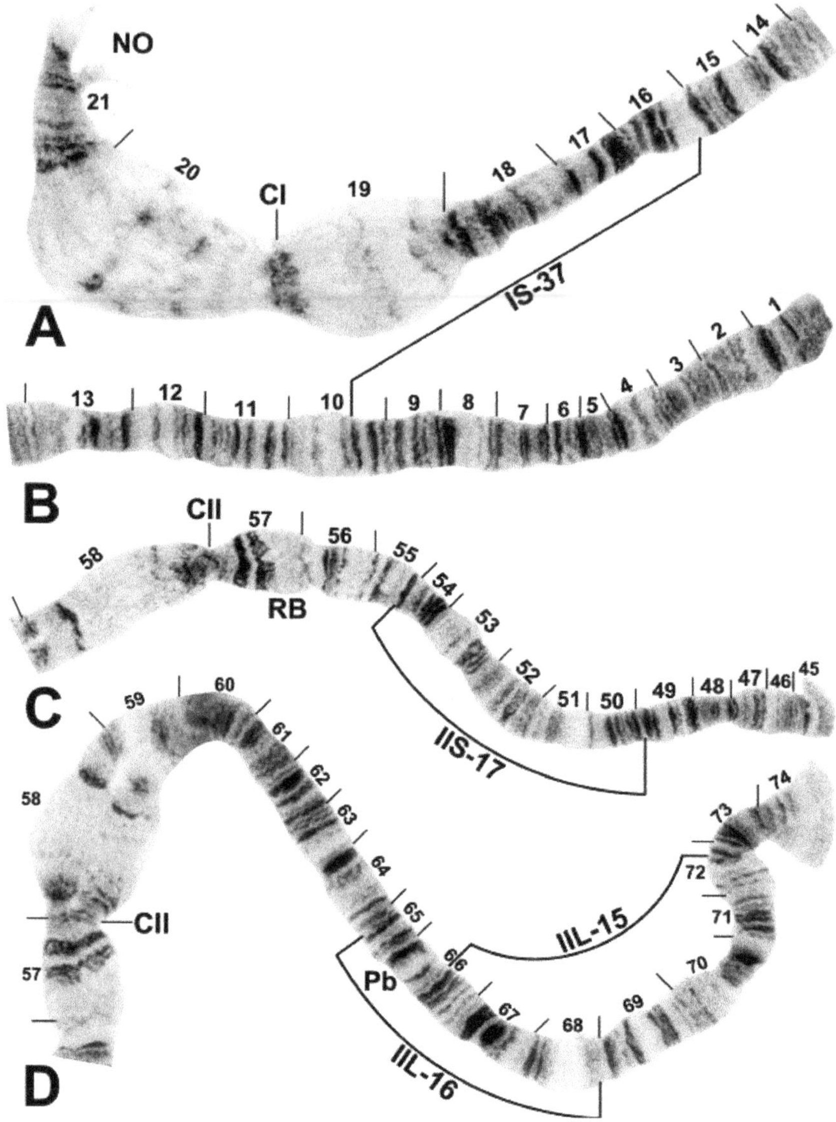

Figure 2. Chromosomes of *Prosimulium supernum* n. sp. from Colorado (male larvae), showing standard sequence. (**A**). IS basal half, including standard (non-transformed) centromere region; bracket indicates breakpoints of autosomal inversion IS-37; CI, centromere of chromosome I; NO, nucleolar organizer. (**B**). IS distal half. (**C**). IIS; bracket indicates breakpoints of autosomal inversion IIS-17; CII, centromere of chromosome II; RB, ring of Balbiani. (**D**). IIL; brackets indicate breakpoints of autosomal inversions IIL-15 and IIL-16; Pb, parabalbiani.

2.3. Morphology

Selected specimens for light microscopy were dissected in 80% ethanol. Larval structures and adult legs were temporarily slide-mounted in a drop of 50% acetic acid. Pupal gills were removed and moved to depression slides in glycerin. Adult heads and terminalia were heated for ca. 2 min in 85% lactic acid, transferred to a drop of glycerin in a depression slide, further dissected into component parts, and oriented for interpretation and imaging.

Structures were photographed at multiple focal planes with a Jenoptik ProgRes® SpeedXT Core 5 digital camera mounted on an Olympus BX40 light microscope. Helicon Focus (version 7.7.5) stacking software was used to form composite images from multiple focal planes. All morphological images were made from specimens collected at the type locality.

Adults in ethanol were chemically dried using hexamethyldisilazane and then pinned through the thorax with a minuten. The pupal exuviae and cocoon were placed in a microvial with glycerin and pinned beneath the associated adult. Dissected parts were pinned in a separate microvial beneath the adult. Descriptions of colors were based on pinned specimens.

Larvae and pupae for scanning electron microscopy were dehydrated through an ethanol series, dried in hexamethyldisilazane, mounted on conductive stubs with double-sided adhesive tape, sputter-coated for 3 min with platinum, and imaged with a Hitachi TM-3000 Scanning Electron Microscope (composite mode, 15 kV, and full vacuum).

Terminology for structures follows that of Adler et al. [11] and, for the larval mandible, Chance [16].

2.4. Type Depositories

After larvae were prepared for chromosomal study they were transferred from Carnoy's fixative to 80% ethanol. The holotype and most paratypes were deposited in the United States National Museum (USNM), Washington, DC. Additional paratypes (adults with pupal exuviae) were deposited in the Canadian National Collection (CNC), Ottawa.

3. Results

***Prosimulium supernum* Adler and Reeves, n. sp.**

Prosimulium esselbaughi: [11] (part: Alpine and Mariposa Counties, California, possibly Nevada).

3.1. Chromosomal Description

Chromosomal sequences of all 18 larvae (10 females, 8 males) prepared for analysis were read completely. The chromosomal complement (haploid number = 3) had standard arm associations and the homologues were tightly paired. The nucleolar organizer was in the standard position (chromosome section 22) for the *P. hirtipes* group (Figure 2A). The centromere region of chromosome I was standard (i.e., not transformed) (Figure 2A), and that of chromosome II was not expanded beyond the standard for the genus (Figure 2C,D). A chromocenter was absent.

All chromosome arms had the standard banding sequence for the genus *Prosimulium*, except IIIL, which was fixed for *IIIL-2* (Figure 3). Sex determination was located on the IIIL arm. Fundamentally, the X sequence carried IIIL-3 (Figure 3A). We interpreted the Y sequence as standard for IIIL-3, with differential expression of bands in sections 85 to 87 (Figure 3B,C). We use "differential band expression" as a descriptive term meaning that the bands in one homologue are well-stained and distinct but in the other homologue the same bands are diffuse and weakly stained. Loops or knots were not expressed in the IIIL-3 region of males. Rather, the homologues in sections 85–87 showed consistent repulsion. Additional rearrangements were frequently associated with the sex arm. Half the larvae in the Colorado populations were heterozygous for IIIL-37 (Figure 3A,B). Differential band expression also occurred outside the IIIL-3 region, in chromosome section 83, in Wyoming larvae (Figure 3C).

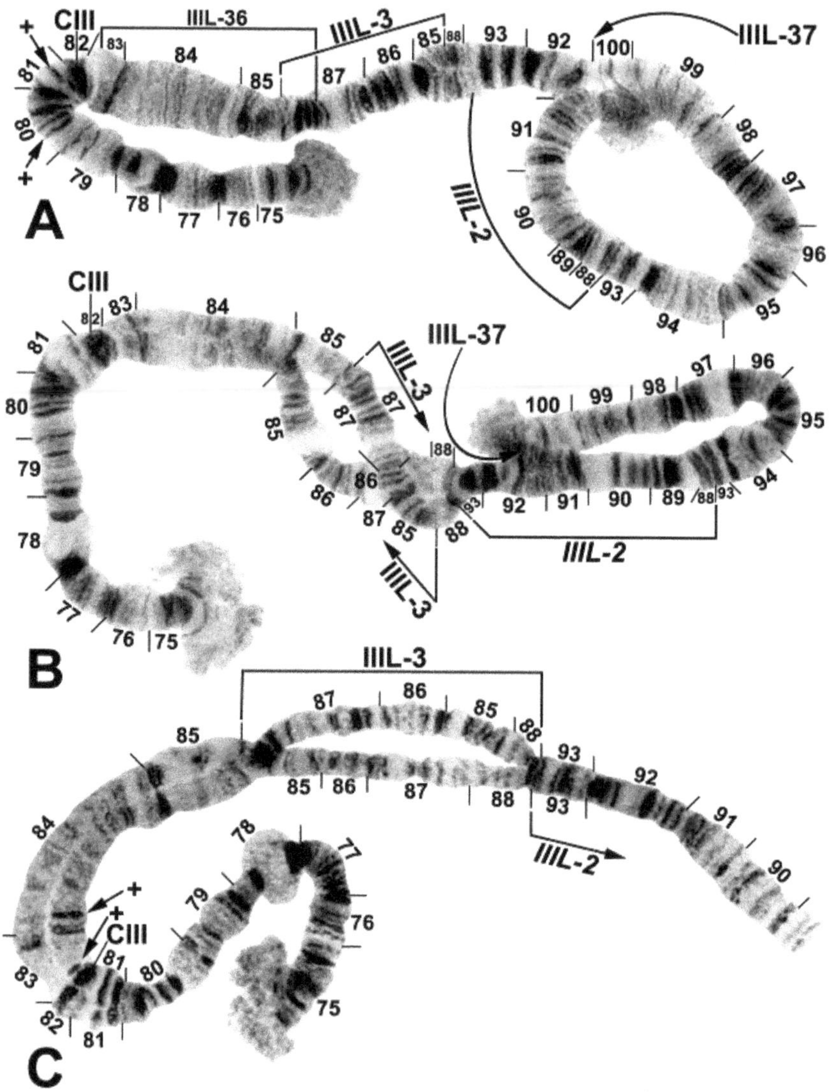

Figure 3. Chromosome III of *Prosimulium supernum* n. sp.; CIII, centromere. (**A**). Sequence of female larva from Colorado, showing the presence of homozygous *IIIL-2* and IIIL-3 and heterozygous IIIL-37 (arrow indicates breakpoints). Limits of IIIL-36 (not present) indicated by brackets; locations of heterobands IIIS hb80 and IIIS hb81 (neither present) indicated by a plus sign (+). (**B**). Sequence of male larva from Colorado, showing the presence of homozygous *IIIL-2* and heterozygous IIIL-3 and IIIL-37 (arrow indicates breakpoints). (**C**). Sequence of male larva from Wyoming, showing the presence of homozygous *IIIL-2*, heterozygous IIIL-3, and heterobands (+) IIIS hb83 and IIIS hb84.

Six autosomal polymorphisms were found (Table 2): IS-37, IIS-17, IIL-15, IIL-16 (Figure 2), IIIS hb80, and IIIS hb81 (Figure 3A).

Table 2. Frequency of homologues with chromosomal rearrangements in larvae of *Prosimulium supernum* n. sp. from the Rocky Mountains of Colorado and Wyoming.

	Site				
	1a [1]	**1b** [1]	**2**	**3**	**4**
Female:Male	3:3	0:2	4:2	2:0	1:1
IS-37	– [2]	–	0.08	–	–
IIS-17	–	–	–	0.25	0.25
IIL-15	0.42	–	–	–	
IIL-16	–	–	–	–	0.25
IIIS hb80	–	–	–	0.25	–
IIIS hb81	1.00	1.00	0.08	–	
IIIL-2 [3]	1.00	1.00	1.00	1.00	1.00
IIIL-3 [4]	*	*	*	*	*
IIIL-36	–	–	0.17	–	0.25
IIIL-37	–	–	0.17	0.50	0.25

[1] Site 1a/1b corresponds to the type locality. [2] A dash (–) indicates a frequency of 0.00. [3] Italics indicate the inversion was fixed. [4] An asterisk (*) indicates that IIIL-3 was linked to the X chromosome; the Y chromosome was standard for IIIL-3.

3.2. Morphological Description

Female. Thorax length 1.4–1.7 mm (mean = 1.5 mm, n = 7). Body grayish brown, pollinose, except pronotum, postpronotal lobes, and scutellum pale yellowish brown. All hair yellowish to golden. Head about 0.6 times as wide as thorax. Frons and clypeus well-haired; frons 0.3 times as wide as head. Labrum slightly shorter than clypeus. Antenna (Figure 4E) brownish, paler basally, with scape, pedicel, and 9 flagellomeres; proportional lengths of pedicel, first flagellomere, and second flagellomere 1.7: 1.6: 1.0. Maxillary palp (Figure 4A) brownish, with proportional lengths of third, fourth, and fifth palpomeres 1.5: 1.0: 1.6; sensory vesicle (Figure 4B) elongated, slender, about 0.4 times length of third palpomere, with short neck and wide mouth about 0.25 times length of vesicle. Lacinia (Figure 4A) with 27–29 teeth. Mandible (Figure 4D) with 23–25 inner teeth and 7 or 8 outer teeth. Cibarium (Figure 4C) at junction with pharynx smooth, unarmed, shallow, broadly U-shaped. Pleural membrane, katepisternum, and postnotum bare. Precoxal bridge incomplete. Legs yellowish, except coxae, trochanters, apices of femora, and tibiae pale brownish; tarsi brownish except basal ¾ of posterior margin of hind basitarsus brownish yellow. Hind leg with basitarsus (Figure 4I) nearly parallel-sided, about 5.3 times as long as wide, and 0.7 times as wide as greatest width of hind tibia; calcipala and pedisulcus absent. Claw (Figure 4J) unarmed or with minute basal tooth. Wing 3.6–4.2 mm long (mean = 4.0 mm, n = 6). Costa, subcosta, and radius with fine setae dorsally and ventrally. Halter grayish white. Segment VIII (Figure 4F) with sclerotized sternal plate; other segments lacking sclerotized sternal plate. Hypogynial valves (ovipositor lobes) (Figure 4F) gently curved toward midline, obliquely truncated posteromedially, membranous except inner margin of each valve sclerotized; inner margins concave, creating teardrop-shaped space. Genital fork (Figure 4H) with stem and arms slender, well sclerotized; space between arms mitre-shaped; each arm expanded into slender triangular lateral plate directed posteromedially. Anal lobe in lateral view (Figure 4G) narrow anteriorly, expanded posteroventrally as broadly rounded lobe extended to, or slightly beyond, anterior margin of cercus. Cercus in lateral view (Figure 4G) short, subrectangular, rounded posteriorly, about twice as wide as long. Spermatheca (Figure 4H) broadly tapered apically, about as long as basal width, wrinkled, heavily pigmented except broad basal area completely devoid of pigment; spermathecal duct and both accessory ducts unpigmented.

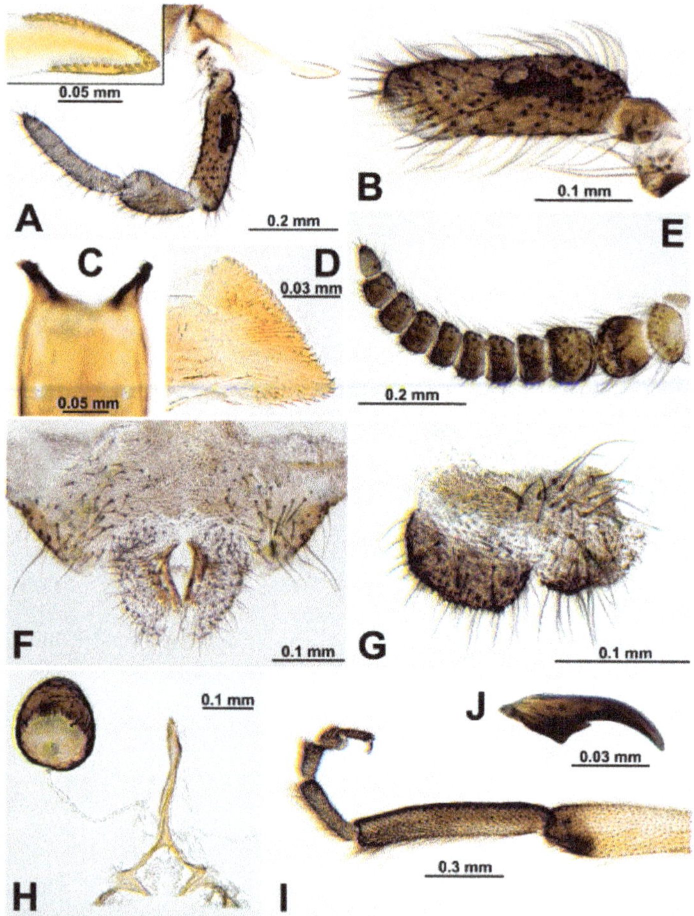

Figure 4. Female of *Prosimulium supernum* n. sp. (**A**) Maxillary palp with 5 palpomeres and lacinia; inset shows apex of lacinia. (**B**) Maxillary palpomeres I, II, and III, showing sensory vesicle. (**C**) Cibarium, anterior. (**D**) Mandible, apex. (**E**) Antenna. (**F**) Sternite VIII and hypogynial valves. (**G**) Anal lobe and cercus, lateral view. (**H**) Genital fork and spermatheca. (**I**) Hind tibia (apical portion only), tarsus, and acropod with claws. (**J**) Hind claw.

Male. Thorax length 1.3–1.4 mm (n = 2). Body dark brown to matte black, except scutellum pale brown. All hair pale golden. Head 0.8 times as wide as thorax. Antenna (Figure 5L) brownish, with scape, pedicel, and 9 flagellomeres; proportional lengths of pedicel, first flagellomere, and second flagellomere 1.7: 1.5: 1.0. Maxillary palp (Figure 5J) brownish, with proportional lengths of third, fourth, and fifth palpomeres 1.2:1.0:2.0; sensory vesicle (Figure 5I) small, about 0.2 times as long as third palpomere, with small, round mouth. Lacinia (Figure 5G) with small apical set of hairs. Katepisternum, pleural membrane, and postnotum bare. Legs pale brownish, except coxae, trochanters, and apices of femora and tibiae brown. Hind basitarsus (Figure 5K) 3.2 times as long as its greatest width, 0.75 times as wide as greatest width of hind tibia; calcipala and pedisulcus absent. Wing 3.4–3.5 mm long (n = 2). Costa and radius with fine setae; subcosta with fine setae ventrally. Halter grayish brown. Gonocoxite in ventral view (Figure 5A) about 1.2 times longer than gonostylus. Gonostylus in ventral view (Figure 5A) smoothly curved toward midline, gradually tapered, with 2 apical spinules; in inner lateral view (Figure 5F) about 1.6 times as long as its greatest width. Ventral plate in ventral view (Figure 5A) subquadrate, slightly tapered, with minute setae in broad triangular pattern; anterior margin rather straight (although dorsal wall in one specimen with protuberance of dark cuticle) and posterior margin slightly concave; body of plate slenderer and more tapered in progressively tilted views (Figure 5B,C); arms somewhat divergent from each other and forming broad U-shape; in lateral view (Figure 5E) convex posteriorly; in terminal view (Figure 5D) subtriangular, strongly rounded distally. Median sclerite (Figure 5D) short, bifurcated apically. Paramere subquadrate, with slender anterior projection, without spines or setae. Dorsal plate absent. Aedeagal membrane with fine setae. Abdominal tergite X (Figure 5H) minute, subrectangular, with or without anterior and posterior incisions. Cercus (Figure 5H) small, rounded, with 23 or 24 setae.

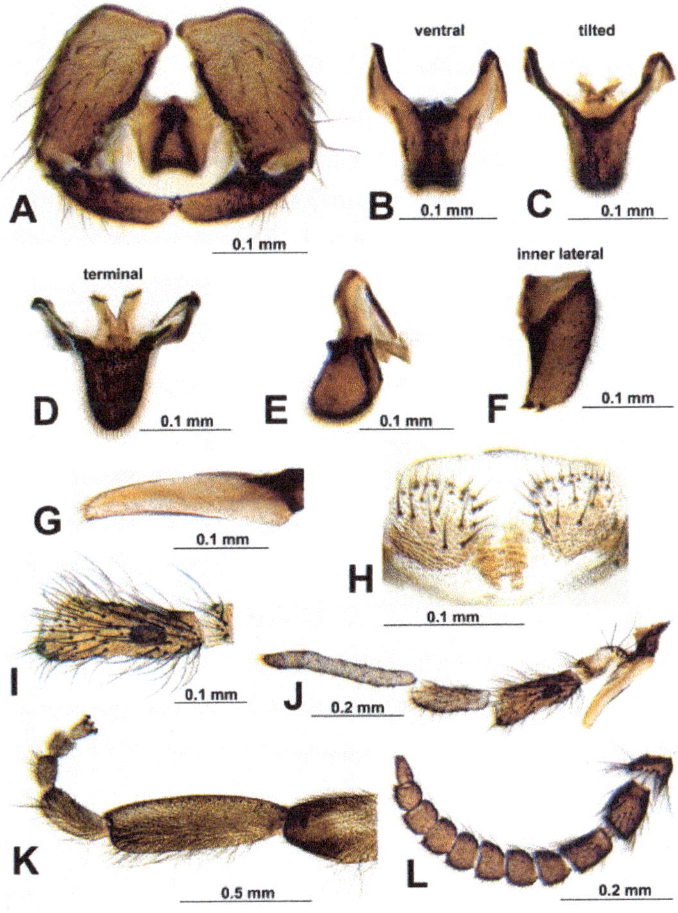

Figure 5. Male of *Prosimulium supernum* n. sp. (**A**) Genitalia, ventral view. (**B**) Ventral plate, ventral view (slightly tilted). (**C**) Ventral plate, ventral view (moderately tilted to show median sclerite). (**D**) Ventral plate and median sclerite, terminal view. (**E**) Ventral plate, lateral view. (**F**) Gonostylus, inner lateral view. (**G**) Lacinia. (**H**) Tergite X and cerci. (**I**) Maxillary palpomere III, showing sensory vesicle. (**J**) Maxillary palp with 5 palpomeres and lacinia. (**K**) Hind tibia (apex only), tarsus, and acropod with claws. (**L**) Antenna.

Pupa. Length (excluding gills) (n = 7) 4.2–5.1 mm, mean = 4.6 mm. Cephalic plate with dense covering of minute, rounded microgranules and 1 pair of unbranched facial trichomes. Thorax (Figure 6D,E) superficially wrinkled (most prominent laterally), densely covered with minute, rounded microgranules; 3 or 4 unbranched dorsal trichomes per side. Gill (Figure 7A,B) about 0.5–0.6 times as long as pupa, with 16 (rarely 15) slender, grayish filaments in 3 groups arising from short basal stalk about as long as wide; stalks of all 3 groups about as long as to 3 times longer than wide; branching pattern: dorsal group with 8 (rarely 7) filaments arranged as [2 + (1 + 2)] + (1 + 2) or [1 + (1 + 2)] + (1 + 2), lateral and ventral groups each with 2 petiolate pairs of filaments; in lateral view, dorsal group often separated from lateral and ventral groups (Figure 7A); filaments furrowed (Figure 6C). Abdomen densely covered with minute, rounded microgranules, dorsally with postscutellar bridge bearing 4 small unbranched setae per side; segment I with 3 or 4 small unbranched setae per side; segment II with 5–7 small unbranched setae per side; segments III and IV each with 4 recurved hooks and 2 or 3 small unbranched setae per side; segments V–IX each with spine comb and 1–5 small unbranched setae per side; segment IX (Figure 6F) with pair of long terminal spines. Abdomen laterally with pleurites each bearing 1 or 2 small unbranched setae per side; striate membrane with 0–3 small unbranched setae per side; segments V and VI each with 1 short, stout seta in tiny sclerite per side. Abdomen ventrally with segment IV bearing pair of slender hooks on each side; segments V–VII each with pair of stout, bifid or trifid hooks per side. Cocoon sac-like, without definitive structure, densely woven, typically covering pupa and part of gill.

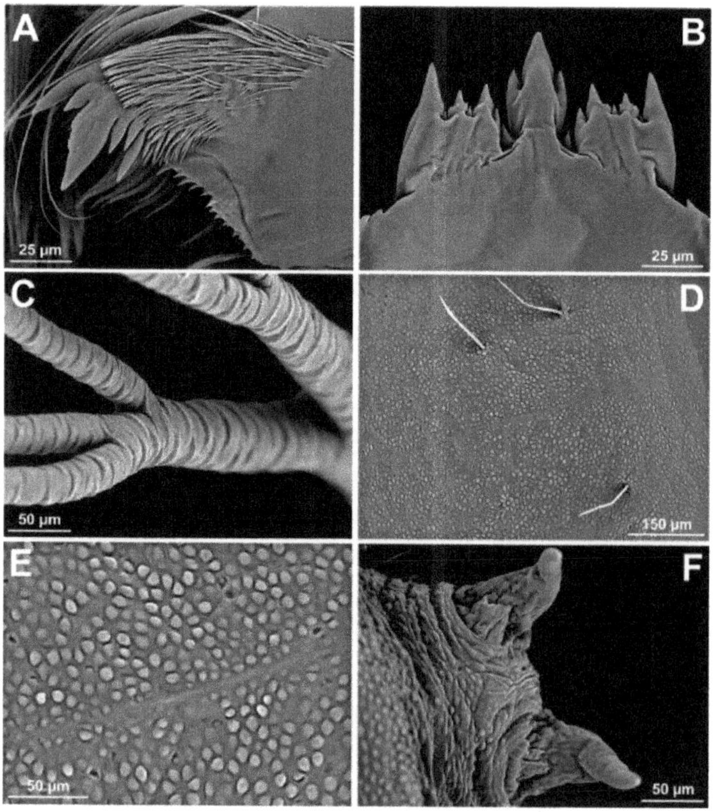

Figure 6. Scanning electron micrographs of *Prosimulium supernum* n. sp. (**A**) Larval mandible, apex,

aboral surface. (**B**) Hypostoma, ventral view (teeth tilted dorsally). (**C**) Pupal gill filaments, showing surface sculpture. (**D**) Pupal thoracic sculpture and trichomes, dorsal view; the ecdysial line is faintly visible, running obliquely from right to left. (**E**) Pupal thoracic sculpture, dorsal view along the ecdysial line. (**F**) Pupal segment IX, showing terminal spines, dorsal view.

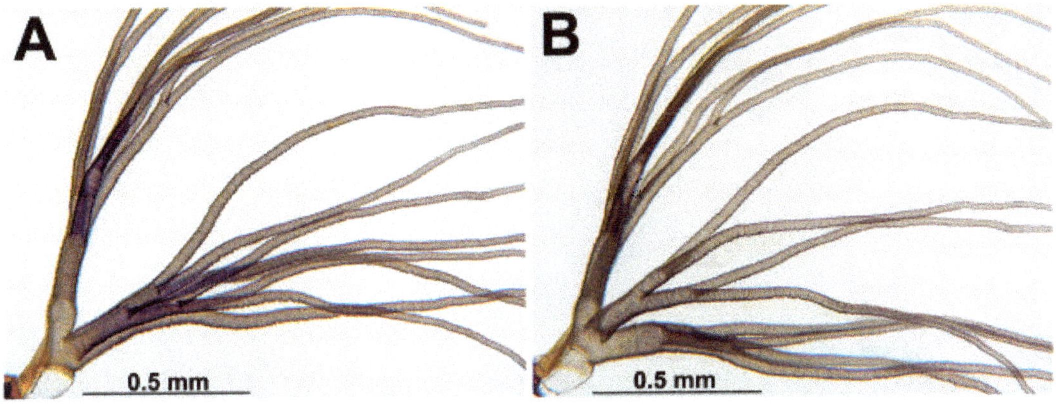

Figure 7. Pupal gill of *Prosimulium supernum* n. sp. (**A**,**B**) Lateral views with slight differences in orientation; apices of filaments not shown.

Larva. Length (n = 4) 6.0–7.6 mm, mean = 6.6 mm. Body (in Carnoy's fixative) grayish brown to pale brownish. Head capsule (Figure 8A) yellowish brown to chestnut brown, palest anteriorly; head spots pale brown, often obscure or faint; anteromedial and sometimes first anterolateral spots typically most conspicuous. Venter of head capsule (Figure 8B) brown, with horizontal long spot and round spot on each side of postgenal cleft brownish. Antenna about as long as, or slightly shorter than, labral fan stalk, with basal and medial articles hyaline and distal article dark brown; proportional lengths of proximal, medial, and distal articles 1.0: 1.7: 1.2. Labral fan (n = 15) with 17–21 (mean = 19.5) primary rays (25–28 for California larvae). Mandible (Figure 6A) with 5 apical teeth, numerous spinous teeth, and 14 or 15 marginal teeth. Maxillary palp about 2 times as long as basal width, with fine, colorless setae along its length (Figure 8D,E). Hypostoma (Figure 8C) with median tooth extended anteriorly beyond all other teeth; sublateral teeth posterior to lateral teeth and extended to same level as tines of median tooth (or beyond if teeth are tilted dorsally; Figure 6B); with 3–5 lateral serrations and 3 or 4 sublateral setae per side. Postgenal cleft (Figure 8B) short, with anterior margin truncate or slightly arched, about 0.3 times as long as postgenal bridge (measured from anterior margin of anterior tentorial pits to hypostomal groove). Cervical sclerites (Figure 8A) minute, enclosed within occiput. Gill histoblast of 16 long, thread-like filaments. Lateral plate of thoracic proleg well-sclerotized, slender, L-shaped. Abdominal cuticle with short, colorless, unbranched setae. Rectal papillae of 3 finger-like lobes. Anal sclerite rectangular, with anterior arms about 1.3–1.4 times as long as posterior arms. Posterior circlet with 67–74 rows of 11–13 hooklets per row.

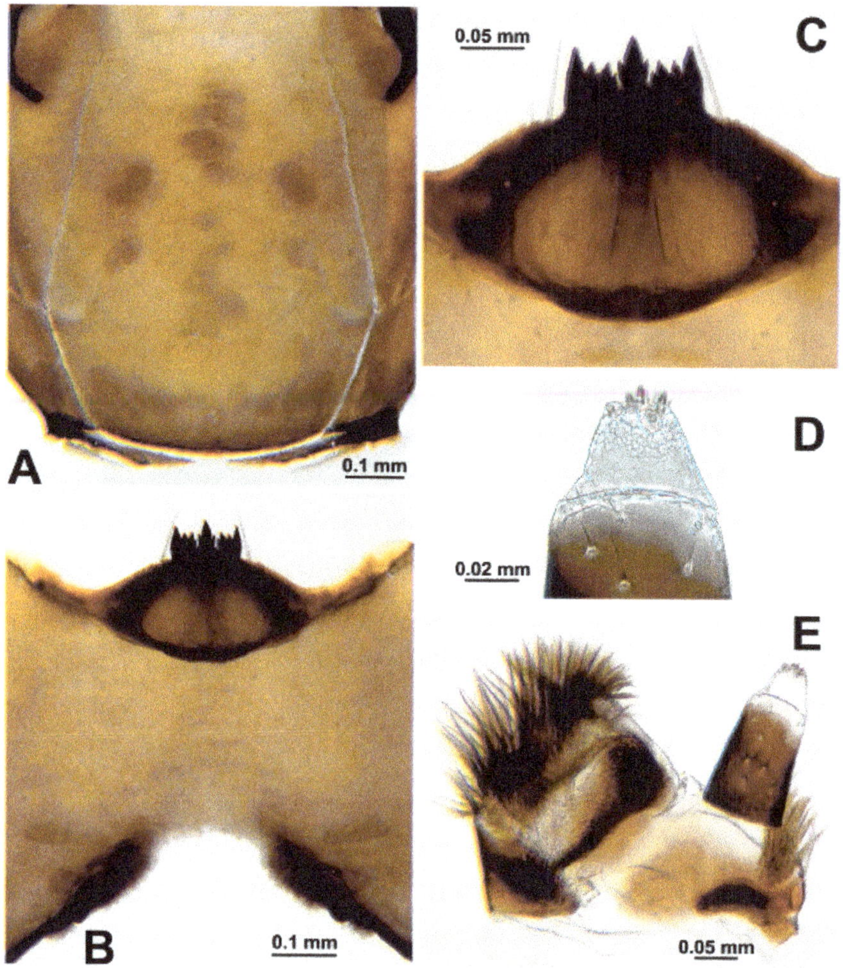

Figure 8. Larva of *Prosimulium supernum* n. sp. (**A**) Frontoclypeal apotome. (**B**) Head capsule, ventral view. (**C**) Hypostoma, ventral view. (**D**) Maxillary palp, apex. (**E**) Maxilla with maxillary palp, lateral view.

3.3. Diagnosis

The following characters place *P. supernum* n. sp. in the *P. hirtipes* group: chromosomes fixed for inversion *IIIL-2* (Figure 3); female with laciniae and mandibles toothed and spermatheca about as long as, or longer than, wide; male with gonostylus bearing 2 apical spinules; pupa with 10–16 filaments; and larva with abdomen rather abruptly expanded at segment V.

Chromosomally, *P. supernum* n. sp. can be distinguished from all other species of the *P. hirtipes* group in western North America by the centromere region of chromosome I in standard (not transformed) configuration (Figure 2A), X-linkage of IIIL-3 (Figure 3), absence of a chromocenter, and absence of fixed inversions other than *IIIL-2*; however, females of *P. supernum* n. sp. and *P. esselbaughi* both carry *IIIL-2* and IIIL-3 and cannot be distinguished from one another.

Morphologically, the female of *P. supernum* n. sp. cannot be reliably distinguished from the females of most other western species in the *P. hirtipes* group, such as *P. esselbaughi* and *P. daviesi* Peterson and Defoliart. The male of *P. supernum* n. sp. can be distinguished from those of 5 of the 13 species in the *P. hirtipes* group with known males (that of *P. idemai* Adler, Currie and Wood is unknown) by having a dark brown to black (rather than orange) scutum, a ventral plate (Figure 5A) with a weakly concave posterior margin and long basal arms (in ventral view), and yellowish femora. *Prosimulium supernum* n. sp. is not reliably distinguished from the remaining 7 species (*P. daviesi*, *P. doveri* Sommerman, *P. esselbaughi*, *P. fulvithorax* Shewell, *P. minifulvum*, *P. rusticum* Adler, Currie and Wood, and *P. travisi* Stone). *Prosimulium opleri* Peterson and Kondratieff, known from a single specimen (male) collected in Rocky Mountain National Park, Colorado, within 14 km of our nearest collection site for *P. supernum* n. sp., is currently in synonymy with *P. shewelli* Peterson and DeFoliart. It differs most prominently by the following characters: first flagellomere distinctly longer than pedicel (rather than about as long as), sensory vesicle slightly less than half as long palpomere III (rather than only about 0.2 times as long as), costal base and stem vein with black (rather than pale golden) setae, and terminalia "unusually small" *sensu* [17] for the *P. hirtipes* group (rather than of typical size). The pupa of *P. supernum* n. sp. with its 16 (rarely 15) filaments can be separated from those of other 16-filamented western species in the *P. hirtipes* group by the clustering of the lateral and ventral branches apart from the dorsal branch (Figure 7A); however, if this configuration is not expressed (Figure 7B), the pupa becomes inseparable from those of species with the branches rather evenly spaced. The larva of *P. supernum* n. sp. can be distinguished from all western species of *Prosimulium*, except *P. esselbaughi*, *P. idemai*, and *P. minifulvum* by the middle hypostomal tooth extended well beyond the lateral teeth (Figures 6B and 8C). The head spots of *P. supernum* n. sp. are generally paler than those of *P. idemai*, which has 13 or 14 filaments (rather than 15 or 16) in its gill histoblast.

Overall, *P. supernum* n. sp. is structurally and chromosomally most like *P. esselbaughi*. The most consistent and reliable diagnostic structural difference between the two species is the configuration of the pupal gill, best appreciated in lateral view. In *P. supernum* n. sp., the lateral and ventral branches either run parallel, thus appearing as a single cluster separate from the dorsal branch (Figure 7A), or all three branches appear separate (Figure 7B), whereas in *P. esselbaughi*, the dorsal and lateral branches typically run parallel and present a cluster separate from the ventral branch [18,19].

3.4. Type Material

Holotype (USNM): Male (pinned) with dissected genitalia (in associated glycerin vial) and pupal exuviae and cocoon (in associated glycerin vial), Wyoming, Albany County, Snowy Range Pass, Libby Creek, 41°21′07″ N 106°17′01″ W, 3233 m asl, 7 July 2008, collected by W. K. Reeves. Paratypes (USNM and CNC): Same location and collector as holotype, 11 June 2007, 13 larvae; 12 June 2008, 2 larvae; 7 July 2008, 7 pupae, 1 male and 5 females (pinned with pupal exuviae in glycerin vials); 17 July 2008, 4 pupae, 2 females (pinned with pupal exuviae in glycerin vials). Colorado, Jackson County, near Cameron Pass, snowmelt tributary of Michigan River, 40°30′31″ N 105°53′05″ W, 3089 m asl, 11 June 2022, W. K. Reeves, 6 larvae; tributary of Michigan River, 40°30′56″ N 105°53′11″ W, 3139 m asl, 20 June 2022, W. K. Reeves, 2 larvae; Cameron Pass, Michigan Ditch, 40°31′13″ N 105°53′32″ W, 3135 m asl, 20 June 2022, W. K. Reeves, 2 larvae.

3.5. Additional Specimens Examined

California, Alpine Co., Rt. 4, trickle 1.2 mi. W of Raymond Meadow Creek bridge 24 June 1991, P. H. Adler, 2 larvae; Mariposa County, Rt. 41, Rail Creek, 12 May 1997, P. H. Adler, 7 larvae; Rt. 41, 1 mi. E of Big Meadow Overlook, 14 May 1997, P. H. Adler, 3 larvae; Rt. 41, 2 mi. W of 5000′ elev. marker, 11 May 1997, P. H. Adler, 3 larvae; Rt. 41, 1.7 mi. N of Avalanche Creek, 12 May 1997, P. H. Adler, 5 larvae; Rt. 41, 4 mi. N of tunnel, 14 May 1997, P. H. Adler, 1 larva; Mono Co., Rt. 108, ca. 2 mi. east of border between Mono County and Tuolumne County, 11 June 1990, P. H. Adler, 1 larva.

3.6. Distribution

Prosimulium supernum n. sp. is confirmed from the Rocky Mountains of Colorado and Wyoming. We also ascribe populations from the Sierra Nevada of California to this species, based on chromosomal and morphological diagnostic characters. A sample of three larvae from Nevada (Pine County, Lehman Creek, 25 August 1966) with chromosomes matching *P. esselbaughi* are possibly those of *P. supernum* n. sp., but the larvae were females; the two species are chromosomally indistinguishable as females. Our map (Figure 1A) shows only localities for which we had both morphological and chromosomal data and could, therefore, make accurate identifications of the new species and *P. esselbaughi* sensu stricto. We suspect that all previous literature records of *P. esselbaughi* sensu lato from Alaska and Canada (i.e., Alberta, British Columbia, and the Yukon) [11,19] pertain to *P. esselbaughi* in the strict sense, based on the distributions and morphological characters (particularly of the pupal gill) that accompanied the records. Literature records of *P. esselbaughi* sensu lato from Montana, Oregon, and Washington [11] might apply to *P. esselbaughi* sensu stricto, *P. supernum* new species, or both; this area possibly represents the transition of the ranges of the two species.

3.7. Bionomics

Larvae of *P. supernum* n. sp., like those of the closely related *P. esselbaughi* [18], are found in fast, cold streams about a meter or more wide. Whereas *P. esselbaughi* has been found from sea level to above the timberline in the area around Anchorage, Alaska [18], *P. supernum* n. sp. is a specialist of high elevations, having been found only at elevations above 3000 m in the Rocky Mountains and above 1500 m in the Sierra Nevada. The larvae of *P. supernum* n. sp. have been collected in association with *Helodon susanae* (Peterson), *P. daviesi*, *Simulium carbunculum* Adler, Currie and Wood, and the *S. arcticum* complex. Like all species of *Prosimulium*, the new species is univoltine. One larva from Mariposa County, California, was infected with an unidentified mermithid nematode.

Topotypical females emerged with their abdomen replete with fat body, suggesting autogeny, at least in the first gonotrophic cycle, as discovered [18] for *P. esselbaughi*. The females are inferred to feed on mammals, based on their fully functional biting mouthparts and claws adapted for mammal feeding. The inference is bolstered by a few records available for *P. esselbaughi*, indicating that large mammals (e.g., sheep and humans) are hosts [18,20].

3.8. Etymology

The species name *supernum* is from Latin, meaning celestial, high, or lofty, in reference to the high-elevation habitat of the species.

4. Discussion

East–west differentiation of North American taxa is a common, long-recognized pattern; the distinction is usually defined by the Rocky Mountains. In the simuliid genus *Prosimulium*, only one of the 38 nominal species, the trans-arctic *P. ursinum* (Edwards), is shared between eastern and western North America [11]. In contradistinction, most nominal species in the group are continuously distributed along a north–south axis [11]. Yet, given the history of glaciation, availability of refugia, dissected topography, and consequent opportunities for isolation, north–south differentiation should be a prominent feature of organisms in western North America. Accordingly, phylogeographic analyses have revealed north–south differentiation in taxa as evolutionarily diverse as birds [21,22], mammals [23], and plants [24].

During the Last Glacial Maximum, the North American Ice Sheet separated the western half of the continent into two major refugia, Beringia in the north and most of the continent to the south of present-day Canada [25], with smaller hypothesized refugia along the Pacific Northwest coast [26]. The biological implication of this vast ice sheet is that populations to its north and south should express disparities in life history, structure, and genetics. This trend is apparent within all nominal species of western *Prosimulium* that are distributed from Alaska southward into the Rockies and Sierra Nevada [11], each of which consists of genetically different northern (e.g., Alaskan) and southern (e.g., central Rocky Mountain) populations. *Prosimulium frohnei* Sommerman and *P. fulvum* (Coquillett) lack a chromocenter in southern populations but both are chromocentric in Alaska; *P. fulvum* also expresses differential sex-chromosome linkage in northern versus southern populations [11,27]. Alaskan and Yukon populations of *P. neomacropyga* Peterson and *P. travisi* show substantial cytogenetic and molecular divergence from Colorado populations, suggesting cryptic species [11,12,15,28]. Northern and southern populations, formerly treated as cytoforms of *P. doveri* [27], were recently recognized as distinct species: the northern *P. doveri* and southern *P. daviesi* [11]. The taxonomic implication is that new species will need to be described, and in other cases some names currently held in synonymy will need to be revalidated. Further insights into these north–south relationships could be gained from molecular analyses, such as the DNA barcoding that was conducted for *P. travisi* and *P. neomacropyga* [12].

Chromosomally and structurally, *P. supernum* n. sp. most closely resembles the more northern *P. esselbaughi* first described from Alaska. It was originally identified as *P. esselbaughi* [11] and is here considered a separate species, differing largely in the pupal gill configuration and chromosomal rearrangements linked to sex. In addition, the floating rearrangement profiles are unique; none of the 17 total floating rearrangements are shared between the two species. *Prosimulium esselbaughi* and *P. supernum* n. sp., thus, represent the sixth example of north–south differentiation of *Prosimulium* in the Western Cordillera and the second case among western *Prosimulium* species in which northern and southern populations are accorded separate species status.

Other simuliid taxa of the Western Cordillera also express north–south disparity. The sister-species *Metacnephia sommermanae* (Stone) and *M. coloradensis* Peterson and Kondratieff, for example, differ cytogenetically and morphologically [11,29]. Why the trend has not been more frequently observed among the Simuliidae might reflect limited north–south genetic investigations and the subtle nature of morphological differences often attributed to intraspecific variation.

Overlain on the large-scale pattern of differentiated populations to the north and south of the North American Ice Sheet is the possibility of further differentiation in northern and southern populations, reflecting isolation, reduced population sizes, and local adaptation in historical microrefugia [30]. *Prosimulium supernum* n. sp. from the Sierra Nevada of California, for instance, might be expected to show different chromosomal rearrangement profiles, although a detailed investigation was not possible for the material now stored in ethanol. Genetically unique populations of other taxa, such as birds and mammals, have been found in the Sierra Nevada [21,31]. Post-glacial factors also might be at work in determining current population genetics at high elevations, such as limited dispersal and local adaptation. Even between the Colorado and Wyoming populations of *P. supernum* n. sp., which are separated by 100 km, chromosomal differences in autosomal polymorphism profiles are stark—only one of nine floating rearrangements is shared.

Post-glacial range expansions of populations north and south of the ice sheet would be expected to bring the differentiated populations into closer proximity, eventually including a zone of overlap in some cases. The species pair, *P. doveri* Sommerman to the north and *P. daviesi* to the south, overlap in the Coast Range of northwestern Washington [11]. A zone of overlap has not been found for *P. supernum* n. sp. and *P. esselbaughi*, although genetic sampling has been limited. *Prosimulium esselbaughi*, however, extends in a cytogenetically homogenous north–south band more than 2100 km, from at least the Anchorage area of southern Alaska to Vancouver in southern British Columbia [15] (as *P. hirtipes* "2 (Alaska)"). The Alaskan population might represent Beringian survival, whereas the British Columbian population is perhaps the result of post-glacial movements southward from the Beringian refugium. The Alaskan and British Columbian populations not only share both fixed inversions (*IIIL-2* and *IIIL-3*), but also the sex inversion and three of the five other floating inversions [15].

The influence of glaciation and subsequent divergence on host use, parasites, and potential vector relationships of *P. supernum* n. sp. and *P. esselbaughi* is unknown. The mammalian hosts of each species and any associated pathogens and parasites transmitted during blood-feeding also would have been subjected to the influences of glaciation. Perhaps, too, the non-vector-transmitted symbiotes (e.g., mermithid nematodes) also carry a signature of Pleistocene isolation. Contemporary populations to the north and south of the great North American Ice Sheet provide ideal subjects for examining potential co-differentiation of black flies and their parasites.

Author Contributions: Conceptualization, P.H.A.; Methodology, P.H.A. and W.K.R.; Validation, P.H.A.; Formal Analysis, P.H.A.; Investigation, P.H.A.; Resources, P.H.A. and W.K.R.; Data Curation, P.H.A.; Writing—Original Draft Preparation, P.H.A.; Review, P.H.A. and W.K.R.; Visualization, P.H.A.; Supervision, P.H.A. and W.K.R.; Project Administration, P.H.A.; Funding Acquisition, P.H.A. and W.K.R. All authors have read and agreed to the published version of the manuscript.

Funding: The research by P.H.A. was supported by NIFA/USDA under project number SC-1700596 and is Technical Contribution No. 7112 of the Clemson University Experiment Station. W.K.R was supported by USDA-ARS project 5410-32000-016-00D.

Institutional Review Board Statement: Not applicable.

Informed Consent Statement: Not applicable.

Data Availability Statement: All data supporting reported results are included in the text.

Acknowledgments: We thank A. G. Wheeler for helpful comments on the manuscript and K. Taylor, L. DeBrey, A. Fabian, and B. Yarnell for logistical assistance in collecting, with additional project support by D. A. Strickman.

Conflicts of Interest: The authors declare no conflict of interest. The funders had no role in the design of the study; in the collection, analyses, or interpretation of data; in the writing of the manuscript; or in the decision to publish the results.

References

1. Schweikert, R.A.; Bogan, N.L.; Girty, G.H.; Hanson, R.E.; Merguerian, C. Timing and structural expression of the Nevadan Orogeny, Sierra Nevada, California. *Geol. Soc. Am. Bull.* **1984**, *95*, 967–979. [CrossRef]
2. McMillan, M.E.; Heller, P.L.; Wing, S.L. History and causes of post-Laramide relief in the Rocky Mountain orogenic plateau. *Geol. Soc. Am. Bull.* **2006**, *118*, 393–405. [CrossRef]
3. Clark, P.U.; Dyke, A.S.; Shakun, J.D.; Carlson, A.E.; Clark, J.; Wohlfarth, B.; Mitrovica, J.X.; Hostetler, S.W.; McCabe, A.M. The Last Glacial Maximum. *Science* **2009**, *325*, 710–714. [CrossRef]
4. Hewitt, G. The genetic legacy of the Quaternary ice ages. *Nature* **2000**, *405*, 907–913. [CrossRef]
5. Brunsfeld, S.J.; Sullivan, J.; Soltis, D.E.; Soltis, P.S. A comparative phylogeography of northwestern North America: A synthesis. In *Integrating Ecology and Evolution in a Spatial Context*; Silvertown, J., Antonovics, J., Eds.; Blackwell Science: Oxford, UK, 2001; pp. 319–339.
6. Ehlers, J.; Gibbard, P. Quaternary glaciation. In *Encyclopedia of Snow, Ice and Glaciers*; Singh, V.P., Singh, P., Haritashya, U.K., Eds.; Springer: New York, NY, USA, 2011; pp. 873–882. [CrossRef]
7. Adler, P.H.; Cheke, R.A.; Post, R.J. Evolution, epidemiology, and population genetics of black flies (Diptera: Simuliidae). *Infect. Genet. Evol.* **2010**, *10*, 846–865. [CrossRef] [PubMed]
8. Adler, P.H. World blackflies (Diptera: Simuliidae): A comprehensive revision of the taxonomic and geographical inventory [2022]. 2022, p. 145. Available online: http://biomia.sites.clemson.edu/pdfs/blackflyinventory.pdf (accessed on 11 November 2022).
9. Weinmann, C.J.; Anderson, J.R.; Longhurst, W.M.; Connolly, G. Filarial worms of Columbian black-tailed deer in California. 1. Observations in the vertebrate host. *J. Wildl. Dis.* **1973**, *9*, 213–220. [CrossRef] [PubMed]
10. Schulz-Key, H.; Wenk, P. The transmission of *Onchocerca tarsicola* (Filarioidea: Onchocercidae) by *Odagmia ornata* and *Prosimulium nigripes* (Diptera: Simuliidae). *J. Helminthol.* **1981**, *55*, 161–166. [CrossRef] [PubMed]
11. Adler, P.H.; Currie, D.C.; Wood, D.M. *The Black Flies (Simuliidae) of North America*; Cornell University Press: Ithaca, NY, USA, 2004.
12. Rivera, J.; Currie, D.C. Identification of Nearctic black flies using DNA barcodes (Diptera: Simuliidae). *Mol. Ecol. Resour.* **2009**, *s1*, 224–236. [CrossRef] [PubMed]
13. Adler, P.H.; Kúdelová, T.; Kúdela, M.; Seitz, G.; Ignjatović-Ćupina, A. Cryptic biodiversity and the origins of pest status revealed in the macrogenome of *Simulium colombaschense* (Diptera: Simuliidae), history's most destructive black fly. *PLoS ONE* **2016**, *11*, e0147673. [CrossRef]

14. Basrur, P.K. The salivary gland chromosomes of seven segregates of *Prosimulium* (Diptera: Simuliidae) with a transformed centromere. *Can. J. Zool.* **1959**, *37*, 527–570. [CrossRef]
15. Basrur, P.K. The salivary gland chromosomes of seven species of *Prosimulium* (Diptera: Simuliidae) from Alaska and British Columbia. *Can. J. Zool.* **1962**, *40*, 1019–1033. [CrossRef]
16. Chance, M.M. The functional morphology of the mouthparts of blackfly larvae (Diptera: Simuliidae). *Quaest. Entomol.* **1970**, *6*, 254–284.
17. Peterson, B.V.; Kondratieff, B.C. The black flies (Diptera: Simuliidae) of Colorado: An annotated list with keys, illustrations and descriptions of three new species. *Mem. Am. Entomol. Soc.* **1995**, *42*, 1–121.
18. Sommerman, K.M. *Prosimulium esselbaughi* n. sp., the Alaskan *P. hirtipes* 2 (Diptera: Simuliidae). *Proc. Entomol. Soc. Wash.* **1964**, *66*, 141–145.
19. Peterson, B.V. The *Prosimulium* of Canada and Alaska. *Mem. Entomol. Soc. Can.* **1970**, *69*, 1–216.
20. Mason, P.G.; Shemanchuk, J.A. Black flies. *Agric. Can. Publ.* **1990**, *1499/E*, 1–19.
21. Hull, J.M.; Keane, J.J.; Savage, W.K.; Godwin, S.A.; Shafer, J.A.; Jepsen, E.P.; Gerhardt, R.; Stermer, C.; Ernest, H.B. Range-wide genetic differentiation among North American great gray owls (*Strix nebulosa*) reveals a distinct lineage restricted to the Sierra Nevada, California. *Mol. Phylogen. Evol.* **2010**, *56*, 212–221. [CrossRef]
22. Bayard de Volo, S.; Reynolds, R.T.; Sonsthagen, S.A.; Talbot, S.L.; Antolin, M.F. Phylogeography, postglacial gene flow, and population history of North American Northern Goshawks (*Accipiter gentilis*). *Auk* **2013**, *130*, 342–354. [CrossRef]
23. Krejsa, D.M.; Talbot, S.L.; Sage, G.K.; Sonsthagen, S.A.; Jung, T.S.; Magoun, A.J.; Cook, J.A. Dynamic landscapes in northwestern North America structured populations of wolverines (*Gulo gulo*). *J. Mammal.* **2021**, *102*, 891–908. [CrossRef]
24. Nadeau, S.; Godbout, J.; Lamothe, M.; Gros-Louis, M.-C.; Isabel, N.; Ritland, K. Contrasting patterns of genetic diversity across the ranges of *Pinus monticola* and *P. strobus*: A comparison between eastern and western North American postglacial colonization histories. *Am. J. Bot.* **2015**, *102*, 1342–1355. [CrossRef]
25. Dulfer, H.E.; Margold, M.; Darvill, C.M.; Stroeven, A.P. Reconstructing the advance and retreat dynamics of the central sector of the last Cordilleran Ice Sheet. *Quatern. Sci. Rev.* **2022**, *284*, 107465. [CrossRef]
26. Colella, J.P.; Lan, T.; Talbot, S.L.; Lindqvist, C.; Cook, J.A. Whole-genome resequencing reveals persistence of forest-associated mammals in Late Pleistocene refugia along North America's North Pacific Coast. *J. Biogeogr.* **2021**, *48*, 1153–1169. [CrossRef]
27. Rothfels, K.H. Cytotaxonomy of black flies (Simuliidae). *Annu. Rev. Entomol.* **1979**, *24*, 507–539. [CrossRef]
28. Madahar, D.P. Cytogenetic study in triploidy in *Prosimulium ursinum* (Simuliidae: Diptera). *Genetics* **1973**, *74* (Suppl. S2), 170–171.
29. Finn, D.; Adler, P.H. Population genetic structure of a rare high-elevation black fly, *Metacnephia coloradensis*, occupying Colorado lake outlet streams. *Freshwat. Biol.* **2006**, *51*, 2240–2251. [CrossRef]
30. Mee, J.A.; Moore, J.-S. The ecological and evolutionary implications of microrefugia. *J. Biogeogr.* **2014**, *41*, 837–841. [CrossRef]
31. Latch, E.K.; Heffelfinger, J.R.; Fike, J.A.; Rhodes, O.E., Jr. Species-wide phylogeography of North American mule deer (*Odocoileus hemionus*): Cryptic glacial refugia and postglacial recolonization. *Mol. Ecol.* **2009**, *18*, 1730–1745. [CrossRef] [PubMed]

diversity

MDPI

Article

DNA Barcoding of Black Flies (Diptera: Simuliidae) in Slovakia and Its Utility for Species Identification

Tatiana Kúdelová [1], Samuel Krčmárik [1,*], Ivona Lužáková [1], Bibiana Bujačková [2], Karin Matická [1] and Matúš Kúdela [1]

[1] Department of Zoology, Faculty of Natural Sciences, Comenius University in Bratislava, Ilkovičova 6, SK84215 Bratislava, Slovakia

[2] Institute of Molecular Biomedicine, Faculty of Medicine, Comenius University in Bratislava, Sasinkova 4, SK81108 Bratislava, Slovakia

* Correspondence: samuelkrcmarik@gmail.com

Abstract: DNA barcoding based on the cytochrome oxidase I gene is increasingly used in black flies (Diptera: Simuliidae), but extensive data for larger areas are still rare. Slovakia, with well-explored black fly fauna, was chosen to verify the reliability of DNA barcoding for species identification. The DNA barcoding region of the COI gene of 235 individuals of 25 black fly species from Slovakia was sequenced. Among them, 30 sequence clusters with assigned Barcode Index Numbers (BINs) were identified, and 5 of them were recorded for the first time. The average intraspecific genetic divergence was 0–3.24%, whereas the average interspecific divergence was 12.3–17.8%. Based on the barcode sequence, 14 species could be identified unambiguously, and 3 of them (*Prosimulium latimucro, Simulium costatum, S. degrangei*) are split into two or more barcode clusters. In eleven species, some degree of barcode sharing occurred, often combined with barcode splitting. The results showed hidden diversity as well as cases of shared barcode sequences among the studied species. Further investigation using other molecular markers is necessary due to the overlap of intraspecific and interspecific variability.

Keywords: black flies; Simuliidae; DNA barcoding; genetic distance; cryptic diversity; Slovakia

Citation: Kúdelová, T.; Krčmárik, S.; Lužáková, I.; Bujačková, B.; Matická, K.; Kúdela, M. DNA Barcoding of Black Flies (Diptera: Simuliidae) in Slovakia and Its Utility for Species Identification. *Diversity* **2023**, *15*, 661. https://doi.org/10.3390/d15050661

Academic Editors: Carlos Barceló, Ignacio Ruiz-Arrondo and Michael Wink

Received: 30 March 2023
Revised: 28 April 2023
Accepted: 8 May 2023
Published: 13 May 2023

1. Introduction

Black flies (Simuliidae) are mainly known as pests and vectors of parasitic diseases. More than 2000 out of 2398 known species feed on the blood of birds or mammals [1]. About 28 species of vertebrate parasites have been recorded in black flies [2]. The main scientific interest has been focused on the vectors of human onchocerciasis. According to WHO estimates from 2017, 220 million people in the world need preventive drugs against onchocerciasis, 14.6 million have skin symptoms of the disease, and 1.15 million have impaired vision [3]. The most common diseases of veterinary importance transmitted by Simuliidae are bovine onchocerciasis [4,5] and avian leukocytozoonosis [6]. In addition to transmitting parasitic diseases, blackfly outbreaks are also problematic. The saliva of black flies contains anesthetic and anticoagulant substances, which can cause allergic reactions; numerous attacks can cause fever (so-called black fly fever), toxic shock, and in some cases, even death [7]. Grazing farm animals on pastures in the vicinity of black fly breeding sites are often stressed by constant attacks, which also causes a decrease in production and associated economic losses [8,9]. In addition to their great importance as a blood-sucking group, the larvae and pupae of black flies also represent an important part of flowing water communities, which also contributes to the need for better knowledge of their diversity. Cytogenetic studies covered about one quarter of known black fly species and revealed that the described morphospecies are often complexes of two or more sibling species. These species might have different habitat preferences and often differ in their ability to transmit parasitic diseases [10,11].

Although cytotaxonomy provides valuable information supplementing the classical taxonomy, molecular methods are being widely used as well. The chromosomal analysis requires a lot of experience, and it can be performed only on the last instar larvae, which were specifically fixed in an ad hoc solution of ethanol and acetic acid (3:1). Therefore, reliable molecular methods for species identification are needed. The most used molecular marker is the barcoding region of the mitochondrial COI gene. Comprehensive studies on the barcoding of black flies were performed for example in North America [12,13], Central America [14], and Thailand [15–18]. Several studies showed the applicability of barcoding in resolving a particular problem with the identity or identification of European black fly species. Barcoding helped to resolve the problem of the identity of *Simulium reptans* and confirmed the presence of a second similar species, *Simulium reptantoides* [19,20]. Used together with cytotaxonomy, the DNA barcoding unambiguously identified all four species of the *Simulium aureum* group in Great Britain [21] and it confirmed that the Italian endemic species *Prosimulium italicum* is not a synonym of the widely distributed and morphologically very similar *Prosimulium hirtipes* [22]. In Europe, larger studies on the barcoding of black flies are still missing, except for Spain, with a barcoding project covering 22 species and 199 individuals [23]. In this study, Slovakia, with its relatively well-explored black fly fauna consisting of 46 species [24], is used as an example area for verifying the reliability of the use of barcoding for black fly species identification.

2. Material and Methods

The larvae and pupae of black flies were collected from branches, rocks, and vegetation in streams and rivers at 26 locations in Slovakia (Table 1, Figure 1). Collected individuals were fixed in 96% ethanol. Larvae and pupae were identified with Zeiss SteREO Discovery.V12 stereoscopic microscope and Zeiss AxioLab microscope using multiple identification keys for European black fly taxa [25–30]. All specimens were identified to the species level based on the morphological characters; the only exceptions are two pupae of the subgenus *Eusimulium*, which were identified to the species level as *Simulium angustipes* Edwards, 1915 and *Simulium rubzovianum* (Sherban, 1961) based on their barcode sequence because no reliable morphological characters are known for identification of the pupae. Nomenclature follows the recent inventory of black flies [1]. Material was deposited at the Department of Zoology of Comenius University in Bratislava, Slovakia.

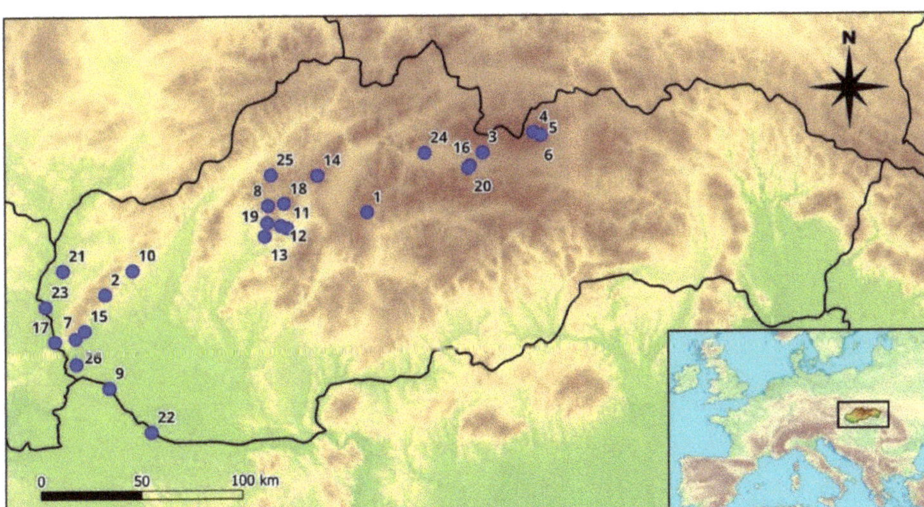

Figure 1. Sampling sites for analyzed specimens. The localities are listed and numbered in Table 1.

Table 1. Species and BIN classification of the studied dataset.

Species	BIN	N	Locality Waterbody, Municipality	Coordinates N, E	Date	No. of Specimen (Per Date)
***Prosimulium hirtipes* (Fries, 1824)**	AER9302	7	[1] Čierny potok, Dolný Harmanec	48.8201, 19.0342	1 May 2021 11 May 2021 10 June 2021	4 2 1
		1	[2] Kamenný potok, Častá	48.3940, 17.3037	24 April 2021	1
		7	[3] Potok Račková, Pribylina	49.1055, 19.8069	11 May 2021 10 June 2021	2 5
	AAB9204	2	[1] Čierny potok, Dolný Harmanec	48.8201, 19.0342	11 May 2021 10 June 2021	1 1
***Prosimulium latimucro* (Enderlein, 1925)**	AEF3934	1	[4] Malé Žabie Javorové pleso, Tatranská Javorina	49.2023, 20.1500	14 September 2009	1
		1	[5] Prostredné Spišské pleso, Vysoké Tatry	49.1910, 20.1972	12 August 2009	1
		2	[6] Malé Spišské pleso, Vysoké Tatry	49.1901, 20.2004	12 August 2009	[illegible]
	AEI9525 *	2	[6] Malé Spišské pleso, Vysoké Tatry	49.1901, 20.2004	12 August, 2009	2
	AEF3642	2	[6] Malé Spišské pleso, Vysoké Tatry	49.1901, 20.2004	12 August 2009	2
***Prosimulium rufipes* (Meigen, 1830)**	AER9302	11	[1] Čierny potok, Dolný Harmanec	48.8201, 19.0342	1 May 2021 11 May 2021 10 June 2021	3 6 2
		9	[3] Potok Račková, Pribylina	49.1055, 19.8069	11 May 2021 10 June 2021	2 7
***Prosimulium tomosvaryi* (Enderlein, 1921)**	AEA2402	2	[1] Čierny potok, Dolný Harmanec	48.8201, 19.0342	1 May 2021	2
		4	[7] Drieňovka, Bratislava-Nové Mesto	48.1877, 17.1234	5 May 2021	4
		3	[2] Kamenný potok, Častá	48.3940, 17.3037	24 April 2021	3
		4	[8] Závada, Kšinná	48.8302, 18.3631	1 April 2017	4
***Simulium erythrocephalum* (De Geer, 1776)**	AAJ6649	2	[9] Vojčianske rameno, Kyselica	47.9751, 17.3702	16 September 2020	2
***Simulium angustipes* Edwards, 1915**	AAF4267	1	[10] Bíňovce	48.5092, 17.4772	6 March 2022	1
***Simulium rubzovianum* (Sherban, 1961)**	AAP9556	1	[13] Hydina, Uhrovec (Látkovce)	48.6929, 18.3499	1 April 2017	5
***Simulium costatum* Friederichs, 1920**	AEH7122 *	6	[12] Sučiansky potok, Nitrianske Sučany	48.7431, 18.4564	1 April 2017	6
	AEH4753	1	[12] Sučiansky potok, Nitrianske Sučany	48.7431, 18.4564	1 April 2017	1
***Simulium cryophilum* (Rubtsov, 1959)**	ACU9243	5	[13] Hydina, Uhrovec (Látkovce)	48.6929, 18.3499	1 April 2017	5
		3	[14] Vríca, Vrícko	48.9764, 18.6875	12 September 2020	3
***Simulium vernum* Macquart, 1826**	AAB8624	1	[7] Drieňovka, Bratislava-Nové Mesto	48.1877, 17.1234	21 March 2022	1
		1	[13] Hydina, Uhrovec (Látkovce)	48.6929, 18.3499	1 April 2017	1
		4	[15] Struha, Bratislava-Vajnory	48.2258, 17.1815	21 March 2022	4
		6	[12] Sučiansky potok, Nitrianske Sučany	48.7431, 18.4564	19 May 2018	6
	AET1431	1	[15] Struha, Bratislava-Vajnory	48.2258, 17.1815	21 March 2022	1
***Simulium argenteostriatum* Strobl, 1898**	AEH6008 *	8	[3] Potok Račková, Pribylina	49.1055, 19.8069	17 July 2019	8

Table 1. *Cont.*

Species	BIN	N	Locality Waterbody, Municipality	Coordinates N, E	Date	No. of Specimen (Per Date)
Simulium degrangei Dorier & Grenier, 1960	ACD5131	3	[16] Belá, Liptovský Hrádok	49.0448, 19.7227	13 June 2013	3
	ACQ6722	1	[16] Belá, Liptovský Hrádok	49.0448, 19.7227	13 June 2013	1
Simuium ornatum Meigen, 1818	AEN0363	1	[17] Dunaj, Bratislava-Devín	48.1681, 16.9873	18 March 2022	1
		3	[11] Nitrica, Diviacka Nová Ves	48.7339, 18.4993	27 July 2017	3
		2	[18] Nitrica, Liešťany (Lomnica)	48.8436, 18.4701	27 July 2017	2
		5	[19] Striebornica, Uhrovec	48.7530, 18.3625	20 September 2017	5
		2	[20] Váh, Liptovský Hrádok	49.0349, 19.7112	13 June 2013	2
		3	[14] Vríca, Vrícko	48.9764, 18.6875	12 September 2020	3
	AEW0869	5	[21] Rudava, Veľké Leváre	48.4888, 17.0096	25 September 2020	5
		1	[19] Striebornica, Uhrovec	48.7530, 18.3625	20 September 2017	1
	AAV2392	1	[11] Nitrica, Diviacka Nová Ves	48.7339, 18.4993	27 July 2017	1
		1	[20] Váh, Liptovský Hrádok	49.0349, 19.7112	13 June 2013	1
	AAN3313	1	[20] Váh, Liptovský Hrádok	49.0349, 19.7112	13 June 2013	1
Simulium trifasciatum Curtis, 1839	AEN0363	2	[19] Striebornica, Uhrovec	48.7530, 18.3625	20 September 2017	2
	AEW0869	1	[19] Striebornica, Uhrovec	48.7530, 18.3625	20 September 2017	1
Simulium colombaschense (Scopoli, 1780)	ADZ9523 *	5	[22] Dunaj, Medveďov	47.7887, 17.6651	16 May 2013	5
Simulium reptans (Linnaeus, 1758)	AAA9951	9	[23] Morava, Vysoká pri Morave	48.3219, 16.9081	30 April 2013	9
Simulium reptantoides Carlsson, 1962	AAA9950	7	[24] Váh, Ivachnová	49.0960, 19.4118	19 July 2014	7
		3	[16] Belá, Liptovský Hrádok	49.0448, 19.7227	19 July 2014	3
Simulium argyreatum Meigen, 1838	AAB8783	7	[1] Čierny potok, Dolný Harmanec	48.8201, 19.0342	1 May 2021	7
		3	[25] Podhradský potok, Zliechov	48.9699, 18.3714	30 June 2018	3
		3	[3] Potok Račková, Pribylina	49.1055, 19.8069	17 July 2019	3
	ADK2119	2	[1] Čierny potok, Dolný Harmanec	48.8201, 19.0342	1 May 2021	2
Simulium maximum (Knoz, 1961)	ACV0745	4	[3] Potok Račková, Pribylina	49.1055, 19.8069	10 June 2021	4
	AES0919 *	4	[3] Potok Račková, Pribylina	49.1055, 19.8069	10 June 2021	4
Simulium monticola Friederichs, 1920	AAB8783	14	[1] Čierny potok, Dolný Harmanec	48.8201, 19.0342	24 September 2020 1 May 2021	6 8
Simulium sp. aff. monticola	ACV0745	14	[1] Čierny potok, Dolný Harmanec	48.8201, 19.0342	24 September 2020 1 May 2021	4 10
Simulium variegatum Meigen, 1818	AAB8783	6	[25] Podhradský potok, Zliechov	48.9699, 18.3714	30 June 2018	6
		9	[3] Potok Račková, Pribylina	49.1055, 19.8069	17 July 2019	9
Simulium balcanicum (Enderlein, 1924)	AAM4036	4	[26] Priesakový kanál, Bratislava-Rusovce	48.0729, 17.1395	23 March 2019	4

Table 1. *Cont.*

Species	BIN	N	Locality Waterbody, Municipality	Coordinates N, E	Date	No. of Specimen (Per Date)
Simulium equinum (Linnaeus, 1758)	AAM3554	3	[11] Nitrica, Diviacka Nová Ves	48.7339, 18.4993	27 July 2017	3
Simulium lineatum (Meigen, 1804)	AAM4036	1	[17] Dunaj, Bratislava-Devín	48.1681, 16.9873	18 March 2022	1
		7	[11] Nitrica, Diviacka Nová Ves	48.7339, 18.4993	27 July 2017	7

*—unique (new) BIN in BOLD database. N—number of specimens per BIN and locality. [1–26]—numbers of localities used in Figure 1.

A small piece of larval or pupal muscle tissue from each specimen was used for DNA extraction. The rest of each specimen was stored in 96% ethanol at −20 °C for further analysis. The DNA was extracted using commercial kit prepGEM Insect (Zygem), following the instructions of the manufacturer.

The barcoding region of the cytochrome *c* oxidase subunit I (COI) fragment was amplified using the primers HCO1490 (5′-GGTCAACAAATCATAAAGATATTGG-3′) and LCO2198 (5′-TAAACTTCAGGGTGACCAAAAAATCA-3′) [31]. The PCR reaction was performed in a total volume of 25 µL: DNA or 2.5 µL Buffer Dream (Fermentas), 2.5 µL $MgCl_2$, 2.0 µL dNTPs, 0.5 µL of each primer, and 0.4 µL DNA polymerase DreamTaq (Fermentas), or 10 µL Red Taq 2X Master Mix, (1.5 mM $MgCl_2$ included); 0.4 µL of each primer; 12.2 µL of Nuclease-Free water, and 2 µL of extracted DNA. The PCR program for COI gene fragment consisted of initial denaturation at 94 °C for 5 min; 35 cycles at 94 °C for 30 s, 45 °C for 30 s, and 72 °C for 30 s; the last polymerization was at 72 °C for 5 min.

The quality of PCR product was checked on 1% agarose gel stained with GoldView. PCR products were purified and sequenced by Macrogen Europe, Amsterdam, The Netherlands.

Sequences were aligned and modified using Geneious 6.1.8 [32]. Genetic distances were calculated using the program MEGA11 [33] using Kimura 2-parameter (K2P) with bootstrap support values estimated 500 replicates. The most suitable evolutionary models were also calculated in the jModelTest2 [34]. Phylogenetic trees were created using the maximum parsimony (MP) method in PAUP 4.0 [35] with 100 random additions, nearest neighbor interchange (NNI) algorithm, and heuristic search approach. Bootstrap support values were estimated 1000 replicates. MrBayes v.3.1.2 [36] program was used for Bayesian phylogenetic inference (BI), with four simulations of Markov chains, 5 M generation, and sampling every 100 generations, with 25% of trees discarded as burn-in. Maximum likelihood (ML) analysis was performed in RAxML [37] through raxmlGUI 2.0 [38] interface.

Two sequences of *Drosophila melanogaster* (Genbank Access Numbers: HM102299.1; HM102298.1) were used as outgroups [39].

All sequences were uploaded to the Barcode of Life Database (BOLD) into dataset DS-SIMSK—Black Fly Barcoding Slovakia (Diptera: Simuliidae), and subsequently to the GenBank database (accession numbers: OQ922995—OQ923229).

Haplotype networks for selected groups were constructed in the PopART (Population Analysis with Reticulate Trees) software [40], using TCS network method [41].

Maps were created in QGIS 3.30.0 [42], using basemaps from https://www.naturalearthdata.com (accessed on 15 February 2023) and https://www.eea.europa.eu/data-and-maps (accessed on 15 February 2023).

3. Results

The DNA barcoding region of 25 black fly species and 235 specimens was successfully amplified. The COI barcodes for three species and one unnamed taxon are published for the first time here (*S. argenteostriatum*, *S. maximum*, *S.* sp. aff. *monticola*, *S. colombaschense*). In 216 specimens of 23 species, we obtained the barcoding region's full length (658 bp). The only exceptions were the species *S. reptans* and *S. reptantoides*, sequenced one-sided in the pilot step of the project, which produced at least 625 bp long sequences. None of the sequences contained stop codons, deletions, or insertions.

3.1. Genetic Distances

In the genus *Prosimulium*, the average intraspecific genetic distances ranged between 0.06% (*P. rufipes*) and 1.81% (*P. latimucro*), and the maximal value of intraspecific genetic distance was found in *P. latimucro* (3.64%). The minimal average interspecific genetic distance was found in *P. hirtipes* (5.08%) and the maximal average interspecific genetic distance was found in *P. tomosvaryi* (10.80%).

In the genus *Simulium*, the average intraspecific genetic distances ranged between 0 (*S.* sp. aff. *monticola*) and 3.24% (*S. ornatum*). The maximal intraspecific genetic distance was found in *S. ornatum* (7.79%), followed by *S. costatum* (7.13%). The average values of interspecific genetic distance varied between 15.85% and 20.28%.

The overlap of intraspecific and interspecific genetic distance occurred in ten species of the genus *Simulium* and two species of the genus *Prosimulium* (Table 2). The genetic distances and possible relations of the haplotypes for selected taxa are visualized in haplotype networks (Figures 2–7).

Table 2. Genetic distances of analyzed species calculated by genera.

Species	n	Intraspecific Genetic Distances				Interspecific Genetic Distances			
		Min	Max	Average	Std	Min	Max	Average	Std
P. hirtipes *	17	0.0000	**0.0195**	0.0047	0.0072	**0.0000**	0.1187	0.0508	0.0491
P. latimucro	8	0.0000	0.0364	0.0181	0.0122	0.0678	0.0997	0.0777	0.0099
P. rufipes *	20	0.0000	**0.0048**	0.0006	0.0011	**0.0000**	0.1147	0.0543	0.0487
P. tomosvaryi	13	0.0000	0.0064	0.0023	0.0016	0.0886	0.1187	0.1080	0.0071
S. erythrocephalum	2	0.0032	0.0032	0.0032	0.0000	0.1184	0.1880	0.1429	0.0175
S. angustipes	1	N/A	N/A	N/A	N/A	0.0765	0.1942	0.1580	0.0149
S. rubzovianum	1	N/A	N/A	N/A	N/A	0.0765	0.1914	0.1645	0.0132
S. costatum *	7	0.0000	**0.0713**	0.0225	0.0336	**0.0586**	0.2028	0.1453	0.0253
S. cryophilum	8	0.0000	0.0048	0.0016	0.0015	0.0586	0.1942	0.1518	0.0279
S. vernum	13	0.0000	0.0312	0.0117	0.0070	0.0877	0.1815	0.1443	0.0196
S. argenteostriatum	8	0.0016	0.0146	0.0058	0.0036	0.0977	0.1814	0.1429	0.0125
S. degrangei	4	0.0000	0.0163	0.0092	0.0065	0.0977	0.1834	0.1458	0.0144
S. ornatum *	25	0.0000	**0.0779**	0.0324	0.0246	**0.0000**	0.1923	0.1275	0.0236
S. trifasciatum *	3	0.0048	**0.0567**	0.0139	0.0268	**0.0000**	0.1858	0.1139	0.0399
S. colombaschense	5	0.0016	0.0032	0.0026	0.0008	0.0670	0.1585	0.1239	0.0205
S. reptans	9	0.0000	0.0130	0.0032	0.0047	0.0600	0.1665	0.1164	0.0219
S. reptantoides	10	0.0000	0.0130	0.0067	0.0044	0.0600	0.1731	0.1234	0.0218
S. argyreatum *	15	0.0000	**0.0415**	0.0143	0.0142	**0.0000**	0.1921	0.1049	0.0530
S. maximum *	8	0.0000	**0.0163**	0.0088	0.0073	**0.0000**	0.2006	0.1106	0.0483
S. monticola *	14	0.0000	**0.0081**	0.0034	0.0023	**0.0000**	0.1984	0.1042	0.0536
S. sp. aff. monticola *	14	0.0000	**0.0000**	0.0000	0.0000	**0.0000**	0.1984	0.1149	0.0447
S. variegatum *	15	0.0000	**0.0113**	0.0055	0.0024	**0.0000**	0.1898	0.1040	0.0531
S. balcanicum *	4	0.0000	**0.0163**	0.0103	0.0059	**0.0081**	0.1808	0.1500	0.0295
S. equinum	3	0.0000	0.0016	0.0011	0.0008	0.1167	0.2028	0.1777	0.0185
S. lineatum *	8	0.0000	**0.0296**	0.0172	0.0088	**0.0081**	0.1813	0.1518	0.0229

*—species with overlapped maximal intraspecific and minimal interspecific genetic distances, overlapping values in bold. Std—standard deviation. n—number of specimens.

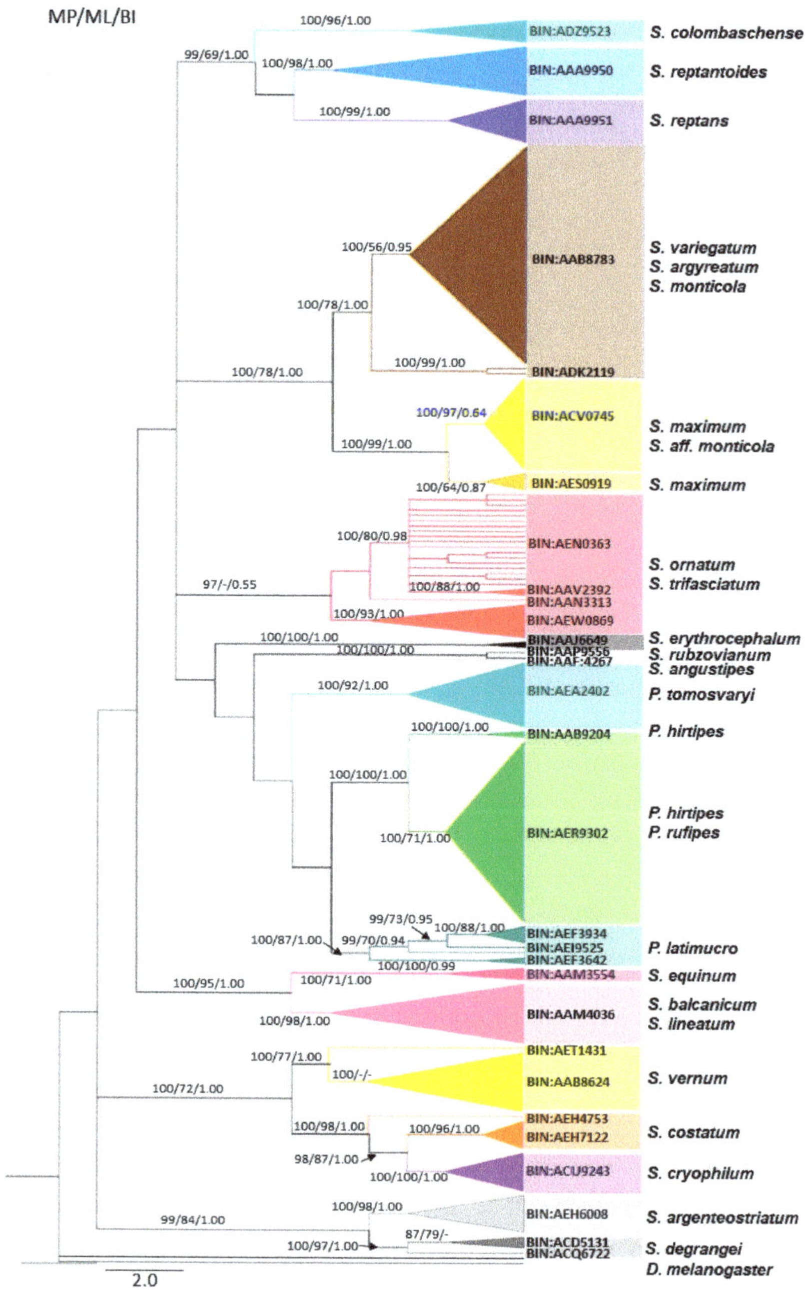

Figure 2. Maximum parsimony (MP) tree based on 235 mitochondrial cytochrome c subunit I haplotypes of 25 black fly species based on 625 bp long alignment. Bootstrap support values for maximum parsimony (MP) and maximum likelihood (ML), and posterior probability values for Bayesian inference (BI) are shown above branches or near branches.

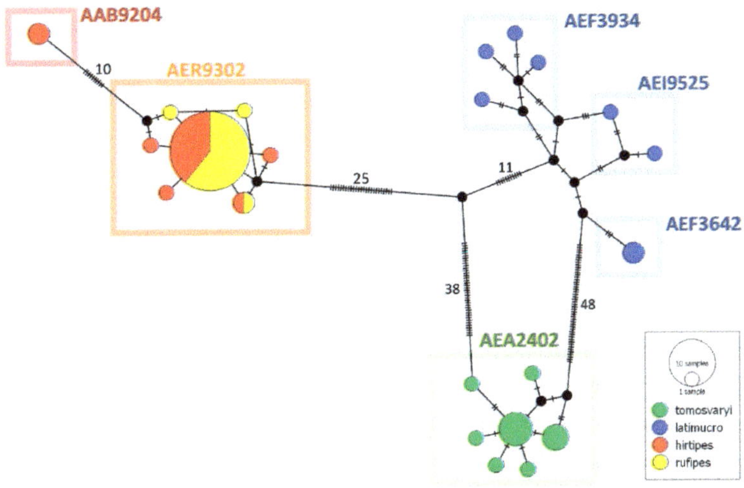

Figure 3. TCS haplotype network of COI 5P haplotypes from 58 individuals of four species of genus *Prosimulium*. Mutational steps are represented by ticks across network connections and also by the number if higher than five.

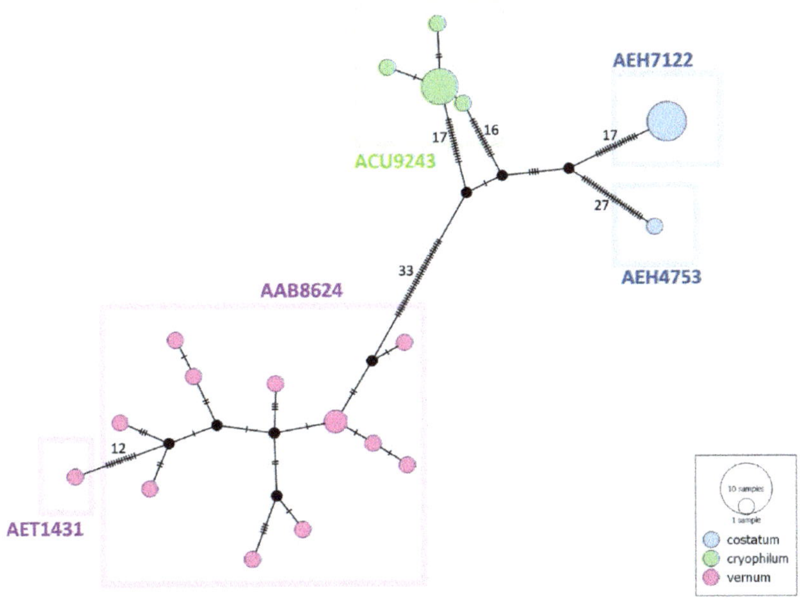

Figure 4. TCS haplotype network of COI 5P haplotypes from 29 individuals of three species of subgenus *Nevermannia*. Mutational steps are represented by ticks across network connections and also by the number if higher than five.

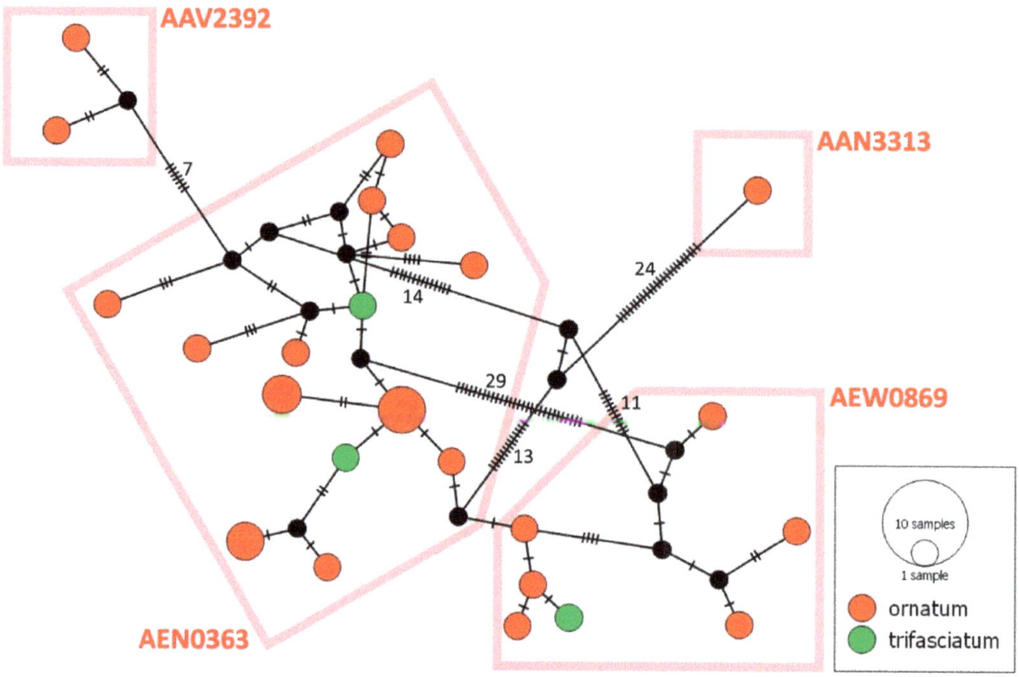

Figure 5. TCS haplotype network of COI 5P haplotypes from 28 individuals of two species of *ornatum* species group. Mutational steps are represented by ticks across network connections and also by the number if higher than five.

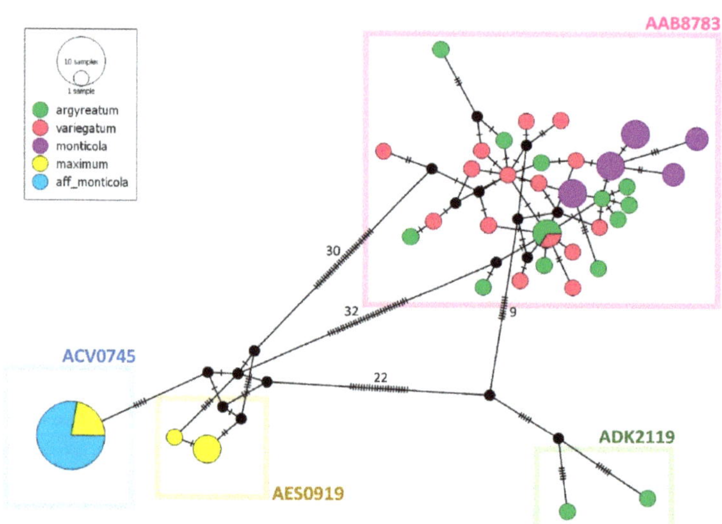

Figure 6. TCS haplotype network of COI 5P haplotypes from 66 individuals of two species of *variegatum* species group. Mutational steps are represented by ticks across network connections and also by the number if higher than five.

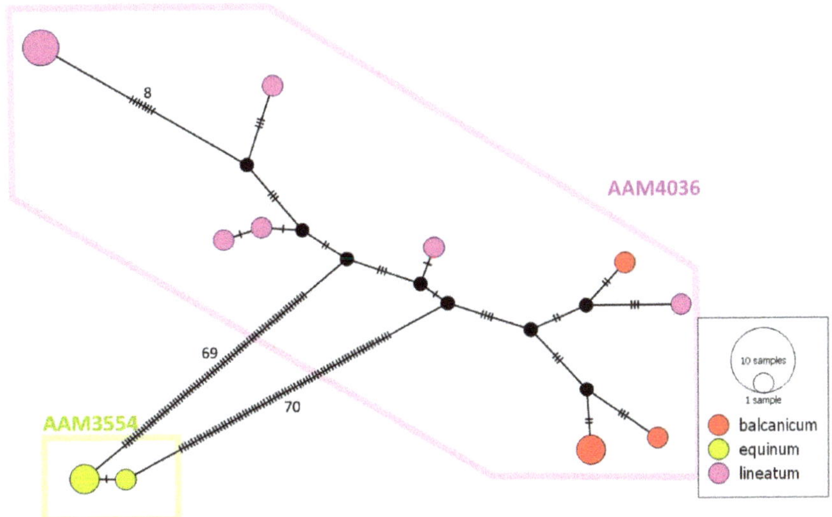

Figure 7. TCS haplotype network of COI 5P haplotypes from 15 individuals of three species of subgenus *Wilhelmia*. Mutational steps are represented by ticks across network connections and also by the number if higher than five.

3.2. *Phylogenetic Trees*

All phylogenetic trees (ML, MP, and BI) yielded similar topologies. All species groups formed well-supported clades (Figure 2), and 14 species formed monophyletic clades with high statistical support. On the other hand, 11 morphologically distinct species did not form monophyletic clades. Non-monophyletic clades formed one species of the genus *Prosimulium* (*P. hirtipes*), all species of *S. variegatum* group (*S. argyreatum, S. monticola, S. variegatum, S. maximum*, and *S.* sp. aff. *monticola*), two species of *S. ornatum* group (*S. ornatum* and *S. trifasciatum*), and two species of subgenus *Wilhelmia* (*S. lineatum* and *S. balcanicum*).

3.3. *Assignment to Barcode Index Numbers (BINs)*

Sequences of all 25 species were classified into 30 BINs. Five BINs were identified as unique (new in the database, Table 1), and three of them were assigned to species for which the COI barcodes are published for the first time. Ten species were assigned a single matching BIN each (*P. tomosvaryi, S. erythrocephalum, S. angustipes, S. rubzovianum, S. cryophilum, S. argenteostiatum, S. reptans, S. reptantoides, S. colombaschense*, and *S. equinum*). Three species (*P. latimucro, S. vernum, S. degrangei*) split into two or more BINs, yet those BINs were species-specific and formed monophyletic clades (Figure 2), thus enabling reliable identification of these taxa. In the case of *S. costatum*, two species-specific BINs were identified but the species appeared to be paraphyletic regarding *S. cryophilum* (Figure 2).

Two species of the subgenus *Wilhelmia* (*S. lineatum, S. balcanicum*) shared the same BIN (merge situation).

Nine of the species were mixtures, i.e., split into two or more BINs and at the same time shared at least one BIN with another species. *P. hirtipes* split into two BINs, one of them shared with the single BIN of *P. rufipes*. The same situation was observed in *S. maximum* with two BINs and one of them shared with *S.* sp. aff. *monticola*. *S. ornatum* split into four BINs and two of them also included specimens of *S. trifasciatum*, split into these two BINs. Three species (*S. argyreatum, S. monticola*, and *S. variegatum*) shared the same BIN, and one additional BIN was found for *S. argyreatum*.

4. Discussion

Since its establishment in 2003, DNA barcoding has become an effective method for specimen identification [43,44]. Thanks to the accessible public databases, the pace of specimen identification to the species level has greatly accelerated. The success rate of specimen identification based on DNA barcodes was nearly 100% in many taxonomic groups [43,45–47]. However, in some groups, the efficiency of DNA barcoding in species identification was lower [48–50].

In black flies, several comprehensive studies showed a high success rate of species identification. A study of 75% of the North American simuliid genera correctly identified nearly 100% of the morphologically distinct species based on DNA barcodes [12]. In Thailand, DNA barcodes provided 96% correct identification of 41 black fly species. Barcodes also successfully differentiated cytoforms of some species complexes; however, in *S. siamense* complex, the levels of success were only 33% [51]. A more recent study from Thailand revealed a 90% level of success in species identification but in several species groups, the efficiency of COI sequences for species identification was very low [18]. In subgenus *Gomphostilbia,* the values of intraspecific and interspecific genetic divergence overlapped in 7 out of 13 species implying that DNA barcoding to identify these species will be ambiguous [52].

Based on COI barcodes, we were able to unambiguously identify only 56% (14 out of 25) of species from our sample. The other 11 species shared a barcode with at least one other species and the values of intraspecific and interspecific K2P genetic distance were overlapping. Sharing of mitochondrial DNA between valid morphological, cytotaxonomic, and biological species has often been recorded across the animal kingdom and represents a serious disadvantage of mitochondrial barcoding [53–55]. The successful identification of species based on DNA barcoding relies on the differences between intraspecific and interspecific genetic divergence and their overlap could lead to errors in species identification. Species non-monophyly (i.e., paraphyly, polyphyly) is the main reason for overlap of intraspecific and interspecific genetic divergence [56].

4.1. Prosimulium, Hirtipes Group

All four species of genus *Prosimulium* reliably present in Slovakia were successfully sampled. Specimens of *Prosimulium tomosvaryi* formed a single well-defined haplotype group, listed under the BIN BOLD:AEA2402. In addition to our samples, the group also includes conspecific specimens from Germany and Romania [57]. Its nearest neighbor BOLD:ADJ9213 represents the British samples of *P. tomosvaryi*.

The alpine species *Prosimulium latimucro* (s.lat.) with intraspecific genetic distance of up to 3.64% shows remarkable mitochondrial diversity, being attributed to three BINs—BOLD:AEI9525, AEF3934, and BOLD:AEF3642. At this point, we consider these groups to be intraspecific diversity, being neighbors of one another and distant to other central European members of the genus. The overall number of BINs for this species, therefore, rises to six public [57]. Two new BINs are currently formed mostly by our specimens from the Tatra Mountains in Slovakia. The exceptions are one single unidentified specimen from Montenegro [57], and one specimen from Spanish Pyrenees (*Prosimulium* sp.) [23,57], both in BIN BOLD:AEF3642. These records indicate a much wider distribution of the clade containing our new BINs. The closest relative of this triplet of BINs as a whole, is *Prosimulium petrosum* Rubtsov, 1955 from Armenia (BOLD:ACQ0837) which is also considered a close relative of *P. latimucro* based on cytotaxonomy and morphology [58].

Two BINs have been identified within samples of *Prosimulium hirtipes* and *P. rufipes* (s.lat.). First of them, BOLD:AAB9204, a widespread haplotype group exclusive to *P. hirtipes* and recorded across its range from the United Kingdom, through Scandinavia to Romania, was present only in two Slovak specimens of *P. hirtipes.* BIN BOLD:AER9302, hitherto including a sole sample of *P. hirtipes* from Spain [23], was present in both, *P. hirtipes* and *P. rufipes* (s.lat.). The BIN was shared almost equally (15 specimens of *P. hirtipes* and 20 of *P. rufipes*), with most specimens carrying a fully identical sequence, not differing in even a single mutation. These unexpected results were consistent in independent DNA isolations and PCR reactions and across localities. The morphological identification of specimens was double-checked by the authors and permanent microscopic slides were made to avoid any ambiguity.

To conclude, in Slovakia the DNA barcoding based on COI 5P region allows reliable identification of *P. tomosvaryi* and *P. latimucro* (s.lat), however, it is not possible to distinguish *P. hirtipes* and *P. rufipes* (s.lat.) sharing a common haplotype group.

4.2. Subgenus Boophtora

Simulium erythrocephalum is the only European member of the subgenus *Boophtora* and has an extremely wide distribution, reaching from the United Kingdom in the west up to the Russian Far East (Adler 2022). Both samples from Slovakia matched the only BIN recorded for this species BOLD:AAJ6649, containing 66 records from various parts of Europe, Armenia, and China [57]. Despite the extant range of the species, so far, there is no indication of higher genetic diversity for this species; however, more detailed sampling is needed.

4.3. Subgenus Eusimulium

We analyzed only two specimens of the subgenus *Eusimulium*, representing two of three species recorded in Slovakia so far. *Simulium angustipes* fits with the only BIN recorded for this species containing 29 samples and distributed from the United Kingdom over Sweden, Finland, Germany, and Armenia up to China. Similarly, *S. rubzovianum* fits with the only BIN recorded for this species containing 50 samples (however, listed under different names) and recorded in various parts of Europe, Morocco, and Turkey. A study of all four species of the subgenus *Eusimulium* in Great Britain showed that the morphologically and chromosomally well-defined populations differed markedly in their COI barcodes [21] and therefore barcoding seems to be suitable for the identification of these species, which are morphologically extremely similar and except for adult males, not possible to identify based on morphological characters only. Therefore, it would be useful to verify if barcoding allows the identification of the *Eusimulium* species also outside of Great Britain.

4.4. Subgenus Nevermannia, Vernum Group

Within subgenus *Nevermannia,* we sampled and sequenced only three species of the *vernum* species group, therefore *Nevermannia* had the smallest representation among all analyzed groups. Two of the analyzed species had a significant divergence of the COI barcode, each consisting of two BINs, with a genetic distance of 3.12% in *S. vernum* and 7.13% in *S. costatum*. The BIN of *S. costatum* BOLD:AEH7122 with six samples from Slovakia is unique, the only sample of *S. costatum* BOLD:AEH4753 fits with five samples from Turkey, additionally *S. costatum* BOLD: AAD1733 was reported from Sweden, United Kingdom, Germany, and Austria but not found in Slovakia. Twelve specimens of *S. vernum* belong to the large group BOLD:AAB8624 consisting of 65 samples distributed across large areas of Europe, and one sample represents a new unique BIN (BOLD:AET1431). In addition, five more BINs were reported where samples identified as *S. vernum* occur in Europe and in southwestern Asia. The eight specimens of *S. cryophilum* showed quite small interspecific genetic distance of 0.48% and all of them are assigned to the BOLD:ACU9243, distributed across Europe.

Higher levels of intraspecific genetic distance usually indicate the presence of species complexes or sibling species [12,14]. In black flies, the value of maximal intraspecific genetic distance around 5% (4.58–6.5%) is typically linked to the species complexes [12]. Therefore, the high level of genetic distance within *S. costatum* (7.13%) indicates the presence of two species.

In the checklist of black flies of Slovakia, 13 species of the subgenus *Nevermannia* are listed [24], and 12 of them belong to the *vernum* species group. It is likely that other species of the *vernum* group occur in Slovakia, as seven of them have been recorded in the surrounding countries and several others in the Alps within Germany, Italy, or France [1].

A more detailed study of the barcodes of the *vernum* group based on a larger dataset is strongly needed.

4.5. Subgenus Simulium, Argenteostriatum Group

Simulium argenteostriatum COI sequences are published for the first time here and they represent a new and unique BIN BOLD:AEH6008, which is quite distant from all the other known barcodes. The most similar BIN is BOLD:ACQ0608 represented by *S. aureofulgens* (Terteryan, 1949) from Armenia with a minimal genetic divergence of 8.66% [57,59]. There is no record of a barcode for the other two species from the group. The closest species within our samples was *S. degrangei* with a minimal genetic divergence of 9.77%.

4.6. Subgenus Simulium, Bukovskii Group

Simulium degrangei, the only representative of this group in Slovakia, showed a high intraspecific variability (up to 1.63%) and consisted of three BINs (Table 1). The COI sequences, including samples from Slovakia, have already been studied [60], and apart from the higher genetic divergence, no indications of the presence of cryptic species were found, therefore we consider the variability to be intraspecific.

4.7. Subgenus Simulium, Ornatum Group

Only two species of the *ornatum* group, *S. ornatum* and *S. trifasciatum*, are known in Slovakia [24]. We morphologically identified both in our samples. The three individuals of *S. trifasciatum* were assigned to the BINs BOLD:AEW0869 and BOLD:AEN0363; both also contained individuals which we determined as *S. ornatum*. The 34 individuals of *S. ornatum* showed very high intraspecific variability up to 7.79% and were split into four BINs. The majority of them (20) belong to BOLD:AEN0363, which is distributed from the United Kingdom up to Iran and Georgia; however, it is missing in northern Europe. Three individuals in our sample were assigned to BOLD:AAN3313, a smaller group recorded only in the United Kingdom and in Spain so far and being close to the previous BIN. The second largest group (10) was assigned to BOLD:AEW0869, a group common in the United Kingdom and northern Europe and reaching up to Romania in the south and east. Finally, one individual is assigned to BOLD:AAN3313, a group containing mostly individuals identified as *S. intermedium* Roubaud, 1906 and with most sequences recorded from Finland, followed by Spain and the United Kingdom. For further analyses, a complex taxonomic revision of the S. *ornatum* complex and the entire *ornatum* group is essential.

4.8. Subgenus Simulium, Variegatum Group

The five species of the *variegatum* species group formed a paraphyletic clade. All species split into four groups corresponded to different BINs, however, the BINs and the species did not match at all. *Simulium maximum* and *S.* sp. aff. *monticola* formed a separate well-supported clade of two BINs (Table 1, Figures 2 and 6), which was quite distant from the other species with a genetic divergence of at least 6.20%. BOLD:AES0919 consisted of five samples of *S. maximum*; BOLD:ACV0745 included all samples of *S.* sp. aff. *monticola* and four samples of *S. maximum*, this BIN showed only minimal diversity, sharing a fully identical haplotype among all samples. *Simulium argyreatum*, *S. monticola*, and *S. variegatum* presented the second well-supported clade within the *variegatum* group (Figure 2) consisting of three BINs. BOLD:ADK2119 contained two specimens of *S. argyreatum*, all the other 42 samples represented the BIN BOLD:AAB8783 and included all three species. The internal variability within this BIN was relatively high. The samples of these three species included pupae only; thus, the morphological identification of specimens does not leave any space for misidentification. A similar result was shown for these three species in Spain [23], and the other unpublished samples in BOLD confirm this pattern. In summary, DNA barcoding based on COI 5P is completely unusable for species identification within the *variegatum* species group in Europe.

4.9. Subgenus Simulium, Reptans Group

All three species of the *reptans* group form separate monophyletic clades with high bootstrap support in the phylogenetic trees and one matching BIN was assigned to each species (Figure 2). All species of *reptans* group can be unambiguously identified by the COI barcodes. The COI sequences of *S. reptans* and *S. reptantoides* have been studied across Europe [19,20,61], and both species differ consistently to a similar extent. *Simulium reptantoides* is missing in the Scandinavian and Baltic countries; however, two lineages of *S. reptans* occur in Sweden and Great Britain, and only one of them was recorded in central Europe and Balkan countries [19,20,61]. *Simulium colombaschense* COI sequences are published for the first time here. Previous chromosomal studies revealed the presence of five cytoforms with different geographical distributions, at least some of which represent reproductively isolated species. According to the place of origin of the sequenced *S. colombaschense* samples located in the Danube river, we assume that they belong to the cytoform 'A', which probably represents the nominal species [62].

4.10. Subgenus Wilhelmia, Lineatum Group

All three specimens of *S. equinum* represented a well-supported sister branch to *S. balcanicum* and *S. lineatum* in the phylogenetic trees (Figure 2), and they were assigned to BOLD:AAM3554, a barcode recorded in 89 samples distributed across Europe and reaching Turkey but missing in northern Europe. *Simulium equinum* in northern Europe is assigned to BOLD:AAP9428 (five specimens in Finland), and this BIN is recorded also for Turkey (three specimens). The other two species of *Wilhelmia*, *S. balcanicum* and *S. lineatum*, formed one well-supported branch in the phylogenetic trees. Despite a few differences between the species in the phylogenetic trees and haplotype network, we could not distinguish between them based solely on the COI barcode sequence. Both species shared one BIN (BOLD:AAM4036), although small differentiation between the species could be found in the phylogenetic trees and haplotype network (Figure 7). This result agrees with the previous study of these two species in larger areas of Europe and Turkey [63]. According to the chromosomal study [64] and the small but consistent morphological difference between the pupae of both species, they can be considered closely related sister species, and the overlap in their barcodes could be the consequence of incomplete lineage sorting.

Mitochondrial haplotypes mixed and shared between valid morphological and cytotaxonomic species have been repeatedly recorded across the black fly family [23,65–67] hindering the ability of DNA barcoding based on COI-5P region to reliably identify species. Our study increases the number of such cases.

In some cases, e.g., *S. degrangei*, despite high genetic diversity and multiple BINs, there are no indications of the existence of several taxa [60]. In others, such as *S. ornatum* group and *S. vernum* group, multiple cryptic species are expected to exist, which may correspond to separate haplotype groups after resolving. The most intriguing situations are those where haplotype groups are partially or fully shared between well-defined species, presumably due to phenomena such as retention of ancestral polymorphism and introgression [68,69].

In the future, it will be necessary to increase the sampling of taxa poorly represented and missing in this study. Nuclear markers may be required for problematic groups to resolve taxonomy and allow molecular identification.

Author Contributions: Conceptualization, T.K.; material collection, T.K., M.K., S.K. and I.L.; material identification, M.K., T.K., K.M. and S.K.; molecular methods, T.K., B.B., I.L. and S.K.; data processing and analyses, T.K., M.K. S.K. and I.L.; figures preparation, T.K., S.K. and I.L.; writing and editing manuscript, M.K., T.K., S.K. and I.L. All authors have read and agreed to the published version of the manuscript.

Funding: This work was supported by the Slovak Research and Development Agency under the Contract no. APVV-19-0076 and by the VEGA grant No. 1/0704/20 of the Ministry of Education, Science, Research and Sport of the Slovak Republic.

Data Availability Statement: Data are publicly available in BOLD (Dataset DS-SIMSK) and GenBank Databases.

Acknowledgments: We are grateful to Ladislav Jedlička for his comments on the preliminary results of this study and to Zuzana Čiamporová-Zaťovičová for providing the material of *Prosimulium latimucro* from three sites in the Tatra Mountains. We thank the three anonymous reviewers, who helped to improve the manuscript.

Conflicts of Interest: The authors declare no conflict of interest.

References

1. Adler, P.H. World Blackflies (Diptera:Simuliidae): A Comprehensive Revision of the Taxonomic and Geographical Inventory. 2022. Available online: https://biomia.sites.clemson.edu/pdfs/blackflyinventory.pdf (accessed on 9 March 2023).
2. Adler, P.H.; McCreadie, J.W. Black flies (Simuliidae). In *Medical and Veterinary Entomology*, 3rd ed.; Mullen, G.R., Durden, L.A., Eds.; Elsevier: London, UK, 2019; pp. 237–259.
3. World Health Organization. Available online: https://www.who.int/news-room/fact-sheets/detail/onchocerciasis (accessed on 30 March 2023).
4. Cheema, A.H.; Ivoghli, B. Bovine onchocerciasis caused by *Onchocerca armillata* and *O. gutturosa*. *Vet. Pathol.* **1978**, *15*, 495–505. [CrossRef]
5. McCall, P.J.; Trees, A.J. Onchocerciasis in British cattle: A study of the transmission of *Onchocerca* sp. in north Wales. *J. helminthol.* **1993**, *67*, 123–135. [CrossRef]
6. Herman, C.M.; Barrow, J.H., Jr.; Tarshis, I.B. Leucocytozoonosis in Canada geese at the Seney national wildlife refuge. *J. Wildl. Dis.* **1975**, *11*, 404–411. [CrossRef]
7. Crosskey, R.W. *The Natural History of Blackflies*; John Wiley & Sons: Chichester, UK, 1990; p. 711.
8. Car, M.; Kutzer, E.; Tauber, R. The blackflies in the autonomic region of South Tyrol-Trentino and their veterinary medical importance. *Wien. Tierartzl. Monatsschr.* **2001**, *88*, 11–17.
9. Zanin, E.; Rivosecchi, L. Attacco Massivo e Ruolo Patogeno di Simulidi del Gruppo reptans (Diptera, Nematocera) sul Bestiame in Provincia di Trento. *Atti Soc. Ital. Sci. Vet.* **1975**, *28*, 865–868.
10. Adler, P.H.; Cheke, R.A.; Post, R.J. Evolution, epidemiology, and population genetics of black flies (Diptera: Simuliidae). *Infect. Genet. Evol.* **2010**, *10*, 846–865. [CrossRef] [PubMed]
11. Adler, P.H.; Crosskey, R.W. Cytotaxonomy of the Simuliidae (Diptera): A systematic and bibliographic conspectus. *Zootaxa* **2015**, *3975*, 1–139. [CrossRef] [PubMed]
12. Rivera, J.; Currie, D.C. Identification of Nearctic black flies using DNA barcodes (Diptera: Simuliidae). *Mol. Ecol. Resour.* **2009**, *9*, 224–236. [CrossRef] [PubMed]
13. Conflitti, I.M.; Pruess, K.P.; Cywinska, A.; Powers, T.O.; Currie, D.C. DNA barcoding distinguishes pest species of the black fly genus Cnephia (Diptera: Simuliidae). *J. Med. Entomol.* **2013**, *50*, 1250–1260. [CrossRef]
14. Hernandez-Triana, L.M.; Chaverri, L.G.; Rodriguez-Perez, M.A.; Prosser, S.W.; Hebert, P.D.; Gregory, T.R.; Johnson, N. DNA barcoding of Neotropical black flies (Diptera: Simuliidae): Species identification and discovery of cryptic diversity in Mesoamerica. *Zootaxa* **2015**, *3936*, 93–114. [CrossRef]
15. Pramual, P.; Wongpakam, K.; Adler, P.H. Cryptic biodiversity and phylogenetic relationships revealed by DNA barcoding of Oriental black flies in the subgenus Gomphostilbia (Diptera: Simuliidae). *Genome* **2011**, *54*, 1–9. [CrossRef] [PubMed]
16. Pramual, P.; Wongpakam, K. Association of black fly (Diptera: Simuliidae) life stages using DNA barcode. *J. Asia Pac. Entomol.* **2014**, *17*, 549–554. [CrossRef]
17. Low, V.L.; Srisuka, W.; Saeung, A.; Tan, T.K.; Ya'cob, Z.; Yeong, Y.S.; Takaoka, H. DNA Barcoding of Simulium asakoae (Diptera: Simuliidae) From Northern Thailand. *J. Med. Entomol.* **2020**, *57*, 1675–1678. [CrossRef] [PubMed]
18. Pramual, P.; Jomkumsing, P.; Wongpakam, K.; Wongwian, P. DNA barcoding of tropical black flies (Diptera: Simuliidae) in Thailand: One decade of progress. *Acta Trop.* **2021**, *224*, 106116. [CrossRef]
19. Day, J.C.; Goodall, T.I.; Post, R.J. Confirmation of the species status of the blackfly Simulium galeratum in Britain using molecular taxonomy. *Med. Vet. Entomol.* **2008**, *22*, 55–61. [CrossRef] [PubMed]
20. Kúdela, M.; Brúderová, T.; Jedlička, L.; Bernotienė, R.; Celec, P.; Szemes, T. The identity and genetic characterization of Simulium reptans (Diptera: Simuliidae) from central and northern Europe. *Zootaxa* **2014**, *3802*, 301–317. [CrossRef]
21. Day, J.C.; Mustapha, M.; Post, R.J. The subgenus Eusimulium (Diptera: Simuliidae: Simulium) in Britain. *Aquat. Insects* **2010**, *32*, 281–292. [CrossRef]
22. Kúdela, M.; Adler, P.H.; Kudelova, T. Taxonomic status of the black fly Prosimulium italicum Rivosecchi (Diptera: Simuliidae) based on genetic evidence. *Zootaxa* **2018**, *4377*, 280–290. [CrossRef]
23. Ruiz-Arrondo, I.; Hernández-Triana, L.M.; Ignjatović-Ćupina, A.; Nikolova, N.; Garza-Hernández, J.A.; Rodríguez-Pérez, M.A.; Oteo, J.A.; Fooks, A.R.; Lucientes Curdi, J. DNA barcoding of blackflies (Diptera: Simuliidae) as a tool for species identification and detection of hidden diversity in the eastern regions of Spain. *Parasit. Vectors* **2018**, *11*, 463. [CrossRef]

24. Jedlička, L.; Knoz, J. Simuliidae Newman, 1834. In *Checklist of Diptera of the Czech Republic and Slovakia*; Electronic Version, 2; Jedlička, L., Kúdela, M., Stloukalová, V., Eds.; Comenius University in Bratislava: Bratislava, Slovakia, 2009; Available online: http://www.edvis.sk/diptera2009/families/simuliidae.html (accessed on 9 March 2023).
25. Jedlička, L.; Kúdela, M.; Stloukalová, V. Key to the identification of blackfly pupae (Diptera: Simuliidae) of Central Europe. *Biol.-Sect. Zool.* **2004**, *59*, 157–178.
26. Bass, J. *Last-Instar Larvae and Pupae of the Simuliidae of Britain and Ireland: A Key with Brief Ecological Notes*; Freshwater Biological Association Scientific Publication 55; Freshwater Biology Association: Ambleside, UK, 1998; p. 101.
27. Rubzov, I.A. Moshki (sem. Simuliidae). In *Fauna SSSR, Nasekomyie Dvukrylyie*; Izdatel'stvo Akademii Nauk SSSR: Moskva–Leningrad, Russia, 1956; p. 860.
28. Rivosecchi, L. Simuliidae: Diptera. Nematocera. In *Fauna d'Italia*; Edizioni Calderini: Bologna, Italy, 1978; Volume XIII, p. 538.
29. Rivosecchi, L.; Addonisio, M.; Maiolini, B. *I Ditteri Simulidi, Nuove Chiavi Dicotomiche per l 'Identificazione Delle Specie Italiane Con Brevi Note Bio-Tassonomiche*; Quaderni del Museo Tridentino di Scienze Naturali: Trento, Italy, 2007; p. 149.
30. Knoz, J. To Identification of Czechoslovakian Black-Flies (Diptera, Simuliidae). *Folia Fac. Sci. Nat. Univ. Purkyn. Brun.* **1965**, *6*, 1–54.
31. Folmer, O.; Black, M.; Hoeh, W.; Lutz, R.; Vrijenhoek, R. DNA primers for amplification of mitochondrial cytochrome c oxidase subunit I from diverse metazoan invertebrates. *Mol. Mar. Biol. Biotechnol.* **1994**, *3*, 294–299. [PubMed]
32. Kearse, M.; Moir, R.; Wilson, A.; Stones-Havas, S.; Cheung, M.; Sturrock, S.; Buxton, S.; Cooper, A.; Markowitz, S.; Duran, C.; et al. Geneious Basic: An integrated and extendable desktop software platform for the organization and analysis of sequence data. *Bioinformatics* **2012**, *28*, 1647–1649. [CrossRef] [PubMed]
33. Tamura, K.; Stecher, G.; Kumar, S. MEGA11: Molecular Evolutionary Genetics Analysis Version 11. *Mol. Biol. Evol.* **2021**, *38*, 3022–3027. [CrossRef]
34. Darriba, D.; Taboada, G.L.; Doallo, R.; Posada, D. jModelTest 2: More models, new heuristics and parallel computing. *Nat. Methods* **2012**, *9*, 772. [CrossRef]
35. Swofford, D.L. PAUP: Phylogenetic Analysis Using Parsimony (and Other Methods), Version 4.0 Beta. 2002. Available online: http://paup.csit.fsu.edu/ (accessed on 16 January 2023).
36. Ronquist, F.; Huelsenbeck, J.P. MrBayes 3: Bayesian phylogenetic inference under mixed models. *Bioinformatics* **2003**, *19*, 1572–1574. [CrossRef]
37. Stamatakis, A. Raxml version 8: A tool for phylogenetic analysis andpost-analysis of large phylogenies. *Bioinformatics* **2014**, *30*, 1312–1313. [CrossRef]
38. Edler, D.; Klein, J.; Antonelli, A.; Silvestro, D. raxmlGUI 2.0: A graphical interface and toolkit for phylogenetic analyses using RAxML. *Methods Ecol. Evol.* **2021**, *12*, 373–377. [CrossRef]
39. Cooper, J.K.; Sykes, G.; King, S.; Cottrill, K.; Ivanova, N.V.; Hanner, R.; Ikonomi, P. Species identification in cell culture: A two-pronged molecular approach. *Vitr. Cell. Dev. Biol.-Anim.* **2007**, *43*, 344–351. [CrossRef]
40. Leigh, J.W.; Bryant, D. PopART: Full-feature software for haplotype network construction. *Methods Ecol. Evol.* **2015**, *6*, 1110–1116. [CrossRef]
41. Clement, M.; Snell, Q.; Walke, P.; Posada, D.; Crandall, K. TCS: Estimating gene genealogies. *Proc. 16th Int. Parallel Distrib. Process. Symp.* **2002**, *2*, 184.
42. QGIS.org. QGIS Geographic Information System. QGIS Association. 2023. Available online: http://www.qgis.org (accessed on 15 February 2023).
43. Hebert, P.D.; Cywinska, A.; Ball, S.L.; DeWaard, J.R. Biological identifications through DNA barcodes. *Proc. R. Soc. Lond. B Biol. Sci. Ser. B–Biol. Sci.* **2003**, *270*, 313–321. [CrossRef] [PubMed]
44. Savolainen, V.; Cowan, R.S.; Vogler, A.P.; Roderick, G.K.; Lane, R. Towards writing the encyclopaedia of life: An introduction to DNA barcoding. *Philos. Trans. R. Soc. B Biol. Sci.* **2005**, *360*, 1805–1811. [CrossRef] [PubMed]
45. Ward, R.D.; Zemlak, T.S.; Innes, B.H.; Last, P.R.; Hebert, P.D. DNA barcoding Australia's fish species. *Philos. Trans. R. Soc. B Biol. Sci.* **2005**, *360*, 1847–1857. [CrossRef] [PubMed]
46. Pentinsaari, M.; Hebert, P.D.; Mutanen, M. Barcoding beetles: A regional survey of 1872 species reveals high identification success and unusually deep interspecific divergences. *PLoS ONE* **2014**, *9*, e108651. [CrossRef]
47. Dincă, V.; Lukhtanov, V.A.; Talavera, G.; Vila, R. Unexpected layers of cryptic diversity in wood white Leptidea butterflies. *Nat. Commun.* **2011**, *2*, 324. [CrossRef]
48. Galimberti, A.; Assandri, G.; Maggioni, D.; Ramazzotti, F.; Baroni, D.; Bazzi, G.; Chiandetti, I.; Corso, A.; Ferri, V.; Galuppi, M.; et al. Italian odonates in the Pandora's box: A comprehensive DNA barcoding inventory shows taxonomic warnings at the Holarctic scale. *Mol. Ecol. Resour.* **2021**, *21*, 183–200. [CrossRef]
49. Elias, M.; Hill, R.I.; Willmott, K.R.; Dasmahapatra, K.K.; Brower, A.V.; Mallet, J.; Jiggins, C.D. Limited performance of DNA barcoding in a diverse community of tropical butterflies. *Proc. R. Soc. B Biol. Sci.* **2007**, *274*, 2881–2889. [CrossRef]
50. Wiemers, M.; Fiedler, K. Does the DNA barcoding gap exist?—A case study in blue butterflies (Lepidoptera: Lycaenidae). *Front. Zool.* **2007**, *4*, 8. [CrossRef]
51. Pramual, P.; Adler, P.H. DNA barcoding of tropical black flies (Diptera: Simuliidae) of Thailand. *Mol. Ecol. Resour.* **2014**, *14*, 262–271. [CrossRef]
52. Phayuhasena, S.; Colgan, D.J.; Kuvangkadilok, C.; Pramual, P.; Baimai, V. Phylogenetic relationships among the black fly species (Diptera: Simuliidae) of Thailand based on multiple gene sequences. *Genetica* **2010**, *138*, 633–648. [CrossRef] [PubMed]

53. Ballard, J.W.O. When one is not enough: Introgression of mitochondrial DNA in Drosophila. *Mol. Biol. Evol.* **2000**, *17*, 1126–1130. [CrossRef] [PubMed]
54. Bachtrog, D.; Thornton, K.; Clark, A.; Andolfatto, P. Extensive introgression of mitochondrial DNA relative to nuclear genes in the Drosophila yakuba species group. *Evolution* **2006**, *60*, 292–302.
55. Seixas, F.A.; Boursot, P.; Melo-Ferreira, J. The genomic impact of historical hybridization with massive mitochondrial DNA introgression. *Genome Biol.* **2018**, *19*, 91. [CrossRef]
56. Meyer, C.P.; Paulay, G. DNA barcoding: Error rates based on comprehensive sampling. *PLoS Boil.* **2005**, *3*, e422. [CrossRef]
57. BOLDSYSTEMS. Available online: http://www.boldsystems.org/index.php (accessed on 27 April 2023).
58. Vlasov, S.; Harutyunova, M.; Harutyunova, K.; Adler, P.H. Chromosomal evidence of species status and evolutionary relationships of the black fly Prosimulium petrosum (Diptera, Simuliidae) in Armenia. *Comp. Cytogenet.* **2016**, *10*, 33. [CrossRef] [PubMed]
59. Andrianov, B.V.; Goryacheva, I.I.; Vlasov, S.V.; Gorelova, T.V.; Harutyunova, M.V.; Harutyunova, K.V.; Mayilyan, K.R.; Zakharov, I.A. Identification of potentially invasive species of black flies [Diptera: Simuliidae] from Armenia based on an analysis of variability in the mtDNA barcode of the cox1 gene and chromosomal polymorphism. *Rus. J. Genet.* **2015**, *51*, 289–299. [CrossRef]
60. Jedlička, L.; Kúdela, M.; Szemes, T.; Celec, P. Population genetic structure of Simulium degrangei (Diptera: Simuliidae) from Western Carpathians. *Biologia* **2012**, *67*, 777–787. [CrossRef]
61. Đuknić, J.; Jovanović, V.M.; Atlagić, J.Č.; Andjus, S.; Paunović, M.; Živić, I.; Popović, N. Simulium reptans (Linnaeus, 1758) and Simulium reptantoides Carlsson, 1962 from the Balkan Peninsula. *ZooKeys* **2020**, *922*, 141–155. [CrossRef]
62. Adler, P.H.; Kúdelová, T.; Kúdela, M.; Seitz, G.; Ignjatović-Ćupina, A. Cryptic biodiversity and the origins of pest status revealed in the macrogenome of Simulium colombaschense (Diptera: Simuliidae), history's most destructive black fly. *PLoS ONE* **2016**, *11*, e0147673. [CrossRef]
63. Inci, A.; Yildirim, A.; Duzlu, O.; Onder, Z.; Ciloglu, A.; Seitz, G.; Adler, P.H. Genetic diversity and identification of Palearctic black flies in the subgenus Wilhelmia (Diptera: Simuliidae). *J. Med. Entomol.* **2017**, *54*, 888–894. [CrossRef]
64. Adler, P.H.; Inci, A.; Yildirim, A.; Duzlu, O.; McCreadie, J.W.; Kúdela, M.; Khazeni, A.; Brúderová, T.; Seitz, G.; Takaoka, H.; et al. Are black flies of the subgenus Wilhelmia (Diptera: Simuliidae) multiple species or a single geographical generalist? Insights from the macrogenome. *Biol. J. Linn. Soc.* **2015**, *114*, 163–183. [CrossRef]
65. LaRue, B.; Gaudreau, C.; Bagre, H.O.; Charpentier, G. Generalized structure and evolution of ITS1 and ITS2 rDNA in black flies (Diptera: Simuliidae). *Mol. Phylogenet. Evol.* **2009**, *53*, 749–757. [CrossRef] [PubMed]
66. Ilmonen, J.; Adler, P.H.; Malmqvist, B.; Cywinska, A. The Simulium vernum group (Diptera: Simuliidae) in Europe: Multiple character sets for assessing species status. *Zool. J. Linn. Soc.* **2009**, *156*, 847–863. [CrossRef]
67. Conflitti, I.M.; Kratochvil, M.J.; Spironello, M.; Shields, G.F.; Currie, D.C. Good species behaving badly: Non-monophyly of black fly sibling species in the Simulium arcticum complex (Diptera: Simuliidae). *Mol. Phylogenet. Evol.* **2010**, *57*, 245–257. [CrossRef]
68. Moritz, C.; Cicero, C. DNA barcoding: Promises and pitfalls. *PLoS Biol.* **2004**, *2*, 1529–1531. [CrossRef] [PubMed]
69. Rubinoff, D.; Cameron, S.; Kipling, W. A genomic perspective on the shortcomings of mitochondrial DNA for barcoding and DNA taxonomy. *J. Hered.* **2006**, *97*, 581–594. [CrossRef]

diversity

MDPI

Communication

First Record of *Phlebotomus* (*Larroussius*) *perfiliewi* (Diptera: Psychodidae), Vector of *Leishmania infantum* and Phleboviruses, in Spain

Mikel A. González [1], Ignacio Ruiz-Arrondo [2,*], Rafael Gutiérrez-López [1,3], Carlos Barceló [1] and Miguel Á. Miranda [1]

[1] Applied Zoology and Animal Conservation Research Group, University of the Balearic Islands, Crta. de Valldemossa km 7.5, 07122 Palma, Spain

[2] Center of Rickettsiosis and Arthropod-Borne Diseases (CRETAV), Infectious Diseases Department, San Pedro University Hospital-Center for Biomedical Research from La Rioja (CIBIR), C/Piqueras 98, 3ª Planta, 26006 La Rioja, Spain

[3] Animal Health Research Center, National Food and Agriculture Research and Technology Institute (INIA-CISA-CSIC), 28040 Valdeolmos, Spain

* Correspondence: irarrondo@riojasalud.es

Abstract: Phlebotomine sand flies are vectors of several diseases of importance for public health, including leishmaniosis, bartonellosis, and sand fly fevers. An entomological survey on blood-feeding Diptera was conducted in June–November 2020–2021 to know the diversity of insect vectors in Mallorca (Balearic Islands, Spain). Among the vectors collected, *Phlebotomus* (*Larroussius*) *perfiliewi* Parrot, 1930 was found being the first record of this species in Spain. *Phlebotomus perfiliewi* s.l. is one of the main vectors of *Leishmania infantum* in the Mediterranean Basin and Central Asia. The identification of this species was confirmed by both morphological features and DNA barcoding. Phylogenetic analyses showed that the specimens captured were *Ph. perfiliewi* s.s. (99.85–100% homologues from Italy and Algeria specimens), with a sequence divergence of 0.17%. The cytochrome c oxidase subunit I gene clearly separates the three species that make up the *Ph. perfiliewi* species complex. In addition, we also provide a brief discussion about their identification remarks, phylogenetic relationships, and vector status.

Keywords: Balearic Islands; barcoding; Mallorca; Phlebotominae; sand fly; taxonomy

Citation: González, M.A.; Ruiz-Arrondo, I.; Gutiérrez-López, R.; Barceló, C.; Miranda, M.Á. First Record of *Phlebotomus* (*Larroussius*) *perfiliewi* (Diptera: Psychodidae), Vector of *Leishmania infantum* and Phleboviruses, in Spain. *Diversity* **2023**, *15*, 400. https://doi.org/10.3390/d15030400

Academic Editor: Luc Legal

Received: 23 January 2023
Revised: 3 March 2023
Accepted: 7 March 2023
Published: 10 March 2023

1. Introduction

Phlebotomine sand flies (Diptera: Psychodidae: Phlebotominae) are blood-feeding insects of major medical–veterinary significance as they are vectors of a large number of pathogens to animals and humans, including protozoa, bacteria, and viruses [1]. The parasitic protozoan *Leishmania* sp. (Kinetoplastida: Trypanosomatidae) is responsible for two clinical major forms, cutaneous leishmaniosis and visceral leishmaniosis, which are endemic in several countries of the European Union, including the Iberian Peninsula, as well as the Balearic Islands [2]. In Europe, the geographic distribution of sand flies increased considerably in the last years, spreading into new areas [3] and causing progressively more autochthonous outbreaks of phlebotomine-borne diseases [4]. Therefore, studies on sand flies are increasing attention along most European countries.

Twenty-two species of sand flies were described in Europe [5], thirteen of them are present in Spain [6], and five in the Balearic Islands [7,8] [*Phlebotomus perniciosus* Newstead, 1911; *Phlebotomus sergenti* (Parrot, 1917); *Phlebotomus papatasi* (Scopoli, 1786); *Sergentomyia minuta* (Rondani, 1843); and *Phlebotomus ariasi* (Tonnoir, 1921)]. Based on their geographical distribution and abundance, *Ph. perniciosus* and *S. minuta* are the most widespread and abundant species in the Spanish territory [6]. Regarding their vector status, *Ph. ariasi*, *Phlebotomus langeroni* Nitzulescu, 1930, *Phlebotomus mascittii* Grassi, 1908, and primarily *Ph.*

perniciosus are highlighted to be proven or suspected vectors of *Leishmania infantum* Nicolle, 1908 in Spain [6]. However, other sand fly species were also highlighted in other European countries to be involved in the transmission of *L. infantum*, such as *Ph. tobbii* Adler, Theodor and Lourie, 1930 and *Ph. neglectus* Tonnoir, 1921, as well as *Phlebotomus papatasi* (Scolopi, 1976) and *Phlebotomus perfiliewi* s.l. Parrot, 1930, which are also vectors of other Phlebovirus. The latter two species were pointed out to be changing their distribution range [3].

The taxonomic status of *Ph. perfiliewi* s.l. still remains undefined and some discrepancies were reported between morphological analysis and molecular markers [9]. Overall, at least three specific or subspecific value names are proposed into the *Ph. perfiliewi* species complex, depending on the authors: *Phebotomus perfiliewi* s.s. Parrot, 1930, *Phlebotomus galilaeus* Theodor, 1958, and *Phlebotomus transcaucasicus* Perfiliev, 1937, the former being the taxa that occurs in western Mediterranean regions.

The current study provides the finding of a new phlebotomine record from Spain, and in particular in Mallorca, the largest island of the Balearic Islands.

2. Materials and Methods

As part of a project focused on the collection of blood-feeding arthropods, a multi-trapping entomological survey was conducted in eight farms (Formatges Burguera: 39.366579 and 3.02172891, 3 masl; Son Ajaume nou: 39.6448596 and 2.65217317, 89 masl; Can Cosme: 39.5222862 and 3.10583271, 80 masl; Son Simó: 39.8173189 and 3.05957789, 14 masl; Ranxo Ses Roques: 39.8331397 and 3.10518693, 3 masl; Centre Hipic Son Reus: 39.6377295 and 2.66639607, 76 masl; Sa Teulera: 39.5840583, 3.1387411, 69 masl; and Son Feliu: 39.5300347 and 3.0338470, 142 masl) in the island of Mallorca (Spain), between June and November 2020–2021. Two types of downdraft traps were located close to animal barns (composed mostly of domestic animals such as pigs, rabbits, cows, sheep, equines, and dogs) in each of the sampling sites. The first type was hand-made traps equipped with incandescent light (12 V, 0.3 amps) as attractant. The second type was mini CDC traps (6 V, model 512; Bioquip, Compton, CA, USA) with CO_2 as bait. Both traps operated on a weekly basis with batteries and specimens were retained in collection cups with fine mesh in the bottom to prevent from escaping. A subsample of the total 520 sand fly female collections (ca. 35%) and all males (n = 841) were analyzed from the eight sampling sites. Head and terminal segments of the abdomen of each female and male terminalia were dissected and mounted on a microscope slide with Hoyer's medium. The rest of the body was retained for molecular characterization. Morphological identification was based on features of the male and genitalia, and pharyngeal armature of females, following the available phlebotomine sand fly keys [10,11]. Diagnostic features of the specimens were photographed and measured (mean ± SD) under a compound microscope (Carl Zeiss 37081, Jena, Germany, 40× magnification) coupled with a camera (AxioCam ICc 1), and the images were processed by ZEN 2.3 lite software.

Four males and four females showing morphological features compatible with *Ph. perfiliewi* s.l. were selected for molecular characterization of the barcoding region. The DNeasy Blood and Tissue Kit (Qiagen, Hilden, Germany) was used for genomic DNA extraction of sand flies following the manufacturer's instructions. Reactions were performed using the universal DNA primers for polymerase chain reaction (PCR) amplification of a 658-bp fragment of the mitochondrial cytochrome c oxidase subunit I (COI) gene: LCO1490: 5′-GGTCAACAAATCATAAAGATATTGG-3′ and HCO2198: 5′-TAAACTTCAGGGTGACCAAAAAATCA-3′ [12] following thermocycling conditions, agarose gel electrophoresis, and PCR product purification methods described by Ruiz-Arrondo et al. [13]. In addition, the four *Ph. perfiliewi* s.l. female specimens previously selected were also screened for *Leishmania* sp. parasites following the procedures described elsewhere [14] with few modifications. Amplicons of COI were sequenced in both directions in the Genomics Platform of the CIBIR (La Rioja, Spain). The sequences were edited using Bioedit software 7.2 and compared with sequences previously deposited in GenBank.

Phylogenetic analyses were performed by constructing multiple alignments of nucleotide sequences including eight high-quality amplicon-length sequences with 658 pb, together with individuals of *Ph. perfiliewi* s.l. (n = 19) from other countries retrieved from GenBank. These countries were selected to hold the three members of the *Ph. perfiliewi* species complex: *Ph. perfiliewi* s.s., *Ph. galilaleus*, and *Ph. transcaucasicus*. In addition, one specimen of *Ph. perniciosus* collected in the present study was included as the outgroup. These analyses were constructed using MAFFT vs. 7 (https://mafft.cbrc.jp/alignment/server/, accessed on 15 January 2023) and subsequently edited with GBlocks (http://molevol.cmima.csic.es/castresana/Gblocks_server.html, accessed on 10 January 2023). The phylogenetic tree was built using the maximum likelihood (ML) method in IQ-tree v.2.2.0 (http://www.iqtree.org/, accessed on 10 January 2023). The best-fitting evolutionary model was TPM2u + F + I. Intraspecific and interspecific genetic divergences were calculated based on the Tamura–Nei model in MEGA X [15]. Sequence similarity searches were carried out through the Barcode of Life Data System (https://www.boldsystems.org/index.php/IDS_OpenIdEngine, accessed on 10 December 2022) and BLASTn (MegaBlast option; https://blast.ncbi.nlm.nih.gov/Blast.cgi, accessed on 10 December 2022). Detailed specimen records and sequence information of sand flies were submitted to the GenBank public database under the following accession numbers: OP824886-OP824894.

3. Results and Discussion

Thirty-seven specimens of *Ph. perfiliewi* s.l. (12 females and 25 males) captured by suction traps, equipped either with light (n = 17) or with CO_2 (n = 20), were recorded in two rural farms (Sa Teulera and Son Ajaume nou) of the island of Mallorca (Balearic Islands) from mid-June to mid-July 2021. The analysis showed that only 4.8% (n = 25) of the total sand fly male collections corresponded with *Ph. perfiliewi* s.l. from Sa Teulera (n = 24) and Son Ajaume nou (n = 1), respectively, and all the females of *Ph. perfiliewi* s.l. were derived from Sa Teulera (n = 12). So far, *Ph. perfiliewi* s.l. is distributed in the Mediterranean Basin (from coast of France and Corsica to eastwards of Turkey and Ukraine) and North Africa [16]. Therefore, the present finding represents the most westerly distribution of this species in Europe, increasing the number of known phlebotomine sand fly species in Spain to fourteen, and in the Balearic Archipelago to six.

Among the sand fly species cited in Europe, female *Ph. perfiliewi* s.l. can be misidentified with *Ph. perniciosus*, as both species share common morphological features. However, various characters were proposed to separate both species attending to morphology. For instance, females can be differentiated by the length and teeth of the pharyngeal armature and features of the spermatheca (number of body segments, shape of the neck, and size of bulges and ducts). In contrast, males are separated based on both the shape and colour of the terminalia of the *aedeagus* [10,11].

The morphological analysis of the sand fly females captured in our study indicated that the pharyngeal armature can be used as a reliable feature to discriminate both species (*Ph. perniciosus* and *Ph. perfiliewi* s.l.). In *Ph. perniciosus* the pharyngeal armature usually occupies more than a quarter length of pharynx and teeth are arranged disorderly (Figure 1A), whereas in *Ph. perfiliewi* s.l. the pharyngeal armature is smaller, and teeth end anteriorly in a clear line of demarcation (Figure 1B). We found that the number of segments of the body of the spermatheca is not a reliable morphological feature to separate both sand fly species, as there is an overlap ranging from 9 to 12 in *Ph. perniciosus* (n = 10 specimens examined from our study) and from 11 to 13 (n = 10 specimens examined) in *Ph. perfiliewi* s.l. (Figure 1C,D). This remark was also noted by other authors [11]. The neck of the spermatheca is longer (ca. 3/4 of the body of the spermatheca), more prominent, and thicker in *Ph. perfiliewi* s.l. (spermatheca body: 48.3 $\pm$ 3.4 μm; neck: ca. 37.8 $\pm$ 1.9 μm length and 3.0 $\pm$ 0.2 μm wide) than in *Ph. perniciosus* (spermatheca body: 40.6 $\pm$ 3.0 μm; neck: 24.6 $\pm$ 3.6 μm length and 2.1 $\pm$ 0.1 μm wide), which usually measures close to 2/4 the length of the spermatheca body. Bending next to the base of the neck is generally more apparent in *Ph. perniciosus* than in *Ph. perfiliewi* (Figure 1(E_1,E_2,F_1,F_2)). Ducts are not longer than three times (usually shorter) the length of a spermathecal body in *Ph. perniciosus* and about four times in *Ph. perfiliewi* s.l. Males can be easily separated by the terminalia of the *aedeagus* (Figure 2). The distal region of the *aedeagus* of *Ph. perniciosus* is bifid (fork-shaped, usually both ends are asymmetric) (Figure 2A–C), whereas in *Ph. perfiliewi* s.l., the *aedeagus* extremity is rounded (oar-shaped, usually having minuscules subapical denticles) (Figure 2B–D). The distal region of the *aedeagus* has a conspicuous colorless membrane (translucent) in *Ph. perfiliewi* s.l., whereas the *aedeagus* in *Ph. perniciosus* is completely dark or the subdistal region is lighter. The measures of the *Ph. perfiliewi aedeagus* (length: 141.0 ± 5.2 μm; thickness: 13.2 ± 1.4 μm; transparent part: 41.1 ± 3.4 μm) were similar to those reported by Depaquit [9], but overall, slightly higher.

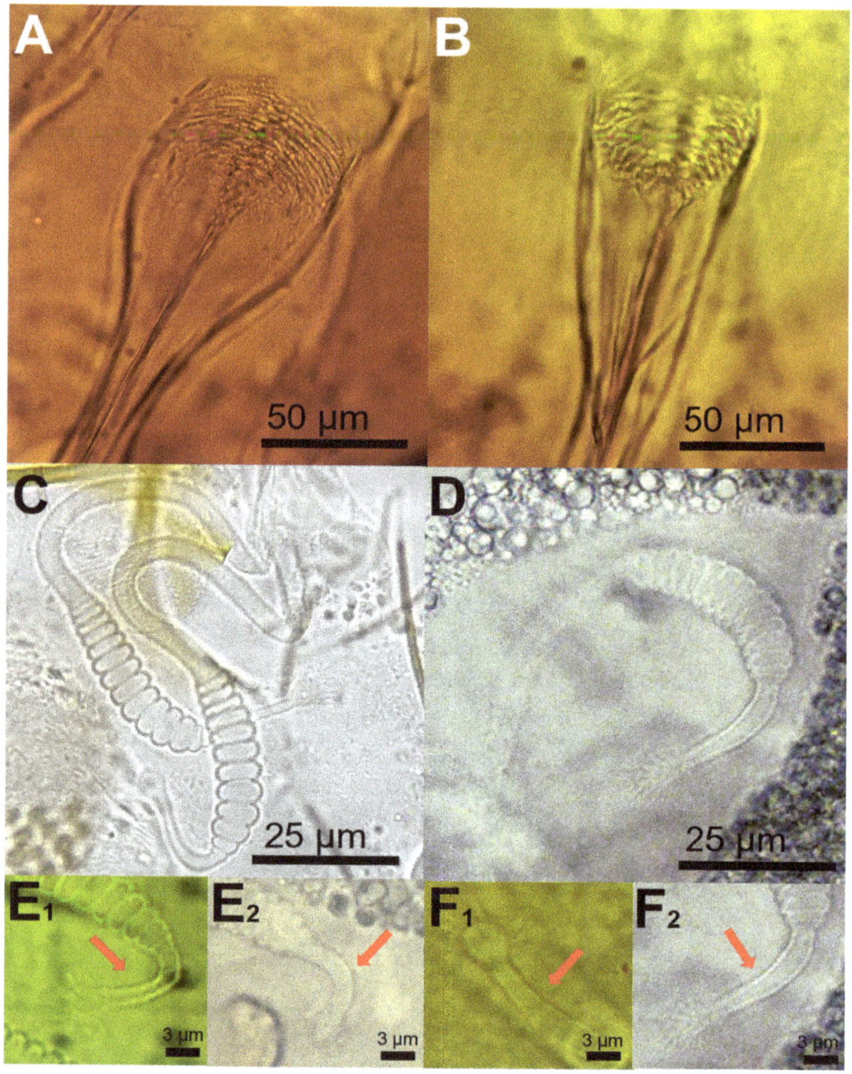

Figure 1. Morphometric characters used to discriminate females of *Ph. perniciosus* (**A**,**C**,**E**) and *Ph. perfiliewi* s.l. (**B**,**D**,**F**). (**A**,**B**) = pharynx) and (**C**,**D**) = spermatheca and ((E_1,E_2) and (F_1,F_2)) = two different examples of spermatheca necks (arrows).

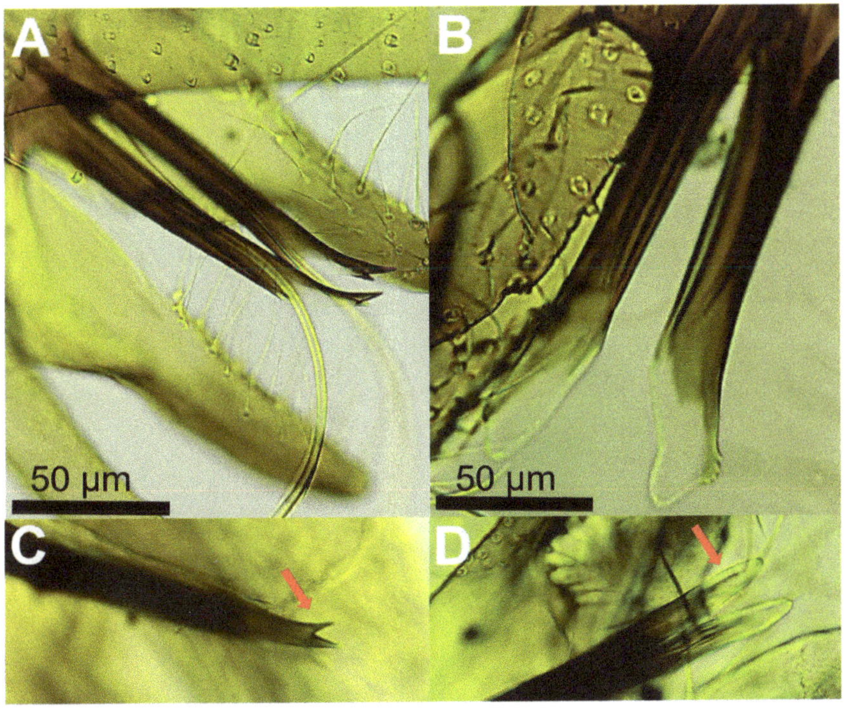

Figure 2. *Aedeagus* (male genitalia) of *Ph. perniciosus* (**A**,**C**) and *Ph. perfiliewi* s.l. (**B**,**D**). Detail of the *aedeagus* tips (arrows).

The sand flies studied here were confirmed molecularly as *Ph. perfiliewi* s.l., showing 99.85–100% identity (query cover of 100%) with homologues from Italy (accession number: KY646194) and Algeria (KJ481177) (Figure 3). The collected *Ph. perfiliewi* s.l. specimens clustered together with the sequences of *Ph. perfiliewi* s.l. from Italy and Algeria and one individual from Jordan.

Individuals of *Ph. perfiliewi* s.l. from other countries (Turkey, Azerbaijan, Greece, Israel, and Jordan) were grouped in a different clade, which is further divided into two subclades (Figure 3). The ML tree showed that *Ph. perfiliewi* s.l. sequences grouped in three clades, each clade consisting of one of the three species forming the *Ph. perfiliewi* species complex. Our results are in agreement with those reported by other authors based on morphological features and both Cytochrome b (Cyt b) and internal transcribed ribosomal spacer 2 (ITS2) genes [9,17,18]. The sequences from Spain, Italy, Algeria, and one from Jordan correspond to *Ph. perfiliewi* s.s., which is a species well-distributed from North Africa to Crimea, including the western Mediterranean. Thus, phylogenetic analyses based on the COI gene showed congruent results on the known geographic distribution of the three species of the *Ph. perfiliewi* s.l. It is interesting to note that Depaquit et al. [9] observed that *Ph. perfiliewi* s.s. was closer to *Ph. galilaeus* than *Ph. transcaucasicus* based on the Cyt b gene, which slightly differs from our ML-tree results, where *Ph. transcaucasicus* was closer to *Ph. galilaeus* than *Ph. perfiliewi* s.s. In contrast, the above-mentioned authors obtained incongruent phylogenetic hypotheses based on the Cyt b and ITS2 genes [9]. A priori, the COI gene could offer a clearer resolution than the two aforementioned genetic markers, but a greater number of sequences from more diverse origins would be needed for further conclusions.

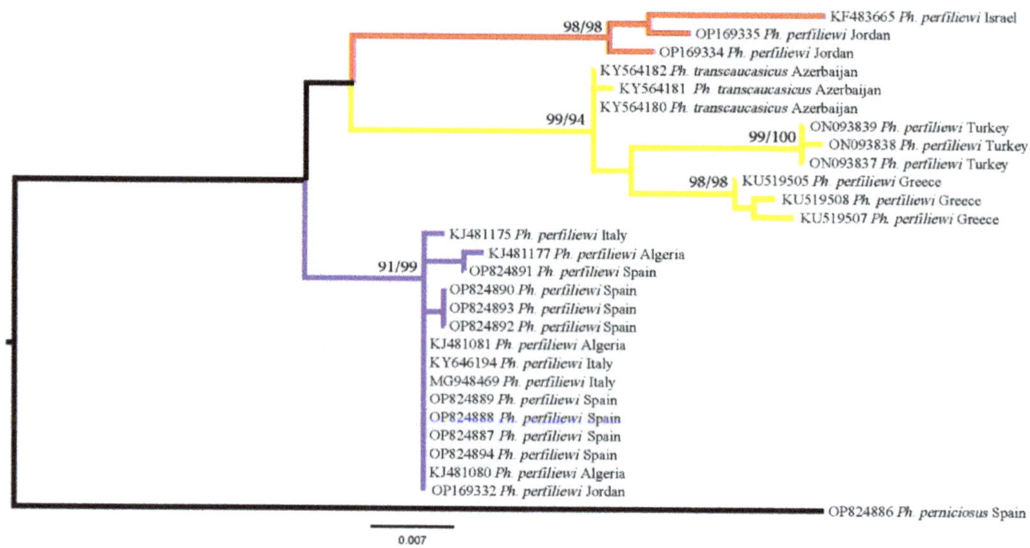

Figure 3. Maximum likelihood phylogenetic tree based on COI sequences in *Ph. perfiliewi* s.l. At specific branches, the first and second values separated by "/" indicate the topological branch support for the ML analysis (aLRT/bootstrap), with values > 75% defining high stability. Red colour = cluster 1, *Ph. galilaeus;* yellow colour = cluster 2, *Ph. transcaucasicus;* and violet colour = cluster 3, *Ph. perfiliewi* s.s.

In our study, the sequence divergence was 0.17% for Spanish *Ph. perfliewi* s.s. and 2.9% for the other *Ph. perfiliewi* s.l. sequences retrieved from GenBank. When the sequences of the individuals were separated according to their geographical origin in the three species of the *Ph. perfiliewi* species s.l., the intraspecific divergence was 0.18% for *Ph. perfiliewi* s.s., 1.6% for *Ph. galilaeus*, and 1.4% for *Ph. transcaucasicus* (Table 1). The interspecific divergence varied between 3.7% and 4.5% among the three species of the *Ph. perfiliewi* s.l. and was higher than 7.0% when compared with *Ph. perniciosus* (Table 1).

Table 1. Interspecific (between groups) pairwise Tamura–Nei model genetic divergence based on COI gene in the three *Ph. perfiliewi* species of the Perfiliewi complex and *Ph. perniciosus*.

	Ph. perfiliewi s.s.	*Ph. galilaeus*	*Ph. transcaucasicus*	*Ph. perniciosus*
Ph. perfiliewi s.s.				
Ph. galilaeus	0.038			
Ph. transcaucasicus	0.037	0.045		
Ph. perniciosus	0.073	0.081	0.083	

In Europe, the main species involved in the transmission of leishmaniosis disease are *Ph. perniciosus*, *Ph. ariasi*, *Ph. papatasi*, and *Ph. perfiliewi* s.l. [19]. *Phebotomus perfiliewi* s.l. is considered a vector with an important role in the transmission dynamics of *L. infantum* in the Mediterranean Basin and in Central Asia [16]. Its vector role for *L. infantum* was recorded several times, particularly in Italy and Iran, but also as a proven or suspected vector of visceral leishmaniosis in Albania, Algeria, Croatia, Greece, Israel, Malta, Morocco, Palestine, Republic of Macedonia, Romania, Tunisia, Turkey, and Hungary [1,16,20–24]. In our study, the four specimens of *Ph. perfiliewi* s.l. individually analyzed for *Leishmania* parasites resulted negative, and therefore, its vector role in the Balearic Islands needs further research.

In countries such as Italy, *Phlebotomus perfiliewi* s.s. was overcome in abundance to *Ph. perniciosus*, particularly in rural areas along the central, south, and north Italy, where it plays a major role in the leishmaniosis sylvatic cycle, with wildlife acting as reservoirs rather than dogs [25–27]. In addition, *Ph. perfiliewi* s.s. was also related to the potential transmission of the Toscana virus (TOSV) and other phleboviruses [28,29], as well as *Trypanosoma theileri* Laveran, 1902 in 2016 in Italy [30]. Other species of the Perfiliewi complex, such as *Ph. transcaucasicus*, are also dominant in Iran, where it is a proven vector of *L. infantum* and *Leishmania donovani* Laveran and Mensil, 1903 [22,31,32].

In the Balearic Islands, leishmaniosis is considered a regionally endemic disease, with an overall rate of human leishmaniosis of ca. 0.7–3.5 cases per year/100,000 inhabitants, and the prevalence of canine leishmaniosis can reach 45% in some areas of the island [7,33]. The collections of *Ph. perfiliewi* s.s. obtained in the current study were in June–July, which is considered the peak of abundance in several Mediterranean countries [34].

4. Conclusions

The combination of identification by morphology and DNA barcoding is of great value in epidemiological studies, as it provides accurate species identification to separate *Ph. perfiliewi* s.l. from *Ph. perniciosus*, as well as to differentiates within the members of the Perfiliewi complex. *Phlebotomus perfiliewi* s.s. is cited here for the first time in Spain and the Balearic Islands, being the most westerly citation of this species in the Mediterranean Basin. This species was associated with farms and was captured either in light or in CO_2 traps. Its epidemiological role remains unknown due to the low number of specimens analyzed in this study for the presence of *L. infantum*; however, based on its proven role in the transmission of *Leishmania* parasites and arboviruses in the Mediterranean Basin, the risk of leishmaniosis transmission exists all over the island.

Further entomological surveys should be conducted to identify the presence of this species in mainland Spain due to its potential role as a vector of diseases with interest for animal and public health. The authors suggest rechecking the material of sand flies previously collected through the Mediterranean coastline of Spain, as it might be possible that female specimens were misidentified with *Ph. perniciosus* due to their closely morphological features. In addition, it would be interesting to deepen whether this species only occurs in restricted areas located in rural areas of Mallorca, or perhaps it was recently introduced.

Author Contributions: Conceptualization, M.A.G., M.Á.M. and I.R.-A.; methodology and formal analysis, C.B. and R.G.-L.; writing—original draft, M.A.G. and I.R.-A.; writing—review, editing and visualization M.A.G., M.Á.M., I.R.-A., C.B. and R.G.-L.; project administration, funding acquisition and supervision, M.Á.M., R.G.-L. and C.B. All authors have read and agreed to the published version of the manuscript.

Funding: This study was funded by the Agriculture, Fisheries and Food council through the project num. BIA08/20-2 from the agricultural and fishing guarantee fund (FOGAIBA) of the Government of the Balearic Islands. Samplings were supported by own resources of ZAP-UIB and FOGAIBA. R.G.L. was initially supported by a Margalida Comas contract from the Government of the Balearic Islands and from the European Regional Development Fund (FEDER) (REF- PD/038/2019) and he is currently funded by a Juan de la Cierva 2019 Formación contract (FJC2019-041291-I) from the Ministry of Science and Innovation.

Data Availability Statement: Not applicable.

Acknowledgments: The authors would like to thank the farm owners for letting set the traps and the sampling procedures. We are also grateful to Marc Oliva Gadea, biology degree student from the Autonomous University of Madrid (UAM) for his support in mounting sand flies.

Conflicts of Interest: The authors declare no conflict of interest.

References

1. Maroli, M.; Feliciangeli, M.D.; Bichaud, L.; Charrel, R.N.; Gradoni, L. Phlebotomine sandflies and the spreading of leishmaniases and other diseases of Public Health concern. *Med. Vet. Entomol.* **2013**, *27*, 123–147. [CrossRef] [PubMed]
2. Herrador, Z.; Gherasim, A.; Jimenez, B.C.; Granados, M.; San Martín, J.V.; Aparicio, P. Epidemiological changes in leishmaniasis in Spain according to hospitalization-based records, 1997–2011: Raising awareness towards leishmaniasis in non-HIV patients. *PLoS Negl. Trop. Dis.* **2015**, *9*, e0003594. [CrossRef] [PubMed]
3. Medlock, J.M.; Hansford, K.M.; Van Bortel, W.; Zeller, H.; Alten, B. A Summary of the evidence for the change in European distribution of Phlebotomine sand flies (Diptera: Psychodidae) of Public Health importance. *J. Vector Ecol.* **2014**, *39*, 72–77. [CrossRef]
4. González, E.; Jiménez, M.; Hernández, S.; Martín-Martín, I.; Molina, R. Phlebotomine sand fly survey in the focus of Leishmaniasis in Madrid, Spain (2012–2014): Seasonal dynamics, *Leishmania Infantum* infection rates and blood meal preferences. *Parasites Vectors* **2017**, *10*, 368. [CrossRef]
5. Wagner, R.; Djong, H. Fauna Europaea: Psychodidae. In *Fauna Europaea: Nematocera*; De Jong, H., Ed.; version 1.3; Museum für Naturkunde Leibniz-Institut für Evolutions und Biodiversitätsforschung: Berlin, Germany, 2004. Available online: http://www.faunaeur.org (accessed on 15 November 2022).
6. Bravo-Barriga, D.; Ruiz-Arrondo, I.; Peña, R.E.; Lucientes, J.; Delacour-Estrella, S. Phlebotomine sand flies (Diptera, Psychodidae) from Spain: An updated checklist and extended distributions. *ZooKeys* **2022**, *1106*, 81–99. [CrossRef] [PubMed]
7. Alcover, M.M.; Ballart, C.; Martín-Sánchez, J.; Serra, T.; Castillejo, S.; Portús, M.; Gállego, M. Factors influencing the presence of sand flies in Majorca (Balearic Islands, Spain) with special reference to *Phlebotomus pernicious*, vector of *Leishmania Infantum*. *Parasites Vectors* **2014**, *7*, 421. [CrossRef]
8. Gil Collado, J.; Morillas-Marquez, F.; Sanchis Marin, M.C. Los flebotomos de España. *Rev. Sanid. Hig. Publica* **1989**, *63*, 15–34.
9. Depaquit, J.Ô.; Bounamous, A.; Akhoundi, M.; Augot, D.; Sauvage, F.; Dvorak, V.; Chaibullinova, A.; Pesson, B.; Volf, P.; Léger, N. A Taxonomic study of *Phlebotomus (Larroussius) perfiliewi* s.L. *Infect. Genet. Evol.* **2013**, *20*, 500–508. [CrossRef]
10. Gatt, P.; Williams, J.; Mifsud, D. New distributional data on sandflies from rubble walls in the Maltese Islands with an illustrated key to the Maltese species (Diptera: Phlebotominae). *Bull. Entomol. Soc. Malta.* **2009**, *2*, 95–110.
11. Dantas-Torres, F.; Tarallo, V.D.; Otranto, D. Morphological keys for the identification of Italian phlebotomine sand flies (Diptera: Psychodidae: Phlebotominae). *Parasites Vectors* **2014**, *7*, 479. [CrossRef]
12. Folmer, O.; Black, M.; Hoeh, W.; Lutz, R.; Vrijenhoek, R. DNA Primers for amplification of mitochondrial Cytochrome c Oxidase Subunit I from diverse metazoan invertebrates. *Mol. Mar. Biol. Biotechnol.* **1994**, *3*, 294–299.
13. Ruiz-Arrondo, I.; Hernández-Triana, L.M.; Ignjatović-Ćupina, A.; Nikolova, N.; Garza-Hernández, J.A.; Rodríguez-Pérez, M.A.; Oteo, J.A.; Fooks, A.R.; Lucientes Curdi, J. DNA barcoding of blackflies (Diptera: Simuliidae) as a tool for species identification and detection of hidden diversity in the eastern regions of Spain. *Parasites Vectors* **2018**, *11*, 463. [CrossRef] [PubMed]
14. Nicolas, L.; Prina, E.; Lang, T.; Milon, G. Real-Time PCR for detection and quantitation of *Leishmania* in mouse tissues. *J. Clin. Microbiol.* **2002**, *40*, 1666–1669. [CrossRef] [PubMed]
15. Tamura, K.; Stecher, G.; Kumar, S. MEGA11: Molecular evolutionary genetics analysis Version 11. *Mol. Biol. Evol.* **2021**, *38*, 3022–3027. [CrossRef]
16. ECDC (European Centre for Disease Prevention and Control). Phlebotomus perfiliewi—Current Known Distribution. March 2022. Available online: https://www.ecdc.europa.eu/en/publications-data/phlebotomus-perfiliewi-current-known-distribution-march-2022 (accessed on 15 November 2022).
17. Artemiev, M.M.; Neronov, V.M. *Distribution and Ecology of Sandflies of the Old World (Genus Phlebotomus)*; Institut Ėvolyutsionnoĭ Morfologii i Ėkologii: Moscow, Russia, 1984; p. 207.
18. Lewis, D.J. *A Taxonomic Review of the Genus Phlebotomus (Diptera: Psychodidae)*; British Museum (Natural History): London, UK, 1982; pp. 121–209.
19. García San Miguel, L.; Sierra, M.J.; Vazquez, A.; Fernandez-Martínez, B.; Molina, R.; Sanchez-Seco, M.P.; Lucientes, J.; Figuerola, J.; de Ory, F.; Monge, S.; et al. Phlebovirus-associated diseases transmitted by phlebotominae in Spain: Are we at risk? *Enferm. Infecc. Microbiol. Clin.* **2021**, *39*, 345–351. [CrossRef]
20. Duport, M.; Lupaşcu, G.H.; Cristescu, A. Contribution à l'étude des phlébotomes des biotopes naturels de Roumanie. *Arch. Roum. Path. Exp. Microbiol.* **1971**, *30*, 387–398.
21. Izri, M.A.; Belazzoug, S. *Phlebotomus* (*Larroussius*) *perfiliewi* naturally infected with dermotropic *Leishmania infantum* at Tenes, Algeria. *Trans. R. Soc. Trop. Med. Hyg.* **1993**, *87*, 399. [CrossRef]
22. Oshaghi, M.A.; Ravasan, N.M.; Hide, M.; Javadian, E.A.; Rassi, Y.; Sadraei, J.; Mohebali, M.; Mehdi Sedaghat, M.; Hajjaran, H.; Zarei, Z.; et al. *Phlebotomus perfiliewi transcaucasicus* is circulating both *Leishmania donovani* and *L. infantum* in northwest Iran. *Exp. Parasitol.* **2009**, *123*, 218–225. [CrossRef]
23. Zivkovic, V. Recherches récentes sur les phlébotomes (Diptera, Psychodidae) dans un foyer endémique de leishmaniose viscérale en Serbie (Yougoslavie). *Acta Parasitol. Lugoslavica* **1975**, *6*, 37–43.
24. Farkas, R.; Tánczos, B.; Bongiorno, G.; Maroli, M.; Dereure, J.; Ready, P.D. First surveys to investigate the presence of canine leishmaniasis and its phlebotomine vectors in Hungary. *Vector-Borne Zoonotic Dis.* **2011**, *11*, 823–834. [CrossRef]

25. Calzolari, M.; Carra, E.; Rugna, G.; Bonilauri, P.; Bergamini, F.; Bellini, R.; Varani, S.; Dottori, M. Isolation and molecular typing of *Leishmania Infantum* from *Phlebotomus perfiliewi* in a reemerging focus of leishmaniasis, Northeastern Italy. *Microorganisms* **2019**, *7*, 644. [CrossRef]
26. Millán, J.; Ferroglio, E.; Solano-Gallego, L. Role of wildlife in the epidemiology of *Leishmania infantum* infection in Europe. *Parasitol. Res.* **2014**, *113*, 2005–2014. [CrossRef]
27. Maresca, C.; Scoccia, E.; Barizzone, F.; Catalano, A.; Mancini, S.; Pagliacci, T.; Porrini, M.; Principato, M.; Venditti, G.; Grelloni, V. A survey on canine leishmaniasis and phlebotomine sand flies in Central Italy. *Res. Vet. Sci.* **2009**, *87*, 36–38. [CrossRef]
28. Calzolari, M.; Angelini, P.; Finarelli, A.C.; Cagarelli, R.; Bellini, R.; Albieri, A.; Bonilauri, P.; Cavrini, F.; Tamba, M.; Dottori, M.; et al. Human and entomological surveillance of Toscana virus in the Emilia-Romagna region, Italy, 2010 to 2012. *Eurosurveillance* **2014**, *19*, 20978. [CrossRef]
29. Calzolari, M.; Chiapponi, C.; Bellini, R.; Bonilauri, P.; Lelli, D.; Moreno, A.; Barbieri, I.; Pongolini, S.; Lavazza, A.; Dottori, M. Isolation of three novel reassortant phleboviruses, Ponticelli I, II, III, and of Toscana virus from field-collected sand flies in Italy. *Parasites Vectors* **2018**, *11*, 84. [CrossRef]
30. Calzolari, M.; Rugna, G.; Clementi, E.; Carra, E.; Pinna, M.; Bergamini, F.; Fabbi, M.; Dottori, M.; Sacchi, L.; Votýpka, J. Isolation of a trypanosome related to *Trypanosoma theileri* (Kinetoplastea: Trypanosomatidae) from *Phlebotomus perfiliewi* (Diptera: Psychodidae). *Biomed Res. Int.* **2018**, *2018*, 2597074. [CrossRef]
31. Rassi, Y.; Javadian, E.; Nadim, A.; Rafizadeh, S.; Zahraii, A.; Azizi, K.; Mohebali, M. *Phlebotomus perfiliewi transcaucasicus*, a vector of *Leishmania infantum* in Northwestern Iran. *J. Med. Entomol.* **2009**, *46*, 1094–1098. [CrossRef]
32. Rassi, Y.; Sanei Dehkordi, A.; Oshaghi, M.A.; Abai, M.R.; Mohtarami, F.; Enayati, A.; Zarei, Z.; Javadian, E. First Report on Natural Infection of the *Phlebotomus tobbi* by *Leishmania infantum* in Northwestern Iran. *Exp. Parasitol.* **2012**, *131*, 344–349. [CrossRef]
33. Epidemiología. Informes y Estadísticas de La Red de Vigilancia Epidemiológica de Les Illes Balears. February 2022. Available online: https://www.caib.es/sites/epidemiologia/es/informes_anuales-11681/ (accessed on 10 December 2022).
34. Cazan, C.D.; Păstrav, I.R.; Györke, A.; Oguz, G.; Alten, B.; Mihalca, A.D. Seasonal dynamics of a population of *Phlebotomus* (*Larroussius*) *perfiliewi* Parrot, 1930 (Diptera: Psychodidae) in North-Eastern Romania. *Parasitol. Res.* **2019**, *118*, 1371–1384. [CrossRef]

diversity

MDPI

Review

Global Diversity, Distribution, and Genetic Studies of Stable Flies (*Stomoxys* sp.)

Gérard Duvallet [1,*] and Jerome A. Hogsette [2]

1 CEFE, Université Montpellier, CNRS, EPHE, IRD, Université Paul Valéry Montpellier 3, F-34199 Montpellier, France
2 USDA–ARS, Center for Medical, Agricultural and Veterinary Entomology, 1600 SW 23rd Drive, Gainesville, FL 32608, USA; jerry.hogsette@usda.gov
* Correspondence: gerard.duvallet@univ-montp3.fr

Abstract: In the subfamily Stomoxyinae there are currently 18 recognized *Stomoxys* species, plus two subspecies. Most *Stomoxys* knowledge was gained through studies with *S. calcitrans*, a cosmopolitan species, economic pest, and vector. Other *Stomoxys* spp. are known only from a few trapped adult specimens. Herein, the *Stomoxys* spp. have been grouped by their ecological diversity, global distribution, and phylogeny and phylogeography. Seven species are dependent to some degree on humans and their activities, particularly animal production. Eleven species are dependent on wildlife to some degree for their development, and little is known about their biology in many cases. Global distributions include one cosmopolitan species (*S. calcitrans*), twelve species found only in Africa, four species only in Asia, and one species (*S. sitiens*) in Africa and Asia. Most genetic studies on *Stomoxys calcitrans* showed little variation in North America, possibly due to the adults' long range flight capability. Phylogeographic analysis of *S. calcitrans* showed a differentiation between Oriental populations (first lineage) and populations from Afrotropical, Palearctic, Nearctic, Neotropical and Oceanian regions (second lineage). Genetic studies were followed by sequencing of the *Stomoxys calcitrans* genome and phylogenetic studies of the *Stomoxys* genus using 10 of the known species. Phylogenetic relationships were established.

Keywords: synanthropic species; genetic sequencing; Afrotropical origin; two subspecies

Citation: Duvallet, G.; Hogsette, J.A. Global Diversity, Distribution, and Genetic Studies of Stable Flies (*Stomoxys* sp.). *Diversity* **2023**, *15*, 600. https://doi.org/10.3390/d15050600

Academic Editors: Carlos Barceló, Ignacio Ruiz-Arrondo and Luc Legal

Received: 14 March 2023
Revised: 21 April 2023
Accepted: 27 April 2023
Published: 28 April 2023

1. Introduction

The genus *Stomoxys* Geoffroy (1762) belongs to the family Muscidae. Even if for a long time the tribe of Stomoxyini was included within the subfamily Muscinae, today it seems that most entomologists accept the proposal to group the 18 recognized species of the genus *Stomoxys* within the subfamily Stomoxyinae [1]. The *Stomoxys* flies are biting flies, 3–10 mm long, hematophagous, with the appearance of a housefly (*Musca domestica* Linnaeus, 1758) but with an adapted biting mouth apparatus, the proboscis, directed forward in the axis of the body and capable of piercing the skin. The subfamily Stomoxyinae includes ten genera, of which the most important are *Haematobosca* Bezzi, 1907, *Haematobia* Lepeletier et Serville, 1828, and *Stomoxys* Geoffroy, 1762 [1]. Diagnosis between the different species of the genus *Stomoxys* is achieved by different morphological characters, in particular the width of frons in males, the colour of thorax and legs, the nervation of wings, the presence and number of hairs and bristles, and the variability of abdominal patterns. Entomologists use the keys of Zumpt [1]. For example, *S. calcitrans* is 4–7 mm long, and has an abdomen punctuated with black spots of rounded shape and a discontinuous median black line. As keys for males and females are separated, the distinction of sexes is based on the posterior part of the abdomen, modified for copulation and oviposition, and on eye spacing at the apex of the head, smaller in males than in females. The rigid biting apparatus, the proboscis, is composed of three long, strongly sclerotized, nonretractile parts: a labium (lower lip), a labrum (upper lip), and a hypopharynx forming two tubes. Saliva is injected into the host dermis through the thinner tube represented by the hypopharynx, while blood is drawn through the larger tube formed by the labium and the dorsal part of the hypopharynx. In *Stomoxys* flies, both sexes are hematophagous; blood is necessary for reproduction, but flies can survive by feeding on nectar [2,3].

Knowledge on *Stomoxys* flies has been acquired mainly by studying the only species that is cosmopolitan: *Stomoxys calcitrans* (Linnaeus, 1758), known as the stable fly (Figure 1).

Figure 1. *Stomoxys calcitrans*, known as stable fly, with a cosmopolitan distribution. Photo: Dr Mikel Alexander González, Spain.

This species is the only one of the genus *Stomoxys* that is present on the European and American continents. This knowledge of *S. calcitrans* has been acquired in relation to their impact on the livestock industry. The adults represent a significant nuisance due to the pain of their bite and the possible transmission of pathogens [4]. In the United States Taylor et al. (2012) [5] estimated the economic impact of stable flies on the livestock industry to be USD 2.2 billion per year. Using the same calculation formulas, French livestock farming losses have been estimated to be EUR 145 million per year for the meat industry and EUR 234 million per year for the dairy industry [6]. The health impacts are also very important. *Stomoxys* adults are mechanical vectors of pathogens present in the blood and skin of their hosts, especially livestock, but occasionally humans. Equine infectious anaemia, African swine fever, West Nile, Rift Valley and lumpy skin disease viruses are known to be transmitted by *Stomoxys* flies, while transmission of other viruses is suspected. Rickettsia (*Anaplasma*, *Coxiella*) and other bacteria and parasites (*Trypanosoma* spp., *Besnoitia* spp.) are also transmitted by *Stomoxys* adults. Finally, stable flies were also found to act as an intermediate host of the helminth *Habronema microstoma* and may be involved in the transmission of some *Onchocerca* and *Dirofilaria*. Being cosmopolite, *Stomoxys calcitrans* might have a greater worldwide impact than previously thought on animal and human pathogen transmission [4,7].

The traditional way of controlling these biting flies relies on the use of insecticides, mainly pour-on insecticides on the back of animals, when most of the flies are seen on the lower legs. However, for several years now, breeders have noticed that these products are no longer effective, despite the high frequency of application. This has been confirmed by numerous laboratory studies showing, phenotypically and genetically, that stable flies have become resistant to all available insecticides [8–12]. Other control methods have become necessary.

This review attempts to gather updated information on the diversity, distribution, phylogeny of the genus *Stomoxys* and the phylogeography of the cosmopolitan species *S. calcitrans*.

2. Diversity and Ecological Remarks

The subfamily Stomoxyinae, comprising 10 genera and 49 species, was described as a well-defined and monophyletic taxon by Zumpt (1973) [1]. Four species are divided into two or three subspecies, which represent a total of fifty-four taxa. Since this work, three new species have been described in the genus *Haematobosca* [13–15]. A total of 52 species and 57 taxa are currently recognized (Table 1).

Table 1. List of genera and number of species and taxa known in 2023 in subfamily Stomoxyinae (Diptera: Muscidae).

Genera	Number of Species	Number of Taxa *
Rhinomusca Malloch (1932)	2	2
Neivamyia Pinto & Fonseca (1930)	5	5
Bruceomyia Malloch (1932)	1	1
Parastomoxys Zumpt (1973)	1	1
Prostomoxys Zumpt (1973)	1	1
Stygeromyia Austen (1907)	2	2
Haematobosca Bezzi (1907)	15	15
Haematobia Lepeletier & Serville (1828)	6	9
Haematostoma Malloch (1932)	1	1
Stomoxys Geoffroy (1762)	18	20

* Taxa: species and subspecies.

The 18 species of the genus *Stomoxys* can be easily identified using morphological characters and the key proposed by Zumpt [1]. They can be grouped by their different ecologies:

Seven species are wholly or partly dependent on human activities and on livestock farming [1].

1. *Stomoxys calcitrans* (Linnaeus, 1758), described by Linnaeus from specimens caught in Sweden, has subsequently been found worldwide and redescribed under 32 synonymous names. The first in-depth studies on its biology were conducted in Egypt [16–18]. These authors showed that the developmental media of this species were the dung of cattle or horses, mixed with straw and urine. They do not develop in pure faecal matter. The authors have also shown that they grow very well in decaying plant material. A review on the biology of this species was recently published [7].

2. *Stomoxys niger* Macquart, 1851 was described from specimens from the island of La Réunion (Indian Ocean). Originally described as *Stomoxys nigra*, the name was changed to *S. niger* when it was realized that the name *Stomoxys* was masculine [19]. It was then redescribed under 12 different synonymous names. This species was then divided into two subspecies when Zumpt (1973) [1] suggested that the species *S. bilineata* Grünberg, 1906 should be considered a subspecies of *S. niger*. The two subspecies, henceforth named *S. niger niger* Macquart, 1851 and *S. niger bilineatus* Grunberg, 1906, are identified by morphological characters of tibiae coloration [1]. Moreover, while *S. niger niger* appears to

be well associated with livestock activities, *S. niger bilineatus* often appears in association with wildlife [20,21]. *S. niger niger* is considered the most common species in Africa. It has been described in many countries on the continent, but also in Madagascar and the Mascarene Islands (Reunion, Mauritius, and Rodrigues). Recently, one of us (GD) has received three specimens from Sao Tomé and Principe. Fiasson (1943) [22] (quoted by [1]) had also described it in Venezuela (South America), but this is obviously an error. This species feeds on cattle, horses, donkeys, and sometimes on humans, and their bite is painful.The subspecies *Stomoxys niger bilineatus* Grunberg, 1906 has been separated from *S. niger niger* on morphological characters, of which the most visible is the coloration of the tibiae. Zumpt (1973) [1] studied specimens from Mali, Ethiopia, Tanzania, Mozambique, Zimbabwe, and South Africa. This species has also been captured in Gabon [20,21,23].

3. *Stomoxys sitiens* Rondani, 1873 was described from specimens from Eritrea, and then redescribed by Brunetti in 1910 under the same name, and by other authors under two other synonymous names with specimens from India and Singapore. This species has since been captured in China [24] and in Thailand [1,25–27]. *S sitiens* is indeed present in both the African continent and Asia. Hafez and Gamal-Eddin (1959) [16,17] showed in Egypt that this species has a similar ecology to *S. calcitrans*. The preimaginal forms are found in the dung of cattle, horses, and donkeys when mixed with straw, but also in decaying vegetation. *S. sitiens* appears to prefer drier, semi-arid environments than *S. calcitrans* [1].

4. *Stomoxys indicus* Picard, 1908 was described as *S. indica* from specimens from the Calcutta region of India. It has been redescribed under eight synonymous names from specimens from Southeast Asia (India, Malaysia, Borneo, Vietnam, Taiwan). This species is found mainly in contact with cattle, but also goats, in the same places as *S. calcitrans*. This species has also been captured in Thailand [26,28,29]. However, it seems (Kano, 1953 [30] cited by Zumpt, 1973 [1]; [26]) that this species has a crepuscular or nocturnal activity. The larvae are found in the dungs of cattle and horses.

5. *Stomoxys bengalensis* Picard, 1908 was described from specimens from India. It was redescribed under a synonymous name in 1910, and is known from the following countries: India, Burma, Vietnam, Malaysia, and Indonesia [1]. It has also been captured more recently in Thailand [26,27]. This species is known to feed on cattle, but its biology is very poorly known.

6. *Stomoxys uruma* Shinonaga and Kano, 1966 was described from specimens from Japan (Ryukyu) and China (Hong Kong). It has also been observed in Thailand [31], Vietnam, and Taiwan. The larval stages are not known and the authors who described it indicate that they feed on cattle and water buffalo. Changbunjong et al. (2012) [32] reported large numbers of *S. uruma* collected in Kho Yai National Park (Thailand).

7. *Stomoxys taeniatus* Bigot, 1888 was described as *S. taeniata* from specimens from South Africa. It was later redescribed under three synonymous names from specimens from Cameroon, Tanzania, Sudan, and Congo. The larval forms are not known and Reid (1956) [33] (quoted by [1]) indicated that specimens captured in Sudan were biting cattle.

Eleven species are known to be wildlife-related or of unknown biology.

8. *Stomoxys omega* Newstead, 1907 was described from specimens from the Congo. These specimens had been captured from a wild buffalo that had just been shot [1]. Zumpt (1973) [1] reported that he examined specimens from the following countries: Liberia, Nigeria, Cameroon, Congo, Uganda, Malawi, and South Africa. This species has been captured in abundance in Gabon in a forested area, far from any livestock [23].

9. *Stomoxys xanthomelas* Roubaud, 1937 was described from specimens from Congo. Zumpt (1973) [1] also indicated its presence in Congo, Uganda, and Tanzania. He specified that its biology is unknown, but specimens were captured in Gabon when Vavoua traps were placed in the canopy 30 m above the ground [34]. Blood meal analysis showed that this species fed on and associated with monkeys. This was confirmed by observing the emergence of adults from chimpanzee faecal material placed in an emergence cage [35].

10. *Stomoxys pallidus* Roubaud, 1911 was described as *S. pallida* from specimens from Benin. It was redescribed from Malawi in 1932 under a synonymous name [1]. Very close

morphologically to *S. xanthomelas* because of the yellow coloration of their abdomens, it differs in other characters. It is known from the following countries: Sierra Leone, Liberia, Congo, Uganda, and Malawi [1]. This species, like *S. xanthomelas*, seems to prefer dense forest areas and feeds on antelopes and hippopotami. The biology of this species is very poorly known.

11. *Stomoxys ochrosoma* Speiser, 1940 was described from specimens from Tanzania. Redescribed in 1932 under a synonymous name from specimens from Uganda, this species is easily recognizable by the yellow colour of its entire body. It is also known from Kenya. Concerning its biology, it has been indicated that the females project their eggs on columns of Dorylinae ants. The larvae would thus develop in the anthills of this species [36] (quoted by [1]). Nothing else is known about the biology of this species.

12. *Stomoxys luteolus* Villeneuve, 1934 was previously described as a subspecies of *S. ochrosoma*, but Zumpt (1973) [1] confirms the separation between the two. This species is known only from Congo and Uganda. No information is available on its biology.

13. *Stomoxys stigma* Van Emden, 1939 is a species very close to *S. omega*, described from specimens from Uganda. No information on its biology is available.

14. *Stomoxys transvittatus* Villeneuve, 1916 was described as *S. transvittata* from specimens from South Africa. It is also known from Malawi, Zimbabwe, Kenya, and Congo. *S. transvittatus* was the most abundant species captured in a forested area in Gabon [23]. No information is available on its biology.

15. *Stomoxys pullus* Austen, 1909 was described as *S. pulla* from specimens from India. A very rare species in collections, nothing is known about its biology. This species has been captured more recently in Thailand [32]. In Thailand, most of the specimens were captured in Khao Yai National Park with abundant wildlife, but others were captured in a local beef cattle farm at the border of the National Park.

16. *Stomoxys boueti* Roubaud, 1911 was described from specimens from Benin. This species is morphologically close to the species *S. uruma* from the Oriental Region and is also known from Congo. Some of the known specimens were taken from buffalo, but nothing else is known about its biology [1].

17. *Stomoxys inornatus* Grünberg, 1906 was described as *S. inornata* from specimens from Cameroon. This species, also described under another synonymous name, is widely known from West and Central Africa: Liberia, Nigeria, Uganda, Kenya, Burundi, Rwanda, Congo, and Sudan. This species has been captured more recently in Gabon [37,38]. The biology of this species is unknown.

18. *Stomoxys varipes* Bezzi, 1906 was described from specimens from Eritrea. Very close morphologically to *S. niger*, this species is also known from Ethiopia, Uganda, Kenya, Tanzania, Rwanda, Congo, Zimbabwe, and Malawi. The biology of this species is unknown [1].

3. Distribution

The global distribution of the 18 species of the genus *Stomoxys* can be summarised in the following map (Figure 2).

Additional details to this distribution map are added:

- *Stomoxys calcitrans* is cosmopolitan, present in all the countries where it has been researched. Almost always found in connection with human activities (breeding), it can sometimes be found in littoral tourist destinations where it can feed on people and become an important nuisance [39,40]. This species is thus adapted to all the climates of the planet, from equatorial zones to mountainous zones with altitude or zones close to the poles.

- *Stomoxys sitiens* is present in both the African continent and Asia.

- twelve Stomoxys species are known only from the African continent: *S. omega*, *S. niger*, *S. xanthomelas*, *S. pallidus*, *S. ochrosoma*, *S. luteolus*, *S. stigma*, *S. transvittatus*, *S. boueti*, *S. taeniatus*, *S. inornatus*, and *S. varipes*.

- four *Stomoxys* species are known only from Asia: *S. indicus*, *S. bengalensis*, *S. pullus*, and *S. uruma*.

- *Stomoxys niger*: the indication of its presence in Venezuela by Fiasson (1943) [22] was questioned by Zumpt (1973) [1]. We have contacted entomologists in this country, who have confirmed that they have never seen any species other than *S. calcitrans*; nor have they ever been aware of a publication indicating the presence of *S. niger* in their country.

- *Stomoxys indicus*: known from several Asian countries, its presence on the island of La Reunion (Indian Ocean) has just been recently confirmed (L. Costet, personal communication). The identification of *S. indicus* has been confirmed morphologically by one of us (GD) and genetically (paper in preparation). Only two species (*S. calcitrans* and *S. niger niger*) were known until now on the island of La Reunion [41]. The presence of *S. indicus* confirms the links between India and the Mascarene Islands.

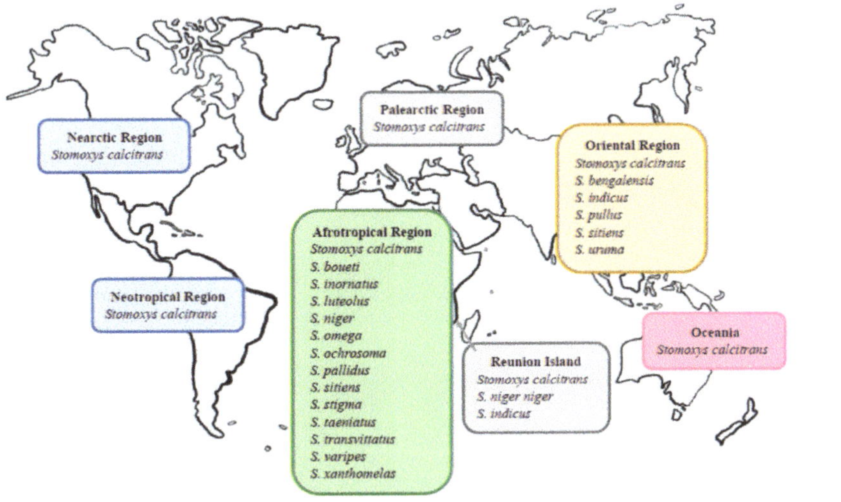

Figure 2. Known distribution of the 18 species of *Stomoxys*.

4. Phylogeography of *Stomoxys calcitrans*

Most studies on *Stomoxys* flies' genetics have been conducted on *Stomoxys calcitrans*. The initial objectives of these studies were to obtain information on population dynamics, the origin of outbreaks, and the identification and geographical distribution of insecticide resistance, to thus develop more effective control strategies. Most studies were conducted at a local level, with only a few at a more global scale. These aspects of *S. calcitrans* genetics were reviewed by Rochon et al. (2021) [7]. Szalanski et al. (1996) [42] observed significant gene flow between populations of *S. calcitrans* from Canada and the USA (Nebraska and Texas). Biochemical and molecular methods showed little variation between these populations, likely related to a bottleneck at the time of the colonization of North America by these flies. This genetic homogeneity is related to the dispersal capacity of these flies [40]. Those results have been confirmed by Kneeland et al. (2013) [43] who, using the amplified fragment length polymorphism (AFLP) method with samples from North America, showed that *S. calcitrans* flies have a high level of gene flow on a continental scale.

Dsouli-Aymes et al. (2011) [44] conducted a phylogeographic analysis to study the population genetic structure of *S. calcitrans* and to trace its global dispersion. They compared twenty populations from the five major zoogeographic regions of the world using mitochondrial (COI-cytochrome c oxidase subunit I, Cyt B-cytochrome B, and ND1-16S-nicotinamide adenine dinucleotide) and nuclear (ITS2-internal transcribed spacer 2) genes. Their results show a differentiation between Oriental populations (first lineage) and populations from Afrotropical, Palearctic, Nearctic, Neotropical and Oceanian regions (second lineage). These two clades were separated by mean genetic distances of 1.7 to 2.3% [44]. The absence of shared haplotypes between them and the genetic distance suggest that they diverged in allopatry, and that the Oriental lineage has been isolated since 0.7–1 million years ago (mid-Pleistocene) [44]. Oriental populations appear in their work as isolated and not participating in the colonization of other regions. Afrotropical populations seem the source of *S. calcitrans* dispersion to other regions. Tsai et al. (2023) [45] also performed a detailed study of the phylogeography of *S. calcitrans*, using 13 populations from Taiwan, 10 specimens from Poland, and all available data in GenBank and BOLD. They confirmed that the likely origin of S. *calcitrans* is in the Afrotropical region. They also suggested that the differences observed between the phylogroups of the Afrotropical, Oriental and Nearctic regions could be due to the isolation of populations in different refugia during the Pleistocene glaciations. They also showed the influence of transnational livestock trade activities by giving the example of the arrival of stable flies from North America to Taiwan with the importation of cattle. They also indicated by examining intra- and inter-population genetic distances, in agreement with [44,46], that the species *S. calcitrans* may in fact contain several cryptic species. Kneeland et al. (2015) [47] also studied the genetic diversity of *S. calcitrans* on a global scale. Specimens from different biogeographic regions were analysed with the AFLP technique. Their results show a lack of genetic differentiation despite geographical barriers. They did not observe a correlation between geographical origin and genetic distance. This is inconsistent with the results of Dsouli-Aymes et al. (2011) [44], but they had no samples from the Oriental region in their work. These studies eventually led to the sequencing of the *Stomoxys calcitrans* genome [48]. This work has allowed the identification of many gene families, allowing a better understanding of the behaviour of this species, but also consideration of new and more effective control methods, e.g., the use of olfactory attractants or repellents.

5. Phylogeny of Genus *Stomoxys*

Dsouli et al. (2011) [49] were the first authors, to our knowledge, to have studied the phylogeny of the genus *Stomoxys* considering 10 of the 18 known different species. Phylogenetic relationships have been established using maximum likelihood and Bayesian methods from DNA fragments from COI, CytB mitochondrial genes, and ITS2 nuclear genes. A phylogenetic tree inferred from the concatenation of the three genes (totalizing 1635 nucleotide sites) and a chronogram resulting from the relaxed clock Bayesian analysis of the same concatenation are available in their publication [49]. The main results of this work were:

- the genus *Stomoxys* appears as paraphyletic because of the inclusion of the species *Prostomoxys saegerae* (Zumpt, 1969) in the analysis. The phylogenetic tree shows *P. saegerae* as a sister group of *S. varipes*. The monospecific genus *Prostomoxys* was created by Zumpt [1] on a simple morphological character showing maxillary palpi as long as the proboscis, when it is shorter for the genus *Stomoxys*. The authors proposed to reintegrate the species *P. saegerae* into the genus *Stomoxys* [49].

- a deep molecular divergence was observed between the subspecies *Stomoxys niger niger* and *S. niger bilineatus*. The authors proposed that these taxa should be considered as distinct species.

- three distinct lineages were observed within the genus *Stomoxys*, identified by Bayesian phylogenetic analyses. The first lineage includes the species *Stomoxys indicus*, which appears as the sister species to all other *Stomoxys* species. The second lineage groups strictly African species *S. inornatus*, *S. transvittatus*, *S. omega*, and *S. pallidus*. The third lineage includes the cosmopolitan *S. calcitrans*, African species such as *S. varipes*, Oriental species such as *S. bengalensis*, and *S. sitiens*, which is present in Africa and the Oriental region.

- the chronogram gave estimations of the divergence time between taxa. If the divergence time between the *Stomoxys* genus and its Stomoxyinae sister-clade (*Haematobia* and *Haematobosca*) is estimated around 30.8 Mya, the age estimate for the emergence of *S. indicus* is estimated at about 27 Mya [49]. The placement of *S. indicus* as the sister group to the remaining *Stomoxys* species lets us suggest that this species would be at the origin of the genus *Stomoxys* in the Oriental region.

6. Conclusions

The subfamily Stomoxyinae currently consists of 18 recognized *Stomoxys* species plus two subspecies. When grouped by ecological diversity, seven species are synanthropic to some degree and benefit particularly from animal production. Eleven species are dependent on wildlife to some degree for their development, and little is known about their biology in many cases. Global distributions include one cosmopolitan species (*S. calcitrans*), twelve species found only in Africa, four species found only in Asia, and one species (*S. sitiens*) found in both Africa and Asia.

Little genetic variation is found in North America, possibly because of gene flow from the adults' long range flight capability. Phylogeographic analysis of *S. calcitrans* showed a differentiation between Oriental populations (first lineage) and populations from Afrotropical, Palearctic, Nearctic, Neotropical and Oceanian regions (second lineage). Sequencing of the *Stomoxys calcitrans* genome allows for better fly management opportunities. Phylogenetic studies of the *Stomoxys* genus using 10 of the known species produced phylogenetic relationships among species. These studies let us suppose that the species *S. indicus* would be at the origin of the genus *Stomoxys* in the Oriental region 27 Mya ago.

As most of the knowledge on *Stomoxys* flies has been acquired from the cosmopolitan *Stomoxys calcitrans*, more research on the biology of other species is necessary; not only those species known to be linked with human activities, but also other species which could play a role in the transmission of pathogens between wild fauna and livestock.

At the same time, because of the developing resistance of these flies to available insecticides, more research is also necessary on new and more sustainable control methods: more efficient and specific trapping systems against adult flies, and biocontrols (parasitoids and predators) against preimaginal stages.

Author Contributions: Conceptualisation, G.D. and J.A.H.; methodology, G.D. and J.A.H.; validation G.D. and J.A.H.; writing—original draft preparation, G.D. and J.A.H.; writing—review and editing, G.D. and J.A.H. All authors have read and agreed to the published version of the manuscript.

Funding: This research received no external funding.

Institutional Review Board Statement: Not applicable.

Data Availability Statement: No new data were created or analyzed in this study. Data sharing is not applicable to this article.

Acknowledgments: We would like to thank Ignacio Ruiz Arrondo, Centro de Investigación Biomédica de La Rioja (CIBIR), Spain, who invited us to write this review article.

Conflicts of Interest: The authors declare no conflict of interest.

References

1. Zumpt, F. *The Stomoxyine Biting Flies of the World. Diptera: Muscidae. Taxonomy; Biology; Economic Importance and Control Measures*; Gustav Fisher: Stuttgart, Germany, 1973; pp. 97–137.
2. Foil, L.D.; Hogsette, J.A. Biology and control of tabanids, stable flies and horn flies. *Rev. Sci. Tech. l'Office Int. Epizoot.* **1994**, *13*, 1125–1158. [CrossRef] [PubMed]
3. Jones, C.J.; Milne, D.E.; Patterson, R.S.; Schreiber, E.T.; Milio, J.A. Nectar Feeding by *Stomoxys calcitrans* (Diptera, Muscidae) Effects on Reproduction and Survival. *Environ. Entomol.* **1992**, *21*, 141–147. [CrossRef]
4. Baldacchino, F.; Muenworn, V.; Desquesnes, M.; Desoli, F.; Charoenviriyaphap, T.; Duvallet, G. Transmission of pathogens by *Stomoxys* flies (Diptera; Muscidae): A review. *Parasite* **2013**, *20*, 26. [CrossRef]
5. Taylor, D.B.; Moon, R.D.; Mark, D.R. Economic impact of stable flies (Diptera: Muscidae) on dairy and beef cattle production. *J. Med. Entomol.* **2012**, *49*, 198–209. [CrossRef]
6. Blanc-Debrune, N. Impact Economique des Principales Espèces de Diptères Sur L'élevage Bovin Français et Méthodes de Luttes Associées. Ph.D. Thesis, Université Claude-Bernard-Lyon I, Villeurbanne, France, 2019; pp. 1–140.
7. Rochon, K.; Hogsette, J.A.; Kaufman, P.E.; Olafson, P.U.; Swiger, S.L.; Taylor, D.B. Stable Fly (Diptera: Muscidae)—Biology, Management, and Research Needs. *J. Integr. Pest Manag.* **2021**, *12*, 38. [CrossRef]

8. Salem, A.; Bouhsira, E.; Lienard, E.; Bousquet-Melou, A.; Jacquiet, P.; Franc, M. Susceptibility of two European strains of *Stomoxys calcitrans* (L.) to cypermethrin, deltamethrin, fenvalerate, lamda-cyhalothrin, permethrin and phoxim. *Intern. J. Appl. Res. Vet. Med.* **2012**, *10*, 249–257.
9. Tainchum, K.; Shukri, S.; Duvallet, G.; Etienne, L.; Jacquiet, P. Phenotypic susceptibility to pyrethroids and organophosphate of wild *Stomoxys calcitrans* (Diptera: Muscidae) populations in southwestern France. *Parasitol. Res.* **2018**, *117*, 4027–4032. [CrossRef]
10. Olafson, P.U.; Kaufman, P.E.; Duvallet, G.; Solórzano, J.A.; Taylor, D.B.; Trout Fryxell, R. Frequency of kdr and kdr-his Alleles in Stable Fly (Diptera: Muscidae) Populations from the United States, Costa Rica, France, and Thailand. *J. Med. Entomol.* **2019**, *56*, 1145–1149. [CrossRef]
11. Barros, A.T.M.; Rodrigues, V.D.; Cançado, P.H.D.; Domingues, L.N. Resistance of the stable fly, *Stomoxys calcitrans* (Diptera: Muscidae), to cypermethrin in outbreak areas in Midwestern Brazil. *Rev. Bras. Parasitol. Veterinária* **2019**, *28*, 5. [CrossRef]
12. Reissert-Oppermann, S.; Bauer, B.; Steuber, S.; Clausen, P.-H. Insecticide resistance in stable flies (*Stomoxys calcitrans*) on dairy farms in Germany. *Parasitol. Res.* **2019**, *118*, 2499–2507. [CrossRef]
13. Pont, A.C.; Mihok, S. A new species of *Haematobosca* Bezzi from Kenya (Diptera; Muscidae). *Stud. Dipterol.* **2000**, *7*, 25–32.
14. Pont, A.C.; Dsouli, N. A new species of *Haematobosca* Bezzi from Gabon (Diptera; Muscidae). *Stud. Dipterol.* **2009**, *15*, 259–266.
15. Pont, A.C.; Duvallet, G.; Changbunjong, T. A new species of *Haematobosca* Bezzi (Diptera: Muscidae) from Thailand. *Zootaxa* **2020**, *4763*, 538–544. [CrossRef] [PubMed]
16. Hafez, M.; Gamal-Eddin, F.M. Ecological studies on *Stomoxys calcitrans* L. and *sitiens* Rond. in Egypt; with suggestions on their control. *Bull. Soc. Entom. Egypte* **1959**, *43*, 245–283.
17. Hafez, M.; Gamal-Eddin, F.M. On the feeding habits of *Stomoxys calcitrans* L. and *sitiens* Rond., with special reference to their biting cycle in nature. *Bull. Soc. Entom. Egypte* **1959**, *43*, 291–301.
18. Gamal-Eddin, F.M. Experimental studies on the development stages of two blood-sucking flies (*Stomoxys calcitrans* Lin. and *S. sitiens* Rond.) in Egypt (Diptera: Stomoxydinae). *J. Arab. Vet. Med. Assoc.* **1963**, *23*, 309–338.
19. Steyskal, G.C. *The Gender of the Genus-Name Stomoxys Geoffroy, 1762 (Diptera; Muscidae)*; Entomological Society of Washington: Washington, DC, USA, 1975; p. 163.
20. Mavoungou, J.F.; Nguema, R.M.; Acapovi, G.L.; Koumba, R.Z.; Mounioko, F.; Lendzele Sevidzem, S.; Kindzi Bakakas, I.; Gilles, J.; Duvallet, G.; M'Batchi, B.; et al. Breeding Sites of Stomoxys spp. (Diptera: Muscidae), a Preliminary Study in the Makokou Region (North-East-Gabon). *Vector Biol. J.* **2017**, *2*, 1. [CrossRef]
21. Lendzele Sevidzem, S.; Mavoungou, J.F.; Zinga-Koumba, C.R.; Koumba, A.A.; Duvallet, G. Factors Influencing Seasonal and Daily Dynamics of the Genus *Stomoxys* Geoffroy; 1762 (Diptera: Muscidae) in the Adamawa Plateau, Cameroon. *Int. J. Zool.* **2019**, *2019*, 3636943. [CrossRef]
22. Fiasson, R. Note sur les parasites animaux du haut-Apure {Venezuela). *Rev. Sci. Médicales Pharm. Vétérinaires l'Afrique Française Libre* **1943**, *2*, 125–151.
23. Mavoungou, J.F.; Picard, N.; Kohagne, L.T.; M'batchi, B.; Gilles, J.; Duvallet, G. Spatio-temporal variation of biting flies *Stomoxys* spp. (Diptera: Muscidae) along a man-made disturbance gradient, from primary forest to the city of Makokou (North-East; Gabon). *Med. Vet. Entomol.* **2013**, *27*, 339–345. [CrossRef]
24. Guo, Y.; Hogsette, J.A.; Greene, G.L.; Jones, C.J. Survey report on pupal parasites of filth flies in livestock and poultry facilities in China. *Chin. J. Biol. Control* **1997**, *13*, 106–109.
25. Tumrasvin, W.; Shinonaga, S. Studies on medically important flies in Thailand V. On 32 species belonging to the subfamilies Muscinae and Stomoxyinae including the taxonomic keys (Diptera: Muscidae). *Bull. Tokyo Med. Dent. Univ.* **1978**, *25*, 201–227. [PubMed]
26. Masmeatathip, R.; Gilles, J.; Ketavan, C.; Duvallet, G. First survey of seasonal abundance and daily activity of *Stomoxys* spp. (Diptera: Muscidae) in Kamphaengsaen Campus, Nakornpathom Province, Thailand. *Parasite* **2006**, *13*, 245–250. [CrossRef] [PubMed]
27. Phasuk, J.; Prabaripai, A.; Chareonviriyaphap, T. A comparison of attractants for sampling *Stomoxys calcitrans* (Diptera: Muscidae) on dairy farms in Saraburi Province, Thailand. *J. Econ. Entomol.* **2016**, *109*, 942–946. [CrossRef]
28. Masmeatathip, R.; Ketavan, C.; Duvallet, G. Morphological studies of *Stomoxys* spp. (Diptera: Muscidae) in Central Thailand. *Kasetsart J. (Nat. Sci.)* **2006**, *40*, 872–881.
29. Keawrayup, S.; Duvallet, G.; Sukonthabhirom, S.; Chareonviriyaphap, T. Diversity of Stomoxys spp. (Diptera: Muscidae) and diurnal variations of activity of Stomoxys indicus and S. calcitrans in a farm, in Wang Nam Khiao District, Nakhon Ratchasima Province, Thailand. *Parasite* **2012**, *19*, 259–265. [CrossRef]
30. Kano, R. Notes on the flies of medical importance in Japan. *Jap. J. exp. Med.* **1953**, *23*, 185–195.
31. Muenworn, V.; Duvallet, G.; Tainchum, K.; Tuntakom, S.; Akratanakul, P.; Chareonviriyaphap, T. Geographic distribution of *Stomoxys calcitrans* in Thailand. *J. Med. Entomol.* **2010**, *47*, 791–797. [CrossRef]
32. Changbunjong, T.; Weluwanarak, T.; Ratanakorn, P.; Maneeon, P.; Ganpanakngan, M.; Apiwathnasorn, C.; Sungvornyothin, S.; Sriwichai, P.; Sumruayphol, S.; Ruangsittichai, J. Distribution and abundance of Stomoxyini flies (Diptera: Muscidae) in Thailand. *Southeast Asian J. Trop. Med. Public Health* **2012**, *43*, 1400–1410.
33. Reid, E.T.M. Notes on the distribution of Stomoxydinae (Dipt., Muscidae) in the southern Sudan. *Entomologist's Mon. Mag.* **1956**, *92*, 343–346.

34. Mavoungou, J.F.; Gilles, J.; Duvallet, G. *Stomoxys xanthomelas* Roubaud; 1937: Une espèce de la canopée en Afrique équatoriale. Bull. Société Entomol. *Bull. Société Entomol. Fr.* **2007**, *112*, 481–483. [CrossRef]
35. Mavoungou, J.F.; Simo, G.; Gilles, J.; De Stordeur, E.; Duvallet, G. Écologie des stomoxes (Diptera: Muscidae) au Gabon. II- Origine des repas de sang et conséquences épidémiologiques. *Parasite* **2008**, *15*, 611–615. [CrossRef] [PubMed]
36. Thorpe, W.H. Observations on *Stomoxys ochrosoma* Speiser (Diptera Muscidae) as an associate of army ants (Dorylinae) in Est Africa. *Proc. R. Entomol. Soc. Lond. (A)* **1942**, *17*, 38–42.
37. Mavoungou, J.F.; Jay-Robert, P.; Gilles, J.; Atsame, E.; Duvallet, G. Ecology of *Stomoxys* flies (Diptera: Muscidae) in Gabon. I- First survey in different ecological areas. *Parasite* **2008**, *15*, 27–34. [CrossRef]
38. Bitome Essono, P.Y.; Dechaume-Moncharmont, F.-X.; Mavoungou, J.; Obiang Mba, R.; Duvallet, G.; Bretagnolle, F. Distribution and abundance of hematophagous flies (Glossinidae, Stomoxys, and Tabanidae) in two national parks of Gabon. *Parasite* **2015**, *22*, 23. [CrossRef]
39. Hogsette, J.A.; Ruff, J.P. Stable Fly (Diptera: Muscidae) Migration in Northwest Florida. *Environ. Entomol.* **1985**, *14*, 170–175. [CrossRef]
40. Jones, C.J.; Hogsette, J.A.; Patterson, R.S.; Milne, D.E.; Propp, G.D.; Milio, J.F.; Rickard, L.G.; Ruff, J.P. Origin of Stable Flies (Diptera; Muscidae) on West Florida Beaches—Electrophoretic Analysis of Dispersal. *J. Med. Entomol.* **1991**, *28*, 787–795. [CrossRef]
41. Gilles, J.; David, J.-F.; Duvallet, G.; de La Roque, S.; Tillard, E. Efficiency of traps for *Stomoxys calcitrans* and *Stomoxys niger* niger on Reunion Island. *Med. Vet. Entomol.* **2007**, *21*, 65–69. [CrossRef]
42. Szalanski, A.L.; Taylor, D.B.; Peterson, R.D. Population genetics and gene variation of stable fly populations (Diptera: Muscidae) in Nebraska. *J. Med. Entomol.* **1996**, *33*, 413–420. [CrossRef]
43. Kneeland, K.M.; Skoda, S.R.; Foster, J.E. Amplified fragment length polymorphism used to investigate genetic variability of the stable fly (Diptera: Muscidae) across North America. *J. Med. Entomol.* **2013**, *50*, 1025–1030.
44. Dsouli-Aymes, N.; Michaux, J.; De Stordeur, E.; Couloux, A.; Veuille, M.; Duvallet, G. Global population structure of the stable fly (*Stomoxys calcitrans*) inferred by mitochondrial and nuclear sequence data. *Infect. Genet. Evol.* **2011**, *11*, 334–342. [CrossRef] [PubMed]
45. Tsai, C.-L.; Lu, C.-N.; Tzeng, H.-Y.; Krafsur, E.S.; Tu, W.-C.; Yeh, W.-B. Global population genetic structure and lineage differentiation of the stable fly, *Stomoxys calcitrans*. *Med. Vet. Entomol.* **2023**, 1–10. [CrossRef] [PubMed]
46. Changbunjong, T.; Weluwanarak, T.; Samung, Y.; Ruangsittichai, J. Molecular identification and genetic variation of hematophagous flies (Diptera: Muscidae: Stomoxyinae) in Thailand based on cox1 barcodes. *J. Asia-Pac. Entomol.* **2016**, *19*, 1117–1123. [CrossRef]
47. Kneeland, K.M.; Skoda, S.R.; Foster, J.E. Genetic variability of the stable fly assessed on a global scale using amplified fragment length polymorphism. *Insect Sci.* **2016**, *23*, 695–703. [CrossRef]
48. Olafson, P.U.; Aksoy, S.; Attardo, G.M.; Buckmeier, G.; Chen, X.; Coates, C.J.; Davis, M.; Dykema, J.; Emrich, S.J.; Friedrich, M.; et al. The genome of the stable fly, *Stomoxys calcitrans*, reveals potential mechanisms underlying reproduction, host interactions, and novel targets for pest control. *BMC Biol.* **2021**, *19*, 41. [CrossRef]
49. Dsouli, N.; Delsuc, F.; Michaux, J.; De Stordeur, E.; Couloux, A.; Veuille, M.; Duvallet, G. Phylogenetic analyses of mitochondrial and nuclear data in haematophagous flies support the paraphyly of the genus *Stomoxys* (Diptera: Muscidae). *Infect. Genet. Evol.* **2011**, *11*, 663–670. [CrossRef]

MDPI
St. Alban-Anlage 66
4052 Basel
Switzerland
www.mdpi.com

Diversity Editorial Office
E-mail: diversity@mdpi.com
www.mdpi.com/journal/diversity

www.ingramcontent.com/pod-product-compliance
Lightning Source LLC
Chambersburg PA
CBHW061517210926
51850CB00008B/1949
9783725804436